Fire Safety Engineering Design of Structures

Third Edition

Fire Safety Engineering Design of Structures

Third Edition

John A. Purkiss and Long-yuan Li

CRC Press
Taylor & Francis Group
Boca Raton London New York

CRC Press is an imprint of the
Taylor & Francis Group, an **Informa** business

A SPON PRESS BOOK

CRC Press
Taylor & Francis Group
6000 Broken Sound Parkway NW, Suite 300
Boca Raton, FL 33487-2742

First issued in paperback 2017

Version Date: 20130923

ISBN 13: 978-1-4665-8547-8 (hbk)
ISBN 13: 978-1-138-07426-2 (pbk)

Library of Congress Cataloging-in-Publication Data

Purkiss, J. A.
    Fire safety engineering design of structures / John A. Purkiss, Long-Yuan Li. -- Third edition.
        pages cm
    Revised edition of: Fire safety engineering desing of structures / John A. Purkess. 2007.
    Includes bibliographical references and index.
    ISBN 978-1-4665-8547-8 (hardback)
    1. Building, Fireproof. 2. Fire protection engineering. I. Li, Long-Yuan. II. Title.

TH1065.P87 2013
693.8'2--dc23
                                                      2013036865

Visit the Taylor & Francis Web site at
http://www.taylorandfrancis.com

and the CRC Press Web site at
http://www.crcpress.com

Disclaimer

Permission to reproduce extracts from British Standards is granted by the British Standards Institution (BSI). British Standards can be obtained in PDF or hard copy format from the BSI online shop (www.bsigroup.com/Shop) or by contacting customer services for hardcopy only (telephone +44(0)20 8996 9001; email: cservices@bsigroup.com).

Code	BSI reference	Book reference
EN 1991-1-2	Table E5	Table 4.4
	Table F2	Table 4.7
EN 1992-1-2	Table 3.1	Table 5.8
	Figure B.2	Figure 7.1
EN 1993-1-2	Figure 3.1	Figure 5.35, Table 5.3
	Table 3.1	Table 5.6
EN 1994-1-2	Table 3.3	Table 5.8
	Figure D.1	Figure 9.1
	Figure D.2	Figure 9.3
	Figure D.3a and b	Figure 9.4
	Table D.5	Table 9.2
	Table D.6	Table 9.1
EN 1995-1-2	Table 3.1	Table 10.1
	Table E.2	Table 10.3
BS 5950 Part 8	Figure C.1	Figure 8.8
	Figure C.2	Figure 8.8
	Table C.1	Table 8.4
PD 7974-3	Table 4	Table 4.7

Contents

Preface

Since the publication of the second edition of this book, *Structural Fire Safety Engineering* has become a recognized engineering discipline which has prompted the need for a third edition. Advances of the discipline also allowed updating of various sections such as material modelling.

The intention of this text is to provide tools to fire safety engineers to enable them to design structures that can withstand the accidental effects caused by fires occurring within part or the whole of the structure. Other than references to furnace testing (whose data are mainly historical), the authors concentrated on the European Design Codes. Designers should still be aware that any design code is subject to revision or amendment and that it is essential that the most recent edition of any guidance be used. Where this produces a discrepancy between this text and a code, the code must be considered the final arbiter.

The large-scale fire tests carried out at Cardington led to a reassessment of frame behaviour, especially in relation to composite steel–concrete frame structures in that whole structure performance markedly outweighs performances of single elements and connection details significantly affect their behaviours.

In discussions of Eurocodes, National Annexes were not considered directly. Any nationally determined parameters were taken at recommended values and not amended to conform to an individual country. However, the UK National Annexes specific comments especially in relation to parametric fires and time equivalence have been noted.

The user of this text is presumed to have knowledge of structural mechanics and a background in the methods of structural design at ambient conditions, since the design of structures at the fire limit state uses either modified ambient design methods or requires data such as member capacities from ambient designs. Some knowledge of the thermodynamics of heat transfer will also be useful. A series of worked examples has been included to provide a feel for the common types of calculations. To help readers gain a better understanding of the principles and procedures involved in fire safety engineering, an extensive reference section follows Chapter 13.

Fire design must be envisaged as part of the overall design of a structure and not treated as an afterthought. To assist in helping designers obtain a full picture of the types of design decisions required for optimum fire performance of a structure, the first chapter provides an overview of the entire subject.

The usual disclaimer must be made by both authors that in the final event any opinions expressed herein are entirely those of the authors, and they alone are responsible for sins of omission or commission.

Whilst the authors have made every attempt to contact possible copyright holders and authors, they apologize if material has been included for which copyright clearance was for any reason overlooked. The authors would be grateful to receive such information so that such omissions can be corrected in any future editions.

Acknowledgements

John Purkiss would like to thank H.L. (Bill) Malhotra, W.A. Morris, and R.D. Anchor for their valuable support and advice over many years. Appreciation is also extended to Colin Bailey, John Dowling, Tom Lennon, and Gerald Newman for their advice and encouragement. He also acknowledges Long-yuan Li for taking up the mantle of co-author.

Gratitude is also expressed to his post-doctoral research fellow Rosen Tenchev, and his post-graduate research students, Sarah Guise, Bahjat Khalafallah, Raymond Connolly, Kamal Mustapha, Abderahim Bali, and Nick Weeks, who assisted, often unwittingly, by discovering references, providing ideas and stimuli, and commenting on early versions of the text of the first edition.

Dr. Purkiss also expresses many thanks to Dr. J.W. Dougill, without whom this book would not have been possible, as many years ago, John's research supervision was responsible for kindling the flames.

Long-yuan Li thanks his former colleagues at Aston University, Chris Page, Colin Thornton, Neil Short, Roger Kettle, and of course, John Purkiss, for their friendship and support. Much of the work presented in this book was conducted at Aston, although some of the work was published at a later date.

We owe a debt of gratitude to a large number of colleagues in the field, too many to name individually, who have given their time and experience to deal with the authors' queries and concerns.

The authors also wish to thank those organizations that went out of their way to deal with permission requests for material now out of print or belonging to defunct organizations.

List of Acronyms

ACI	American Concrete Institute
ASCE	American Society of Civil Engineers
ASFP	Association for Specialist Fire Protection
BIA	Brick Industry Association (USA)
BRE	Building Research Establishment (UK)
BSI	British Standards Institution
CEB	Comité Euro-International du Béton
CIB	Conseil Internationale du Bâtiment pour la Recherche l'Étude et la Documentation (International Council for Building Research Studies and Documentation)
CIRIA	Construction Industry Research and Information Association (London)
ECCS	European Convention for Construction Steelwork
FIP	Fédération Internationale de la Précontrainte
ISE	Institution of Structural Engineers (UK)
ISO	Organisation Internationale de Normalisation (International Organization for Standardization)
RILEM	Réunion Internationale des Laboratoires d'Essais et de Recherches sur les Matériaux et les Constructions (International Union of Testing and Research Laboratories for Materials and Structures)
SCI	Steel Construction Institute (UK)

Notation

A	Roof pitch parameter, steel creep parameter, stress parameter
A_c	Area of concrete cross section
A_f	Area of floor of a compartment
A_h	Area of horizontal openings
A_m	Area of steel protection per unit length exposed to fire
A_r	Area of reinforcement, area of residual timber section
A_s	Area of tension steel, air space
A_{s1}	Area of tension steel resisting concrete compression
A_{s2}	Area of tension steel resisting steel compression
A_t	Total internal area of compartment
A_v	Shear area, area of vertical openings
a	Axis distance from concrete surface to centroid of reinforcing bar(s), thermal diffusivity
a_a	Thermal diffusivity of steel
a_c	Thermal diffusivity of concrete
a_w	Thermal diffusivity of wood
a_z	Width of damage zone
$a_1 - a_6$	Parameters for determination of protection thicknesses
B	Parametric length, steel creep parameter, stress parameter (Khennane and Baker), width of section
B_e	Width of bottom flange of shelf angle floor
b	Reduced width, thermal inertia, width of concrete compression stress block, flange width
b_{eff}	Effective width of slab on composite beam
b_{fi}	Temperature-modified width of concrete compression stress block
b_v	Parameter dependent on openings ratio (α_v)
b_w	Depth of web

b_1/b_2	Width of bottom/top flange of steel beam
C	Constant, roof pitch parameter, steel creep parameter, temperature-time dependent boundary parameter (Hertz)
C_1, C_2, C_3	Creep parameters (Schneider)
$\mathbf{C}$	Global capacitance matrix
$\mathbf{C}_e$	Element capacitance matrix
C_a	Specific heat of dry air
C_l	Specific heat of water
C_s	Specific heat of porous skeleton
C_v	Specific heat of water vapour
c	Equivalent fire resistance, specific heat, web depth, flange outstand
c_a	Specific heat of steel
c_{al}	Specific heat of aluminium
c_c	Specific heat of concrete
c_p	Specific heat of combustion gasses, specific heat of protection
$c_{p,peak}$	Peak value allowing for moisture of specific heat of concrete
c_{pm}	Specific heat of masonry
c_{pw}	Specific heat of timber
c_v	Volumetric specific heat
D	Depth of compartment, overall section depth, steel creep parameter, temperature–time dependent boundary parameter (Hertz)
D_e	Height to underside of slab (shelf angle floor)
D_g	Diffusion coefficient of component in binary mixture
d	Depth, effective depth, shear stud diameter
d_{char}	Depth of charring
$d_{char,b}$	Depth of charring on bottom of beam (Stiller)
$d_{char,col}$	Depth of charring on column (Stiller)
$d_{char,n}$	Depth of charring for multiface exposure
$d_{char,0}$	Depth of charring for single-face exposure
$d_{char,s}$	Depth of charring on side of beam (Stiller)
d_{ef}	Effective reduced section
d_{fi}	Temperature-modified effective depth
d_p	Protection or insulation thickness
d_{wall}	Wall thickness
d_0	Additional reduction in section dimensions
E	Height of stanchion, Young's modulus, temperature–time dependent boundary parameter (Hertz)

E_{al}	Modulus of elasticity of aluminium at ambient temperature
$E_{al,\theta}$	Modulus of elasticity of aluminium at temperature θ (°C)
E_c	Young's modulus of concrete
$E_{c,\theta}$	Temperature-affected Young's modulus of concrete
E_c^*	Slope of descending branch of concrete stress–strain curve
E_{cm}	Young's modulus for concrete
$E_{cm,\theta}$	Temperature-affected Young's modulus of concrete
E_d	Design effect
$E_{dyn,\theta}$	Dynamic Young's modulus for concrete at elevated temperature
$E_{fi,d}$	Design effect of fire
$E_{fi,d,t}$	Design effect of fire at time t
E_r	Young's modulus of reinforcement
E_s	Young's modulus of steel
$E_{s,\theta}$	Young's modulus of steel at temperature θ
$E_{stat,\theta}$	Static Young's modulus for concrete at elevated temperature
E_t	Slope of linear section of stress–strain curve at time t
$\dot{E}$	Evaporation rate of free water
e	Enhancement factor due to membrane action
e_w	Thickness of web
e_1/e_2	Thickness of bottom/top flange of steel beam
F	Force, steel creep parameter, surface area exposed to fire
F_f	Axial load in fire limit state
F_i	Thermal forces (i = 1, 2, 3)
F_s	Tensile force in reinforcement
f	Overdesign factor for timber members
f_{al}	0.2% proof strength of aluminium at ambient temperature
$f_{al,\theta}$	0.2% proof strength of aluminium at temperature θ (°C)
$f_{amax,\theta cr}$	Steel strength at critical design temperature
$f_{amax,\theta w}$	Temperature-reduced steel strength of web
$f_{amax,\theta 1}/f_{amax,\theta 2}$	Temperature-reduced steel strength of top/bottom flange of steel beam
$f_{ay,20°C}$	Ambient steel design strength
$f_{c,\theta}$	Temperature-dependent concrete strength, stress applied to concrete at start of heating
$f_{c,20°C}$	Ambient cylinder strength of concrete
f_{cd}	Concrete design strength based on cylinder strength
f_{ck}	Characteristic strength of concrete at ambient (also $f_{ck,20}$)
f_k	Characteristic strength, modification factor for compartment boundary conditions or horizontal openings

f_{mean}	Mean strength
$f_{scd,fi}$	Temperature-reduced compression reinforcement strength
$f_{sd,fi}$	Temperature-reduced tension reinforcement strength
f_t	Partial derivative of strain with respect to time
f_u	Ultimate strength of steel
f_y/f_{yk}	Characteristic/yield strength of reinforcement (also $f_{yk,20}$)
f_θ	Partial derivative of strain with respect to temperature
f_σ	Partial derivative of strain with respect to stress
$f_1/f_2/f_3$	Temperature–time coordinate functions (Hertz)
G	Distance between ends of haunches in portal frame, shear modulus, temperature gradient parameter
G_k	Characteristic permanent structural load
g	Temperature-dependent stress function (Schneider), acceleration due to gravity
g_0	Parameter defined by depth of compression block
$g_{2,\theta}$	Parameter for strength reduction of stainless steel
H	Heights of windows or vertical openings, horizontal reaction, steel creep parameter
H_p	Heated perimeter
H_t	Strain hardening parameter at time t (Khennane and Baker)
$H_{u,i}$	Calorific value of compartment fire load
h	Overall section depth, height
h_c	Overall slab depth
h_{cr}	Limiting depth to 250°C isotherm for composite slab
h_{eff}	Effective (equivalent) depth of profiled concrete composite deck
h_{qr}	Combined convective radiation heat transfer coefficient
h_u	Depth of concrete stress block in composite slab
h_w	Overall height of web
h_{wall}	Height of wall
h_1	Depth of trough for profile sheet decking, vertical distance between horizontal opening and mid-height of vertical opening
h_2	Depth of concrete above trough in composite deck
$\dot{h}_B$	Heat stored in gas
$\dot{h}_c$	Rate of heat release from compartment
$\dot{h}_L$	Rate of heat loss due to convection

$\dot{h}_{net}$	Net design heat flow due to convection and radiation at boundary
$\dot{h}_{net,c}$	Net rate of heat transmission at boundary due to convection
$\dot{h}_{net,r}$	Net rate of heat transmission at boundary due to radiation
$\dot{h}_R$	Rate of heat loss through openings
$\dot{h}_w$	Rate of heat loss through walls
$\overline{h}'$	Normalized heat load in compartment
$\overline{h}''$	Normalized heat load in furnace
I	Second moment of area
I_a	Second moment of area of steel section
I_c	Second moment of area of concrete
I_r	Second moment of area of reinforcement
$\mathbf{I}$	Identity matrix
i	Radius of gyration
$J(\theta,\sigma)$	Unit stress compliance function (Schneider)
J_l	Flux of free water at boundary
J_v	Flux of water vapour at boundary
$\mathbf{K}$	Global matrix
$\mathbf{K}_c$	Global conductance matrix
$\mathbf{K}_{ce}$	Element conductance matrix $(i = 1, 2, 3)$
K	Coefficient of overall heat transfer, concrete buckling parameter, constant of proportionality in equivalent fire duration, parameter of Ramburg–Osgood equation, portal frame parameter dependent on number of bays, intrinsic permeability of porous skeleton
K_g	Relative permeability of gaseous mixture
K_{ij}	Matrix coefficients $(i,j = 1, 2, 3)$
K_l	Relative permeability of liquid water
k	Ratio between strength and induced stress, constant of proportionality between transient and free thermal strain, parameter, property ratio, timber buckling parameter
$k_{amax,\theta}$	Temperature modification factor for steel strength
k_b	Factor dependent on compartment thermal boundaries
k_c	Temperature modification factor for concrete strength, correction factor for time equivalence on unprotected steelwork, timber strut buckling parameter
$k_{c,m}$	Reduction factor for mean concrete strength
$k_{E,\theta}$	Temperature reduction factor for Young's modulus of steel

k_{eff}	Thermal conductivity coefficient
k_{LT}	Moment correction factor for lateral-torsional buckling
$k_{mod,fi}$	Modification factors applied to timber strengths and elastic modulus
k_s	Strength reduction factor due to temperature for steel
k_{shadow}	Shadow factor for determination of steel temperatures on bare steelwork
k_y	Moment correction factor
$k_{y,\theta}$	Reduction factor on yield strength due to temperature
k_z	Moment correction factor
k_1	Constant relating loss in UPV and strength
k_2	Constant of proportionality between transient and free thermal strain, constant relating loss in UPV and strength
L	Span, temperature dependent boundary condition parameter for cooling period (Hertz)
$\bar{L}_e$	Effective or system length
$L_{f,d}$	Design fire load in kilograms of wood per unit floor area
$L_{fi,k}$	Total design fire load in kilograms of wood equivalent
L_r	Heated perimeter for profile decking
LITS	Load-induced thermal strain (Khoury)
l_θ	Effective length in fire limit state
l_1, l_2, l_3	Dimensions of profile sheet steel decking
M	Bending moment, moisture content, number of elements in solution domain
M_a	Molar mass of dry air
$M_{b,fi,t,d}$	Design bending strength in fire limit state
$M_{Ed,fi}$	Applied moment in fire limit state
$M_{fi,Rd}$	Design resistance moment in fire limit state
$M_{fi,\theta,d}$	Applied design moment in fire limit state
M_{fr}	Moment capacity in fire limit state
$M_{k,i}$	Characteristic mass
M_{pl}	Plastic moment capacity
M_{Rd}	Moment of resistance
M_{Sd}	Ambient design moment
M_U	Moment capacity in fire limit state of concrete in flexure
M_{U1}	Moment capacity in fire limit state of tensile reinforcement for concrete in flexure
M_{U2}	Moment capacity in fire limit state of compression reinforcement for concrete in flexure

M_y	Bending moment about yy axis (major axis), molar mass of water vapour
$M_{y,fi,Ed}$	Moment applied about yy axis
M_z	Bending moment about zz axis (minor axis)
$M_{z,fi,Ed}$	Moment applied about zz axis
m	Fuel load factor
$m_{p\theta}$	Temperature-reduced moment capacity of slab
m_θ	Strain hardening parameter
$\dot{m}_F$	Rate of mass flow of outflow gasses
N	Interpolation matrix
N	Number of shear studs per half length of beam
$N_{fi,cr}$	Axial design resistance based on buckling
$N_{fi,pl,Rd}$	Axial plastic load capacity in fire limit state
$N_{fi,Rd}$	Axial design resistance
$N_{fi,\theta,d}$	Required axial load capacity in fire limit state
N_i	Interpolation function defined at node i
N_{Rd}	Axial capacity or resistance
N_x	Axial membrane force
n	Exponent, parameter in Popovics equation, parameter in Ramburg–Osgood equation, number of slices, number of nodes in element, normal of surface
n_x/n_y	Distance (coordinate) parameter (Wickström)
n_w	Time-dependent parameter (Wickström)
$\hat{n}$	Normal of element boundary
O	Opening factor
P	Pore pressure in concrete
P_a	Partial pressure related to dry air in gaseous mixture
$P_{fi,Rd}$	Shear stud capacity in fire limit state
P_{or}	Initial concrete porosity
P_{sat}	Saturation vapour pressure
P_v	Water vapour pressure, partial pressure related to water vapour in gaseous mixture
p	Moisture content in percent by weight, perimeter of residual timber section
$p_{0,2proof,\theta}$	Temperature-reduced 0,2% proof strength of stainless steel
$p_{2,\theta}$	Temperature-reduced 2% proof strength of stainless steel
$Q_{fi,k}$	Total characteristic fire load
Q_k	Characteristic load
$Q_{k,i}$	Characteristic variable load

Q_{mean}	Mean load
Q	Heat flow vector
q	Uniformly distributed load, vector of heat flux per unit area
$q_{f,d}$	Design fire load per unit floor area
$q_{f,k}$	Generic fire load
$q_{fi,d}$	Uniformly distributed load in fire limit state
$q_{p\theta,slab}$	Load carried by slab
$q_{p\theta,udl}$	UDL supported by beam
q_t	Fire load per unit area of compartment
$q_{t,d}$	Design fire load per unit area of compartment
q_x	Axial distributed load
q_y/q_z	Transversely distributed load in y/z direction
q_{Sd}	Plastic distributed load in fire limit state
$q_{Sd.el}$	Elastic design distributed load in fire limit state
R	Global nodal vector
R_Q	Global nodal vector of internal heat source
R_{Qe}	Element nodal vector of internal heat source
R_q	Global nodal vector of heat flow
R_{qe}	Element nodal vector of heat flow
R	Rate of burning in kilograms of wood equivalent per second, resistance, gas constant
R_d	Design resistance effect
$R_{fi,d}$	Design resistance effect in fire limit state
R_1	Length of portal frame rafter
r	Radius of arris, fillet radius, ratio of applied load to allowable load (as percentage)
S	Frame spacing, specific gravity of timber, initial water saturation
S_d	Fire effect due to structural loads (actions)
SIG1	Changeover stress for calculating steel creep
T	Tensile force resultant in composite beam
t	Time
t_d	Fire duration based on design fire load
t_{eff}	Effective thickness of masonry wall
$t_{e,d}$	Equivalent fire duration
t_f	Flange thickness
$t_{fi,d}$	Calculated fire resistance based on member performance

$t_{fi,requ}$	Design or required fire performance based on classification system
t_j	Time at step j ($j = k, k + 1, \beta$)
t_{lim}	Limiting value of t_{max} due to fire type
t_{max}	Time to maximum temperature for parametric fire
t_{spall}	Time to spalling
t_v	Delay time to allow for moisture
t_w	Web thickness
t_0	Time to maximum charring under parametric conditions
t^*	Parametric time for determining compartment temperature–time response
t_d^*	Parametric fire duration
t_{max}^*	Maximum value of t^*
$\bar{t}$	Time shift
$\bar{t}_{E,d}$	Mean test result
U	Creep activation energy
$U_{c,\theta}$	Temperature modified ultrasonic pulse velocity
$U_{c,20}$	Ambient ultrasonic pulse velocity
$U_{s,\theta}$	Temperature-reduced ultimate strength of stainless steel
u	Axial displacement
u_x	Correction factor for nonstandard concrete diffusivity (Wickström)
$u_1/u_2/u_3$	Distances to reinforcement from face of trough of steel decking
V	Shear force, vertical reaction, volume of steel
$V_{fi,t,d}$	Design shear capacity in fire limit state
V_g	Migration velocity of gaseous mixture
V_i	Volume of steel per unit length
V_l	Migration velocity of free water
V_{Rd}	Shear resistance
v	Deflection in y direction
W	Compartment width, water content, mass per unit length (steel section)
W_{el}	Elastic section modulus
$W_{fi,d}$	Design structural fire load
W_{pl}	Plastic section modulus
W'	Mass per unit length (steel section and plasterboard)
w	Compartment geometry factor, moisture content percent by weight, half width of section, deflection in z direction

w_f	Ventilation factor
X_1	Distance to end of haunch
X_2	Distance to centroid of loading on haunch
x	Cartesian coordinate, depth of neutral axis (reinforced concrete), distance, factor dependent upon t_{max} and t_{lim}
x_n	Depth to centroidal axis
x_{spall}	Depth of spalling
Y	Height to end of haunch
y	Cartesian coordinate, distance
y_T	Height to line of action of tensile force
Z	State function, Zener–Hollomon parameter
z	Cartesian coordinate, distance parameter, lever arm (tension steel), load factor
z'	Lever arm (compression steel)
α	Angle of rotation of stanchion, coefficient of total heat transfer, interpolation parameter, modification factor for shear stud capacity, ratio of ultimate strength at elevated temperature to ambient strength for timber, coefficient of thermal expansion, aspect ratio (<1,0), angle of decking, isobaric heat of evaporation of water from liquid state
α_{al}	Coefficient of thermal expansion for aluminium
α_c	Coefficient of convective heat transfer
α_{cc}	Concrete load duration factor
α_{eff}	Coefficient of effective heat transfer
α_h	Ratio of horizontal compartment openings to floor area
α_m	Coefficient of thermal expansion of masonry, surface absorption
α_r	Coefficient of radiative heat transfer
α_v	Ratio of area of vertical openings in compartment to floor area
β	Interpolation parameter, statistical acceptance/rejection limit, algorithm parameter, transfer coefficient
β_c	Imperfection factor for timber strut buckling
$\beta_{M,LT}$	Moment gradient correction factor
$\beta_{M,y}$	Moment gradient correction factor
$\beta_{M,z}$	Moment gradient correction factor
β_n	Charring rate including allowance for arris rounding
β_{par}	Charring rate due to parametric exposure
β_0	Uniform charring rate of timber

Γ	Parameter to calculate parametric compartment temperature–time response
Γ_e	Boundary of element
Γ_{max}	Maximum value of Γ due to fire load restrictions
γ	Constant (= 0,1 mm/min)
γ_c	Concrete materials partial safety factor
γ_f	Partial safety factor applied to loads or actions
γ_G	Partial safety factor applied to dead loads (permanent actions)
γ_M	Partial safety factor applied to design strengths
$\gamma_{M,fi}$	Materials partial safety factor in fire limit state
$\gamma_{M,fi,a}$	Partial materials safety factor for structural steel
$\gamma_{M,fi,c}$	Partial materials safety factor for concrete in composite construction
$\gamma_{M,fi,v}$	Partial materials safety factor for shear studs
γ_Q	Partial safety factor applied to imposed loads (variable actions)
γ_s	Partial safety factor applied to steel design strengths
γ_w	Moisture-dependent parameter (Schneider)
γ_0	Constant (Schneider)
δ	Deflection, parametric measure of fuel energy lost through openings
δ_b	Deflection due to flexure
$\delta_{bow,b}$	Thermal bowing calculated for beam
$\delta_{bow,c}$	Thermal bowing calculated for cantilever
δ_n	Factor to allow for presence of active fire protection
δ_{q1}	Partial safety factor dependent upon consequences of failure
δ_{q2}	Partial safety factor dependent upon type of occupancy
ε	Strain, local buckling parameter
$\varepsilon_{al,\theta}$	Aluminium strain
ε_b	Volumetric fraction of bound water in concrete
ε_{bo}	Initial volumetric fraction of bound water in concrete
ε_c	Concrete strain
$\varepsilon_{c,\theta}$	Concrete strain at elevated temperature
$\varepsilon_{c,20}$	Concrete strain at ambient
ε_{cr}	Creep strain
$\varepsilon_{cr,c}$	Concrete creep strain
$\varepsilon_{cr,s}$	Steel creep strain
$\varepsilon_{cr,s,0}$	Creep intercept (Dorn model)
ε_{cu}	Ultimate concrete strain

ε_d	Volumetric fraction of dehydrated bound water in concrete
ε_f	Emissivity of fire or furnace
ε_g	Volumetric fraction of gaseous phase in concrete
ε_l	Volumetric fraction of liquid phase in concrete
$\varepsilon_{l,25}$	Volumetric fraction of saturation water at 25°C
$\varepsilon_{l,0,96}$	Volumetric fraction of free water at $P_v = 0{,}96 P_{sat}$
$\varepsilon_{l,1,04}$	Volumetric fraction of free water at $P_v = 1{,}04 P_{sat}$
ε_{lo}	Initial volumetric fraction of free water
ε_m	Emissivity of surface
$\varepsilon_{p,c,\theta}$	Effective concrete plastic strain (Khennane and Baker)
ε_{res}	Resultant or effective emissivity
ε_s	Steel strain, volumetric fraction of solid phase in concrete
$\varepsilon_{s,fi}$	Reinforcement strain in fire limit state
ε_{ss}	Volumetric fraction of porous skeleton in concrete
ε_{th}	Free thermal strain
$\varepsilon_{th,c}$	Free thermal strain for concrete
$\varepsilon_{th,s}$	Free thermal strain for steel
ε_{tot}	Total strain
$\varepsilon_{tot,c}$	Total strain for concrete
$\varepsilon_{tot,s}$	Total strain for steel
ε_{tr}	Transient strain
$\varepsilon_{tr,c}$	Transient strain for concrete
$\varepsilon_{y,s}$	Yield strain for steel
$\varepsilon_{y,s,\theta}$	Yield strain for steel at temperature θ
ε_σ	Elastic strain
$\varepsilon_{\sigma,c}$	Elastic strain for concrete
$\varepsilon_{\sigma,s}$	Elastic strain for steel
$\varepsilon_{0,c}$	Peak concrete strain
$\varepsilon_{1,c}$	Changeover strain between parabola and linear descending branch (Anderberg and Thelandersson)
η	Compartment geometry factor, imperfection coefficient for timber strut buckling, load ratio, strength reduction factor for rectangular stress block
η_{fi}	Ratio between action effects from fire and ambient limit states
$\eta_{fi,t}$	Ratio between action effects from fire and ambient limit states at time t
Θ	Temperature-compensated time (Dorn model)
Θ_0	Temperature-compensated time changeover point between primary and secondary creep

θ	Elastoplastic redistribution factor, rafter sag angle, temperature
θ_a	Steel temperature, ambient temperature
$\theta_{a,t}$	Structural steel temperature at time t
θ_{al}	Aluminium temperature
θ_c	Concrete temperature
θ_{cr}	Critical steel temperature
$\theta_{cr,d}$	Critical design temperature
θ_{cw}	Temperature at wood–char interface
θ_d	Calculated design temperature
θ_e	Vector of element nodal temperatures
$\theta_{f,max}$	Maximum temperature of fire in compartment related solely to geometry
θ_g	Creep temperature parameter (Schneider), furnace or gas temperature
θ_i	Temperature of internal surface, temperature at node i
θ_{lim}	Limiting temperature
θ_M	Mean temperature
θ_m	Temperature of masonry, temperature of surface exposed to fire
θ_{max}	Maximum temperature reached at duration of fire
θ_R	Reference temperature for shelf angle floor
θ_s	Reinforcing steel temperature
θ_t	Fire temperature at time t
θ_w	Wood temperature
θ_0	Ambient or reference temperature, rafter angle at ambient
$\theta_1/\theta_2/\theta_8/\theta_{64}$	Curve fitting parameters for property reduction curves
$\bar{\theta}$	Prescribed temperature at boundary
κ	Adaptation factor, nonlinearity factor (Schneider)
κ_{xy}	Curvature in the xy plane
κ_{xz}	Curvature in the xz plane
κ_1/κ_2	Adaptation factors
λ	Slenderness ratio, thermal conductivity, concrete strength-dependent stress block depth factor, thermal conductivity tensor
λ_a	Thermal conductivity of steel
λ_{al}	Thermal conductivity of aluminium
λ_c	Thermal conductivity of concrete
λ_d	Specific heat of dehydration of bound water

λ_e	Latent heat of evaporation of free water
λ_i	Thermal conductivity of compartment boundary
λ_m	Largest eigenvalue
λ_p	Thermal conductivity of protection material
λ_{rel}	Normalized slenderness ratio
λ_0	Dry thermal conductivity for masonry
λ'	Moisture modified thermal conductivity for masonry
$\bar{\lambda}$	Normalized strut buckling parameter
$\bar{\lambda}_{LT}$	Normalized lateral-torsional buckling parameter
$\bar{\lambda}_{LT,\theta,com}$	Temperature-dependent normalized lateral-torsional buckling parameter with respect to temperature in compression flange
$\bar{\lambda}_\theta$	Temperature-dependent normalized strut buckling parameter
μ_g	Dynamic viscosity of gaseous mixture
μ_l	Dynamic viscosity of liquid
μ_{LT}	Correction factor dependent upon loading
μ_y	Correction factor dependent upon loading
μ_z	Correction factor dependent upon loading
μ_0	Utilization factor
ξ	Reduction factor applied to permanent loads
ξ_{cm}	Mean strength reduction factor for concrete
$\xi_{s,02}$	Temperature dependent modification factor for steel strength based on 0,2% proof strength
$\xi_{\theta,x}/\xi_{\theta,y}$	Nondimensional uniaxial temperature rise parameters (Hertz)
$\xi(\theta)$	Change in property at elevated temperature
ρ	Density
ρ_a	Density of structural steel, mass concentration of dry air in gaseous phase
ρ_{air}	Density of air
ρ_{cem}	Mass of anhydrous cement per unit volume concrete
ρ_g	Specific weight of gaseous mixture
ρ_l	Specific weight of water
ρ_p	Density of insulation
ρ_v	Mass concentration of water vapour in gaseous phase
ρ'_p	Moisture modified density of insulation
σ	Stefan–Boltzmann constant

$\sigma_{al,\theta}$	Aluminium strength at temperature θ
σ_c	Concrete stress
$\sigma_{c,\theta}$	Concrete strength at a temperature θ
$\sigma_{c,20}$	Concrete strength at ambient conditions
σ_d	Design strength for timber
σ_E	Euler buckling stress
σ_f	Standard deviation of strengths
$\sigma_{h'}$	Standard deviation of normalized heat flow in compartment
$\sigma_{h''}$	Standard deviation of normalized heat flow in furnace
σ_L	Standard deviation of compartment fuel load
σ_{par}	Strength parallel to grain
σ_Q	Standard deviation of applied loading
$\sigma_{s,f}$	Temperature-modified steel strength
σ_t	Standard deviation in results of furnace tests
$\sigma_{w,t}$	Axial compressive strength of timber
$\sigma_{y,s}$	Steel yield strength
$\sigma_{y,s,\theta}$	Temperature-modified steel yield strength
$\sigma_{y,20}$	Yield strength at ambient
$\sigma_{0,c}$	Peak concrete strength
$\sigma_{1,c,\theta}$	Changeover stress between linear and elliptic parts of curve
$\sigma_{1c,\theta}/\sigma_{2c,\theta}$	Concrete stress in biaxial compression
Φ	Configuration factor for radiation, creep function (Schneider), insulation heat capacity factor, ventilation factor related to mass inflow, view factor for profile sheet decking
φ	Creep function (Schneider)
$\varphi_{LT,\theta,com}$	Temperature-dependent lateral-torsional buckling parameter
φ_θ	Temperature-dependent strut buckling parameter
χ_{fi}	Strength reduction factor due to strut buckling in fire limit state
$\chi_{LT,fi}$	Strength reduction factor due to lateral-torsional buckling in fire limit state
ψ	Fire load density factor, end moment ratio
ψ_{fi}	Value of load reduction factor appropriate to fire limit state
$\psi_0/\psi_1/\psi_2$	Load combination factors
Ω	Solution domain
Ω_e	Element domain
ω_k	Mechanical reinforcement ratio

∇	Gradient operator
ΔE_t	Incremental in linear slope of stress–strain curve at time t, incremental in linear slope of strain-hardening parameter at time t
$\Delta H/R$	Activation energy
ΔH_c	Heat of combustion of wood (18,8 J/kg)
Δp	Plastic strain semi axis (Khennane and Baker)
Δt	Time increment
Δt_{cr}	Critical value of time step
Δx	Layer thickness, nodal spacing
Δy	Nodal spacing
$\Delta \varepsilon_{th,c}$	Free thermal strain increment, transient strain increment
$\Delta \varepsilon_{tot,c}$	Total strain increment
$\Delta \theta$	Temperature gradient, temperature increment
$\Delta \theta_{a,t}$	Incremental increase in steel temperature
$\Delta \theta_g$	Increase in furnace (gas) temperature over ambient
$\Delta \theta_{lim}$	Increase in limiting temperature
$\Delta \sigma_c$	Concrete stress increment
$\Delta(.)$	Increment of variable (.)

SUBSCRIPTS

The following subscripts are employed extensively throughout the text:

a	Structural steel (acier)
c	Concrete
fi	Fire
Rd	Resistance
Sd or Ed	Design (applied)
s	Reinforcing steel, general reference to steel
e	Element
el	Elastic
pl	Plastic
–	Hogging moment
+	Sagging moment
d	Design
0	Peak values of stress or strain for concrete
20 or 20°C	Ambient conditions

TEMPERATURE UNITS

The conventional degree Celsius unit for temperature has been used rather than the absolute measure (Kelvin). This generally causes no problem except in heat transfer calculations in which Kelvin must be used for the radiation component. Also, some empirical formulae require the use of Kelvin.

Chapter 1

Fire safety engineering

Before setting the groundwork for the complete subject of fire safety engineering and its influence on the overall planning, design and construction of building structures, it is necessary to attempt to define what is meant by fire safety engineering. There is, as yet, no absolute definition, although the following may be found acceptable:

> *Fire safety engineering* can be defined as the application of scientific and engineering principles to the effects of fire in order to reduce the loss of life and damage to property by quantifying the risks and hazards involved and provide an optimal solution to the application of preventive or protective measures.

The concepts of fire safety engineering may be applied to any situation where fire is a potential hazard. Although this text is mainly concerned with building structures, similar principles are equally applicable to the problems associated with oil and gas installations and other structures such as highway bridges. The additional hazards from gas and oil installations are primarily caused by the far more rapid growth of fire and the associated faster rates of temperature rise. This has been recognized by considering the testing of material responses under heating régimes other than those associated with more conventional cellulosic fires. The design methods used are, however, similar to those for normal cellulosic-based fires.

Most non-building structures face risks of fire damage, but the fact that the risks are extraordinarily low means that such contingencies can normally be ignored. However, in the example of a petrol tanker carrying a highly combustible cargo such as petrol that collides with part of the supporting structure of a highway bridge, the resultant damage from a fire can be large, often necessitating replacement of the original structure (Anon, 1990; Robbins, 1991).

The largest group of structures at risk from fire damage is low-rise domestic housing that generally does not require sophisticated design methods. The spread of smoke and toxic gases that prevents occupants from escaping is a far greater risk than structural collapse (Malhotra, 1987).

Certainly for a long period in the U.K., with the possible exception of World War II from 1939 to 1945, there were few, if any, recorded cases of deaths of occupants in a fire caused directly by the collapse of a structure. There have been unfortunate cases, however, of fire fighters trapped by a collapse of a structure well after the occupants were evacuated. This relatively low incidence of deaths resulting from collapse does not imply that structural integrity (load-bearing response) is unimportant, but is rather a testimony to the soundness of structural design, detailing, and construction. It has already been noted that the common cause of death in a fire is asphyxiation, i.e., being overcome by smoke and gases or being trapped, unable to escape, and exposed to the effects of heat. It is therefore extremely important to consider all the issues that may play a part in ensuring life safety in a fire-affected structure.

1.1 DESIGN CONCERNS

Elements that relate both to life and property safety within the discipline of fire safety engineering can be readily identified. These areas are not mutually exclusive as an action that increases life safety may also increase property safety. The key areas are identified below.

Control of ignition — This involves controlling the flammability of materials within a structure, by maintenance of the structure fabric and finishes or by fire safety management measures such as imposing a ban on smoking or naked flames.

Control of means of escape — This can be achieved either by the imposition of statutory requirements requiring suitable escape facilities or by the education of occupants.

Detection — This covers the installation of methods whereby a fire may be detected, preferably at the earliest possible stage.

Control of spread — Control of the spread of a fire within a building or to adjacent properties may be effected by in-built features (such as compartmentation) or requirement of adequate control between buildings, or by mechanical means such as venting, smoke screens, and sprinklers.

Prevention of structure collapse — This covers load-bearing capacity and structure integrity in whole or in part. Each of these topics can now be considered in greater depth.

1.1.1 Control of ignition

This topic falls into three subheadings; the first two are concerned with spread of flame and the third with management and maintenance of a structure. Ignition can occur through a variety of mechanisms. Generally, these are accidental, e.g., lighted cigarette ends, electrical faults, or overheating of a mechanical or electrical plant. However, deliberate actions or arson cannot be discounted.

1.1.1.1 Control of flammability

There have been too many cases where fire spread rapidly due to the unsuitable nature of the linings of a structure. Any material used in finishes on any part of a structure must limit the spread of flames or flammability. This in general is controlled by the imposition of flammability and fire retardation tests required by national or international standards. In the UK, this issue is covered by the relevant sections of the Fire Test Standard (BS 476: Parts 3, 6 and 7 or the equivalent European standards).

It is also essential to ensure that materials used in a structure reduce fire hazards. It is clearly impractical to insist that the contents of any structure make no contribution to the combustible fire load in a structure, but it is necessary to ensure that the contents present the smallest hazard possible. This means that surface coatings should not be easily ignitable. Furthermore, after several domestic fires involving foam-filled furniture occurred in the UK, the use of certain foams that produce large quantities of highly toxic smoke on ignition is now controlled by legislation.

1.1.1.2 Control of fire growth

One classic means of controlling fire spread is by the use of vertical or horizontal fire compartments. However, these compartments are only satisfactory if there is no possible route for smoke or flame through the compartment boundary. Fire spread can also occur within a room or to a compartment beyond its point of origin if the original fire boundary is incapable of containment because of unsatisfactory closures to the room of origin (Hopkinson, 1984). A more recent case of fire spread attributed to lack of fire stopping after replacement of the original façade was the Torre Windsor Tower in Madrid (Redfern, 2005; Dowling, 2005; Pope, 2006). Additional problems occurred, namely what appeared to be longer than normal response by the fire brigade and the lack of fire protection on steel columns above the 17th floor (Arup, 2005).

An additional problem may arise even if compartment boundaries are satisfactory. The installation of services after the civil (or structural) part of the construction sequence is complete may destroy fire breaks, or replacements of fire breaks may not meet a satisfactory standard. This situation can also arise when subsequent modifications are made, forced either by changes of the use of the structure or by repairs to or replacements of existing services. A further problem can occur from failure to clear away accumulations of combustible rubbish that can either be ignited by fire as at Bradford (Anon, 1985; 1986) or can gradually cause flashover by very slow fire growth, i.e., smouldering (Anon, 1987; 1988).

Such problems can be reduced by ensuring that a fully effective fire safety management policy is in place.

1.1.1.3 Fire safety management

In single occupancy structures, it is relatively easy to set up procedures to ensure that all occupants are aware of proper procedures in the event of a fire and there are suitable people to act as marshals and direct the fire brigade as required. In multiple occupancy situations such as shopping malls where occupancy changes frequently because of a large transient population, setting up procedures is more difficult. It is therefore essential that the owners, often corporate bodies, set up a fire safety management strategy and ensure that a responsible group of persons is on duty at all times to take full control in the case of an outbreak of fire.

Note that this function can be handled by the staff employed for normal day-to-day security provided they are fully and properly trained. It is also essential that full records of the fire detection, fire control, and fire fighting systems are kept and that a full check is made of occupancy to ensure that no action taken will negate any part of those systems. It is essential that where a fire engineering approach to building design is approved and adopted, the measures contained in that design are followed at all times and that financial exigencies are not allowed to compromise fire safety.

1.1.2 Means of escape

There are generally statutory requirements for the provision of escape routing in all except the simplest single-storey structures. Such requirements are based on the concept of the maximum length of escape route to a safe place, whether an external fire door or a protected fire-escape stair well. The maximum lengths are based on type of occupancy and are also dependent on the method of escape, i.e., whether along a corridor or through a fire compartment. For multi-storey structures, it may well be possible to make use of the concept of phased evacuation by which only a reduced number of storeys adjacent to the fire affected zone are cleared initially, with other floors cleared subsequently if needed.

Other requirements relate to the total number of fire escapes, the dimensions of escape routes that are normally functions of the building type, the number of people expected in the building at any one time, and the potential mobility of such persons. The escape routes are sized to allow complete evacuation from the fire compartment into either a protected area or the outside of the structure in some 2,5 min, with a basic travel velocity on staircases of approximately 150 persons per minute per metre width of escape route.

It should, however, be recognized that staircases are built in discrete widths and that doubling staircase width will not double the throughput as an individual person requires finite space. Minimum widths also need to be specified. The above design figures are for able-bodied persons and need modification if it is likely that disabled persons will constitute a segment of building occupancy (Shields, 1993).

The historical background for imposing requirements on escape routes and evacuation is given in Read and Morris (1993). The imposition of requirements followed a series of disastrous fires over some 50 years from 1881, including a theatre fire in Vienna responsible for some 450 deaths and a fire in Coventry in 1931. Much of the background of current legislation in the UK is given in a Ministry of Works Report (1952) based on then-current international practice.

All escape routes must be lined with non-flammable, non-toxic materials. It should be noted that the fire doors opening onto escape routes may have a lower fire resistance performance requirement than the structure because they are required to function in the very early stages of a fire when the major concern is evacuation rather than structural stability. It must be pointed out that fire doors propped open, even by fire extinguishers, are totally ineffective!

It is regrettable that there have been too many cases where the requisite number of escape routes were provided but they were not kept clear. The fire doors at the ends of the escape routes were locked and could not be opened. This occurred at Summerland (Anon, 1973) and the Dublin Stardust disco fire (Anon, 1983).

It is equally important that the occupants of a structure are educated to respond to the warnings of a fire. In domestic situations where the occupants are in a familiar situation, response may be faster than that in an unknown situation. There is still a large amount to be learnt concerning human behaviour in a fire (Canter, 1985; Proulx, 1994). Any warning system must, to use a colloquial phrase, be 'user friendly.' It has still not been determined satisfactorily whether alarm bells or sirens should be implemented by broadcast instructions or graphical displays showing the best manner of exit. It is essential that all escape routes are fully illuminated with self-contained emergency lighting and all signs are also powered by emergency power supplies.

The number of stories, some perhaps apocryphal, about people who totally ignored warnings and continued whatever they were doing before an alarm are legion. One example is a restaurant patron who insisted on continuing to eat the meal that had been paid for in spite of the large quantities of smoke gradually engulfing the diner. Evidence suggests that individuals will continue as long as possible behaving as if the fire did not exist or there were no warnings (Proulx, 1994).

The educational process must also extend to the owners and lessees of any structural complex. This process must form part of the fire safety management policy adopted. For buildings where the occupancy is controlled, part of the educational process can take the form of a fire drill procedure. This, however, must be treated with caution as it is the authors' experience that the more people who know when the drills are to take place, the more likely it is that the drill will be circumvented and its efficacy lost. The authors even noted individuals going in the opposite direction to the flow of evacuees to collect items from offices, and when questioned glibly responded to the effect that "it is only a drill!"

1.1.3 Fire detection and control

In order to ensure life safety through evacuation, it is necessary to ensure that means are available for detection and control of a fire. Control is needed both to reduce the production of smoke to allow more efficient evacuation and to keep temperatures down in the structure to reduce subsequent damage.

1.1.3.1 Fire detection

Systems installed for fire detection may be manual, automatic, or a combination of these.

Manual systems — Manual systems, such as the traditional frangible glass panel that, when broken, automatically sets off the fire alarm system, can be relatively simple. However, they require a human response to realize the existence of a fire and perceive and determine its severity. Thus, such systems may be of only limited use, especially in situations where the presence of individuals cannot be certain.

Automatic systems — These rely either on the monitoring of the presence of excessive amounts of heat or smoke by a sensor that either directly activates a fire-fighting system, such as the fusible head of a sprinkler, or indirectly activates a fire control and evacuation system. Recent developments in automatic systems include the use of low-power lasers and infrared sensors to monitor the presence of smoke.

Many automatic systems rely on combinations of heat and smoke detection sensors. The positioning of either type can be very sensitive to normal ambient conditions and the usage of the building in which they are situated. Kitchens or areas where smoking is allowed can be particularly problematic, although the level of problems formerly associated with such areas has decreased with the advent of computer control.

In all cases, other than for small low-occupancy structures, all detection devices should be linked into a system to indicate either the source of the fire or the point at which the alarm was sounded, to initiate control of the fire by the operation of roller shutter doors to seal off compartments, smoke curtains, or automatic venting systems, and to initiate evacuation procedures and automatically register the outbreak of the fire at the local fire brigade station.

1.1.3.2 Smoke control

It is absolutely essential that during evacuation any build up of smoke is such that evacuees have clear visibility and that the bottom level of smoke is not allowed to fall below a level of about 2,5 to 3 m above floor level during the first 15 min of a fire (Building Research Establishment, 1987; Morgan and Gardner, 1991). There may also be a requirement to keep smoke temperatures below a critical value. The requirement on smoke control is in part due to problems caused by toxic material within smoke and in part due to the totally disorientating effect caused by loss of visibility. In general, either forced venting a fire to control smoke generation will be necessary or, in the early stages, smoke curtains can be used to form reservoirs and contain smoke.

Only in very few cases where natural venting occurs very early in a fire, notably in single-storey construction where roofing material has no specific fire resistance requirement or is designed to fail and collapse early during a fire, does no consideration need to be given to smoke control. However, for warehouses where the contents may emit toxic smoke in the early stages of a fire, such smoke will need to be contained in specific areas before the roof self-vents. Such containment need only be for a sufficient period to ensure full visibility during evacuation. This period is likely to be very short because of the relatively small levels of human occupancy and the general availability of direct level access to fire exits. Smoke control must be employed where compartment volumes are large or escape routes are long.

These problems become much more severe in large open-plan building structures such as shopping malls and atrium structures. The levels of smoke production and the amount of ventilation required can be determined (Morgan and Gardner, 1991; Marshall, 1992; Marshall and Morgan, 1992; Hansell and Morgan, 1994). In such structures, the installation of an automatic smoke venting system that initiates when a fire is detected is a *sine qua non*. Any such venting system must be automatic and can either rely on natural or forced draft ventilation. In both cases, the access points needed by the fire brigade and also the possibility of falling cladding material must be considered. In the case of a forced draft ventilation system, a completely reliable standby power supply must be available.

1.1.3.3 Fire-fighting systems

In sensitive areas, automatic fire-fighting devices initiated either manually or by a fire detection system will be installed. Such automatic devices will vary depending on the type of fire to be expected, but they generally operate by smothering a fire and denying the fire any source of oxygen. Sprinklers effectively act by reducing the temperature of the burning contents. Any fire-fighting system installed as part of the fabric of the structure will be supplemented by the supply of both suitable portable fire extinguishers and possibly by hose reels for local fire fighting.

A large number of structures are also likely to have sprinkler systems installed either at the prompting of an insurance company to reduce property losses or as part of a trade-off between active and passive systems allowed by some regulatory bodies, e.g., England and Wales Building Regulations, Approved Document B (Department of the Environment, 1992a). Such sprinkler systems are operated automatically by the melting of fusible elements or frangible glass in the heads of sprinklers. A drawback of water sprinkler systems is that a substantial amount of damage may be caused on floors other than those where the fire occurs by the seepage of water. A sprinkler system also has the advantage that the amounts of smoke are much reduced giving increased opportunity for evacuation.

In tests carried out after the Woolworth's fire (Anon, 1980; Stirland, 1981), the maximum temperatures at ceiling level with sprinklers would have been 190°C compared to 940°C without sprinklers. The volume of smoke and gases in the first 7 min produced with sprinklers was 1500 m³ and without some 10,000 to 20,000 m³, with only some 10% of the fire load consumed with operative sprinklers compared to the result without them. It was also estimated that with sprinklers an extra minute would have been available for evacuation and that the fire would have been brought under full control in 22 min.

Concern has been expressed in some quarters about the efficiency of sprinkler systems in a fire as there have been cases where they failed to operate. The evidence is not completely clear-cut, Stirland (1981), suggesting that such concern is unnecessary.

There can also be inherent problems caused by the interaction between venting systems and sprinkler systems. The problems identified by Heselden (1984) were caused when water cooled the smoke plume and thereby destroyed its upward buoyancy. A series of tests were carried out as a result (Hinkley and Illingworth, 1990; Hinkley et al., 1992) and design guidance published (Morgan, 1993). The problem caused by the interaction between smoke venting and sprinklers is that the smoke plume does not rise and causes a loss in visibility during evacuation or that the upward velocity due to the vents causes a loss of effect of the water droplets descending from the sprinkler heads. Day (1994) indicates that where both systems are fitted then for storage areas, the sprinklers should operate before smoke vents, but in other areas where evacuation is important, they may operate together.

Unless a fire is small or can be contained within a localized area by in-built fire-fighting systems, it is generally only by the prompt arrival at the scene by the fire brigade that complete evacuation can be checked and control of the fire within the structure and the avoidance of spread to adjacent structures effected. To check evacuation and fight a fire, access must be provided by protected shafts containing either stairs or lifts. It is now generally a statutory requirement to provide adequate access for fire fighting.

1.1.4 Compartmentation

Any large structure should be divided into compartments vertically, horizontally, or both. This requirement is to limit the spread of fire throughout a structure and may also be imposed to allow the phased evacuation of a multi-storey structure (only the floors contained in a fire-affected compartment are evacuated initially, and the remaining floors above and below the fire areas are evacuated later). The rules governing compartmentation are generally vague about why values expressed as maximum floor areas or volumes limiting compartment sizes were selected.

It is probable that most criteria are historically based on long-past experience and may be no longer valid with improved fire-fighting methods (Malhotra, 1993). The issue of fire spread between horizontal compartments due to failure of glazed curtain wall façades needs to be considered. This is partially due to the use of aluminium within such systems. Glazed curtain wall systems need either sprinkler protection or intumescent protection (Morris and Jackman, 2003). It is important to ensure that compartmentation is maintained after repairs, remedial work, or renovations.

1.1.5 Fire spread between structures

There will also be an imposed restriction to limit the spread of fire across boundaries from one structure to another. Limits can be imposed on the lateral spacing of structures, fire resistance requirements for openings in structures, and materials used for cladding (Read, 1991; LPC, ABI, FPA, 2000).

1.1.6 Structure collapse

Obviously, no structure should collapse totally during evacuation or during the fire-fighting phase. Provided the occupants are sufficiently mobile and aware of the situation, evacuation should be relatively fast. Escape routes should be protected staircases or lead directly outside. They are designed to permit complete evacuation of a fire compartment in some 2,5 min. Fire fighting may continue for a substantial period, and thus there should be a sufficiently long period before the structure shows any sign of collapse.

The UK has a statutory requirement to provide safe access to a building to enable the fire services to carry out fire fighting. Collapse of a structure before a given period (conventionally defined as fire resistance) can be avoided by: (1) designing a structure that will be capable of sustaining a reasonable level of applied load for the whole period (passive approach) although weakened and deformable, or (2) designing measures ensuring that a fire will be contained or temperatures will not reach a level that will cause mechanical distress to the structure (active approach).

Unfortunately, the World Trade Center towers collapsed due to fire, and the Pentagon explosion produced a partial collapse in 2001. It should be remembered that in both cases, the fires were deliberately caused when fully laden aircraft were flown into the buildings. In the case of the World Trade Center, current thinking is that the aircraft impact was not a major cause of the collapse. The cause was dislodging of the fire protection on the structure by the impact; the fire load was of a similar magnitude to that of a typical office fire load (Dowling, 2005). The Pentagon suffered only partial collapse because of its *in situ* reinforced concrete frame structure (ISE, 2002; Mlakar et al., 2003).

In reality, a structure is designed to have both approaches operative, although traditionally they were considered separately. Only recently has it been recognized, in the UK at least, that the two systems are interdependent and that one can be used to reduce or modify the needs of the other. This interaction is often referred to as a "trade-off." More information regarding trade-offs is given in BS 9999 (see Section 2.5 for background).

Many of the measures to detect, control or contain a fire within a building are imposed by national or local legislation or imposed by other statutory regulatory bodies. In certain cases, building insurers may impose additional constraints.

1.2 REGULATORY CONTROL

Regulatory control evolved over a long period to protect the public and ensure that a framework can be put into place to ensure that recurrences of disastrous fires are highly unlikely. Such regulatory control can be imposed through national or international standards or by legislation.

In the UK, many of the requirements for structures of different types and occupancies are covered by parts of the Code of Practice for fire safety that dictate the design, management, and use of buildings. BS 9999 covers matters such as fire safety management, compartmentation, fire safety manual, and fire safety training.

Legislative control generally takes the form of national and local building regulations and specific legal requirements. Some degree of control may also be imposed by insurance companies. The current situation in the UK is covered by Read and Morris (1993) to which reference should be made. Guidance is also published by the Department of the Environment (1992b). It has also become apparent that along with completed structures that require consideration to be given to fire safety, we face an increasing need to consider structures under construction or repair.

1.3 FIRE PRECAUTIONS DURING
CONSTRUCTION AND MAINTENANCE

Fire potential during construction (execution) and during maintenance or repair can be inherently more severe than for completed structures since substantial amounts of highly combustible materials may be stored on the site. Some processes involve the application of heat from bare flames. Active or passive fire protection systems may not be completed or operative. The ventilation characteristics of compartments will be different if cladding or walling are not in place. Compartmentation or fire stops may not be complete, and access to certain areas may be hindered by the construction process. It is also necessary to consider the possible requirement of a full security system linked to fire detection measures (Muirhead, 1993).

These problems were highlighted when the severe damage to the Broadgate Centre (Phase 8) (Robbins, 1990), Minster Court (Bishop, 1991), London Underwriting Centre (Rosato, 1992), Expo '92 Pavilion of Discovery (Byrd, 1992a and b) are considered. All these buildings were under construction at the time of fire damage. The problems outlined above for construction become even more important during reconstruction, repair, or renovation of buildings of supreme historical importance such as Windsor Castle (Fowler and Doyle, 1992; Cockcroft, 1993), the Wiener Hofburg (Anon, 1993) or Madrid's Torre Windsor (Pope, 2006). The Windsor Castle situation was exacerbated by the lack of fire stopping in concealed cavities and roof voids.

Fires in buildings undergoing construction or repair are clearly expensive for insurance companies, if indeed the structures are insured. Fire damage led to substantial increases in insurance costs and to the imposition in the UK of a Code of Practice (Building Employers Federation and Loss Prevention Council, 1992). The Department of the Environment issued a similar document relating to Crown works (Department of the Environment, 1991). It should be noted that most of the contents of these documents represent common sense and delineate what should be good site practice. Both documents emphasize the need for fire safety management on every construction site.

Having briefly outlined the areas of fire safety engineering applying to both completed buildings and buildings under construction, it is useful to provide a brief summary before outlining the content of the remainder of this text.

1.4 SUMMARY

This summary classifies the considerations involved in fire safety engineering under the two headings of active and passive provisions (Malhotra, 1986).

1.4.1 Active measures

- Provision of alarm systems
- Provision of smoke control systems
- Provision of in-built fire fighting or fire control systems
- Control of hazardous contents
- Provision of access for external fire-fighting
- Provision of a fire safety management system

1.4.2 Passive measures

- Adequate compartmentation,
- Control of flammability of structure fabric
- Provision of fixed escape routes
- Provision of adequate structural performance

The last item forms the area of concern in this book. The remaining chapters outline the philosophy behind the concepts of structural fire safety engineering, the use of prescriptive methods to satisfy fire resistance requirements, the temperature–time response in a fire compartment, and the basis behind calculation methods to satisfy fire resistance requirements. The next sections deal progressively with the materials data required and the methods of calculation for common structural materials, both as structural elements and complete structures. The final section of the text considers the problems with fire damaged structures.

Chapter 2

Design philosophies

This chapter is concerned with the theoretical justification for the methods that are available to determine the performance of structures or structural elements when subjected to the effects of fire. Much of the material in this chapter is derived from the CIB W14 Workshop Report (1983) that laid down some of the basic principles behind the methods that can be adopted to determine fire resistance. However, before considering the details, it is appropriate to review the concepts of limit state design applied under ambient conditions.

2.1 AMBIENT LIMIT STATE DESIGN

A limit state can be simply defined as the expression of a particular design criterion, e.g., flexural capacity or deflection. When possible, design criteria are considered as a total package. Some are more relevant to determination on the basis of a 'failure' calculation (e.g., flexural capacity). Others (such as deflection) are more relevant to conditions pertaining to the total working or service life of a structure. Thus, the two main categories of limit states are ultimate and serviceability.

The *ultimate limit state* is concerned with the determination of a member or structure capacity at actual or incipient failure. The *serviceability limit state* is concerned with the performance of a structure during its working life time under normal conditions. Other limit states include response to accidental loading or actions.

Serviceability conditions are necessarily checked under the application of working or service loads (actions) on a structure. Service loads are the characteristic loads (actions) multiplied by partial safety factors that can normally be taken as no greater than unity. Thus the failure or ultimate condition must be checked on loading that exceeds the service loads, i.e., load factors greater than unity are applied to the characteristic loads.

Since both loads and material properties are subject to statistical uncertainties in their values, the average value of load or strength properties cannot be used. Instead, characteristic values based on a 5% acceptance limit on loads and a 5% rejection limit on material strengths are used. This means that a design is based on loading that has a 5% probability of being exceeded and strengths that have a 95% chance of being met. Thus, assuming a Gaussian distribution of both load and strength variability, the characteristic loads or strengths may be written in terms of the mean and standard deviation.

For loads:

$$Q_k = Q_{mean} + 1{,}64\sigma_Q \tag{2.1}$$

where Q_k is the characteristic load, Q_{mean} is the mean or average load, σ_Q is the standard deviation of the load, and the 1,64 factor relates to the area under the Gaussian distribution curve to give a 5% limit.

For strengths:

$$f_k = f_{mean} - 1{,}64\sigma_f \tag{2.2}$$

where f_k is the characteristic strength, f_{mean} is the mean strength, and σ_f is the standard deviation.

If the load effect S_d that must be resisted, e.g., bending (flexure), is calculated from the characteristic loading and the resistance effect R_d that must be satisfied, e.g., flexural strength, is calculated from the characteristic strengths, the satisfaction of the ultimate limit state may be written as

$$\gamma_f S_d \leq \frac{R_d}{\gamma_m} \tag{2.3}$$

where γ_f is the partial safety factor applied to the loads or actions, and γ_m is a partial safety factor applied to the strengths. It should be noted that the partial safety factors applied to loads (actions) are generally explicitly stated in design codes. The partial safety factors applied to material strengths will also be explicit except where phenomena can be described only by empirical equations (such as shear in reinforced concrete). Partial safety factors are set to:

- Cover uncertainties of calculation methods in the analyses to determine load effects and resistance effects
- Cover variations other than statistical variations in assumed data
- Ensure that materials in the structure behave in a sensibly linear manner during service conditions and thus ensure that continued removal and application of imposed or variable loading does not cause irreversible deformations
- Yield an acceptable probability against failure

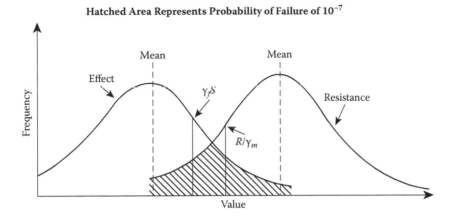

Figure 2.1 Failure envelope for ultimate limit state design.

It is impossible and extremely uneconomic to set partial safety factors to give a zero failure probability. The generally acceptable probability is around 10^{-7} (see Figure 2.1). This low probability of failure and the ensuing choice of partial safety factors also have the effect of ensuring that most designs that satisfy the ultimate limit state will also satisfy most structural service limit states such as deflection.

Where the concern is with accidental loading, there is clearly no requirement to satisfy any service limit states since there is no need to control, for example, deflection. It is only necessary to consider strength. Thus the partial safety factors applied to the loading can essentially be reduced to unity, i.e., to values corresponding to service loading. Equally, the partial safety factors applied to the characteristic strengths can generally also be reduced to unity as there is no concern with straining materials beyond yield or into a nonlinear region. This concept of limit state design can be extended into situations where the effects of fire are concerned as fire can be considered an accidental load.

2.2 FIRE LIMIT STATES

Based on the standard fire or furnace test, a series of failure criteria were identified:

1. The element under test should have sufficient strength (load bearing capacity) to resist the applied loading over the required duration of the test.
2. The temperatures on the unexposed face should be low enough not to cause initiation of combustion of materials stored against that face.

3. There should be no possibility of flame reaching the unexposed face through any weakness or loss of integrity in the construction either inherent in the construction or due to excessive deformation during the test.

These failure criteria can be simply identified as load-bearing capacity (L), insulation (I), and integrity (E). The last item is not amenable to calculation and can be determined only by physical testing and will not therefore be considered further. The other two criteria are capable of assessment by calculation and thus can be expressed mathematically as follows using state functions Z defined as the algebraic difference between calculated response and the minimum required value of that response.

2.2.1 Load-bearing capacity criterion

This can be expressed in one of three ways.
On a time base:

$$Z = t_{fi,d} - t_{fi,requ} \geq 0 \tag{2.4}$$

where $t_{fi,d}$ is the calculated time to failure and $t_{fi,requ}$ is the required time to failure ($t_{fi,requ}$ may be replaced by the equivalent time $t_{e,d}$).
On a strength base:

$$Z = R_{fi,d} - E_{fi,d} \geq 0 \tag{2.5}$$

where $R_{fi,d}$ is the load resistance and $E_{fi,d}$ is the load effect, both evaluated with the application of suitable appropriate partial safety factors over the required time period.
On a temperature base:

$$Z = \theta_{cr,d} - \theta_d \geq 0 \tag{2.6}$$

where $\theta_{cr,d}$ is the critical design temperature and θ_d is the calculated design temperature of the member. This last criterion is applicable only to the load-bearing capacity when a member is exposed to the standard temperature–time curve.

2.2.2 Insulation criterion

This may also be expressed by Equation (2.6) but with the temperatures defined as those on the surface of a member not exposed to the fire.

Table 2.1 Allowable probabilities of failure in fire limit state

Type of occupancy	Load bearing capacity		Insulation and integrity
	Single-storey	*Multi-storey*	
Dwelling	10^{-4}	10^{-6}	10^{-4}
School	10^{-4}	10^{-5}	10^{-4}
Hotel	10^{-6}	10^{-7}	10^{-6}
Hospital	10^{-6}	10^{-7}	10^{-6}
Elderly home	10^{-6}	10^{-7}	10^{-6}
Theatre	10^{-7}	10^{-8}	10^{-7}
Shop	10^{-4}	10^{-5}	10^{-4}
Office	10^{-4}	10^{-5}	10^{-4}
Factory, etc.	10^{-3}	10^{-4}	10^{-3}

Note: The probabilities above are for primary structural members. It may be justifiable to reduce the probabilities for secondary and minor structural elements by factors of 10 and 100, respectively.

Source: Table 6.2/Page 19 of CIB W14 Report (1983) by permission.

2.2.3 Determination of partial safety factors

The partial safety factors used to determine the inequalities of Equations (2.4) to (2.6) must be set to give an acceptable probability of failure. This probability should be a function of the use and size of the structure according to the CIB W14 Report. See Table 2.1.

The partial safety factors are also governed to a certain extent by the fact that fire is seen in an emotive light, in which it is perceived that only a zero risk of fatal human involvement is desirable. This is clearly impossible. It is, therefore, necessary to consider risks that are acceptable to society as a whole. These risks should be functions of the type and usage of the structure under consideration (Rasbash, 1984/5). These partial safety factors can be evaluated by a series of methods of decreasing complexity.

The most complex method is to use a form of Monte Carlo simulation to assess the effect of random variation on all the parameters concerned to evaluate the possible outcomes and the distribution of such outcomes. Whilst this is not a practical proposition for design methods, it is capable of assessing the relative importance of specific variables in a problem and can therefore be used to determine the critical areas of design or for further research (Kordina and Henke, 1987; Purkiss et al., 1989).

The next possible method is a first-order reliability analysis on the mean and standard deviations of the state function Z to determine the reliability index of the state function and set it to an appropriate value. Unfortunately, this method becomes complex unless the state function Z is a simple linear function of two variables. In a fire situation, both the required performance level and the calculated performance level will be nonlinear multivariable functions. Thus, for all practical purposes, the evaluation of the partial safety factors needed in the calculations is set using a certain degree of common sense and the need to link with conventional ambient limit state design in the use of characteristic loads and strengths, for example.

In a full fire safety engineering design approach, structural fire loading is determined as the quasi-permanent portion of a variable load, i.e., the variable load is multiplied by ψ_2 from EN 1990. In the UK, reference should be made to PD 6688-1-2 Table 1, in which values of ψ_2 of 0,3 are taken for domestic, residential and office areas, 0,6 for shopping and congregation areas, and 0,8 for storage.

EN 1991-1-2, however, also allows the load (action) effect required to be resisted in a fire to be set equal to a proportion of the ambient load (action) effect, thus:

$$E_{fi,d} = \eta_{fi} E_d \tag{2.7}$$

where E_d is the design effect at ambient and η_{fi} is the reduction factor, values of which are given in the relevant material design codes. Typically η_{fi} takes values around 0,6 to 0,7 depending upon the type of construction. For fire testing to determine compliance with code data (or to provide code data), the structural load (including self-weight) must be taken as η_{fi} times the total collapse (or ultimate) load, again including self-weight.

It should be noted that the material partial safety factors for use in the fire design sections are generally lower than those specified for accidental damage or actions in the main sections of the design codes. Part of the reason the materials partial safety factors are lower in the fire design case is that an acceptance or rejection limit greater than 5% may be permissible.

All numerical examples in this text have been carried out with the recommended values of partial safety factors or load ratios. It is thus essential that the relevant National Annexes are consulted to verify whether such values have been amended. Having established the mechanisms by which the partial safety factors to be used can be determined, it is necessary to consider the methods whereby a structure or structural element can be assessed when exposed to fire.

2.3 ASSESSMENT MODELS

Three heating and three structural models of increasing complexity are available. They are illustrated in Figure 2.2 (CIP W14 Report, 1983). The assessment method used to evaluate fire performance is related to the heating or temperature exposure model rather than the structural model.

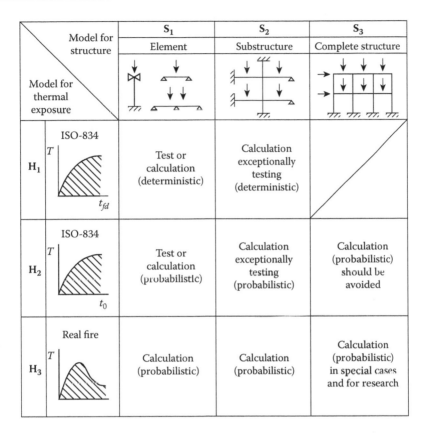

Model for structure	S_1	S_2	S_3
Model for thermal exposure	Element	Substructure	Complete structure
H_1 ISO-834 T ... t_{fd}	Test or calculation (deterministic)	Calculation exceptionally testing (deterministic)	
H_2 ISO-834 T ... t_0	Test or calculation (probabilistic)	Calculation exceptionally testing (probabilistic)	Calculation (probabilistic) should be avoided
H_3 Real fire T ...	Calculation (probabilistic)	Calculation (probabilistic)	Calculation (probabilistic) in special cases and for research

Figure 2.2 Matrix of assessment models for structural fire safety design. (*Source:* CIB W14 Report, 1983. With permission.)

2.3.1 Assessment method: level 1

This method relates the heating model to exposure to the temperature–time relationship generated in a standard fire or furnace test and allows assessment of simple structural elements or subassemblies by test or by calculation with the duration determined from regulations or codes. Historically, this method has been used to determine the fire resistance of structural elements, provide the prescriptive data that have been the cornerstones of most regulatory procedures, and provide the tabular material in various design codes of practice (Chapter 3). This method has also been used to provide a data bank to calibrate some of the available calculation models.

The combination of level 1 heating régime (standard furnace curve) and complete structures (Model type S_3) is not considered by Witteveen (1983) owing to the large discrepancy between the levels of sophistication of the models.

Since the cooling phases in real structures may well be important, as demonstrated in the Cardington tests (see Chapter 12), it is the authors' view that Witteveen's view remains correct.

2.3.2 Assessment method: level 2

At this level, the thermal model demonstrates exposure to the standard furnace curve but the duration of that exposure is determined by the equivalent fire duration time related to the actual fire characteristics of the compartment in which the structural element or subassembly is contained. For a full discussion of equivalent duration of fire exposure, reference should be made to Section 4.5.2.

2.3.3 Assessment method: level 3

In this case, the temperature–time response used in the thermal model is generated from the actual characteristics of the compartment, i.e., its fire load (combustible fuel), the available ventilation sources, and the thermal characteristics of its boundaries. In lieu of a full calculation of such a response, it is permissible to use parametric equations to determine the compartment temperature–time response (Lie, 1974; EN 1991-1-2). The time of exposure is taken as the minimum that causes any of the appropriate limit states of load-bearing capacity or insulation to be no longer satisfied. The relevant failure criteria for various member types are given in Table 2.2.

Although some testing has been carried out with a natural fire exposure, it is intended that this level of assessment should be used only for calculation. The determination of the compartment temperature–time response is dealt with in Chapter 4.

2.3.4 Practical considerations

In general, only levels 1 and 2 will be required for most structures. It is only for innovative or complex structures that recourse is likely to be needed to a

Table 2.1 Design criteria for structural members

Member	Design criteria		
	Load bearing	Insulation	Integrity
Beam	L	I	–
Slab	L	I	E
Column	L	–	–
Wall	L[a]	I	E

[a] A wall requires only insulation and integrity unless it is load bearing.

level 3 assessment. However, in certain circumstances, a level 3 assessment can show the need for a much reduced level of fire protection as the temperatures reached and the actions induced in the fire will not be sufficient to cause loss of load-bearing capacity.

Although EN 1991-1-2 covering the calculation of actions during fire exposure allows any of the three determination levels to be used, most of the codes for calculation of member resistances to applied actions appear only to allow calculation using level 1 or level 2 assessment using equivalent fire durations. The reason is that many of the design equations are related implicitly to the standard furnace temperature–time curve since most data for calibrating such equations were derived from furnace tests. Thus many of the resulting design aids available also correspond to level 1 assessments. This may not be always apparent in the design code.

2.4 APPLICABILITY OF ASSESSMENT LEVELS

Because most regulatory authorities use standard gradings to determine the required fire resistance of a building, structural fire resistance has been based on standard periods or gradings of fire resistance of 30, 60, 90, 120 and 240 min determined from element tests. For ease of application, these gradings have traditionally been related solely to the size of a structure (height, floor area or volume as appropriate) and to its usage (and by implication the combustible fire load) (Ministry of Works, 1946). These parameters were expressed in terms of broad-based categories.

Until recently, no account was generally taken of the existence of automatic active fire protection methods such as sprinkler systems in the determination of the required fire grading (Section 2.5). It should also be pointed out that the required fire gradings in the past have often been based on obsolete data from buildings with outmoded forms of construction. Response in changing these gradings to take account of new methods or materials of construction was often slow. To a certain extent, this innate conservatism of approach can be explained as the emotive response to fire and its effects rather than a rational response. However, this conservatism may hold back development.

Gradually, some countries are coming to terms with the need to consider a more rational approach in determining the required duration of fire resistance for a structure, either by allowing a limited calculation using risk analysis (Kersken-Bradley, 1986) or by acknowledging and specifying active measures as recently occurred in England and Wales (Department of the Environment, 1992a).

Regulatory bodies are moving away from a prescriptive approach to fire resistance performance by telling designers what must be done to provide a specific method of construction with its required fire resistance to a functional approach in which designers are told what must be achieved rather than how. The concept of fire safety engineering is now recognized as a legitimate design tool.

Thus a much more flexible approach in terms of the use of calculation methods on individual elements or the whole structure is available. Under certain circumstances, a designer can adopt a more flexible approach on the use of heating régimes other than the standard furnace curve. It is to be emphasized that a large number of uncertainties surround the application of full calculation methods of which the designer needs to be aware (Purkiss, 1988). However, to a certain extent, these uncertainties are covered by partial safety factors and the inherent conservatism in the design values used for various parameters needed in such calculations.

However, this freedom to use a full level 3 assessment is unlikely to be available for most structures. It may be possible to determine the fire load and ventilation characteristics for a structure as designed and built. However, any change of use during the life of the structure may alter both the fire load and ventilation giving totally different compartment temperature–time responses, and thus a totally different need for fire protection or fire design. It is possible that such a change of use may need not to be notified to any regulatory authorities, or worse, authorities may fail to be notified. Thus the need for change to a fire design may not be realized. This problem may be mitigated by the existence of a proper fire safety management policy.

However, alterations to the fire safety design imposed by change of use may be prohibitively expensive. It is therefore only really possible to design the structure for either exposure to the standard temperature–time curve on a time equivalent basis or to a natural fire response where a structure's use cannot change during its life. Examples of such structures are bus and railway stations, sports stadia, airport terminals, and large shopping complexes (Kirby, 1986; Kirby, undated).

It also should be pointed out that it may not be economic for some structures to consider calculation methods even when the standard furnace curve is used as the thermal exposure model. This will be especially true when the required fire resistance period is low, say 30 min. For reinforced concrete, for example, the cover required to satisfy durability will be far higher than that required to provide adequate fire performance.

2.5 INTERACTION BETWEEN ACTIVE AND PASSIVE MEASURES

Traditionally, structural fire performance has always been achieved by passive measures such as the provision of adequate cover to reinforcing steels or a given thickness of protection on a steel beam to meet a specified regulatory fire resistance. Equally, the insurers of structures have either insisted on or offered financial inducements via reduced insurance costs for the installation of active fire prevention measures commonly in the form of sprinkler systems.

The insurer's motive is clearly to reduce the amount of damage caused to the property and contents in the case of a fire. If an active system successfully controls a fire by either preventing its spread or controlling its temperature to such an extent as to keep temperatures below those needed to cause structural damage, the question can be raised whether the provision of such active measures should not be allowed to contribute to the overall fire resistance requirements. This argument led to discussions about possible trade-offs between the two approaches as to the possibility that active measures can lead to a reduction of the passive requirements (Stirland, 1981; Read, 1985).

The solution to this problem is essentially statistical and has been addressed by Baldwin and Thomas (1973) and Baldwin (1975) who demonstrated that, depending on relative costs and an assessment of the probability of failure of a sprinkler system, the fire resistance grading required for an office block could be reduced by about 1 hr if sprinklers are fitted over that period required were no sprinklers to be fitted.

It is thus seen that there is a benefit from the trade-off although the benefit gained may not be fully utilized, i.e., with a fully operative sprinkler system, a fire may well be contained or indeed extinguished before any structural damage occurs. There is, however, still the need to consider the possibility that a sprinkler system will be inoperative when a fire occurs. Thus, until further guidance becomes available, it may well be prudent only to consider the trade-off in terms of reduced fire resistance periods, recognizing that one of the prime functions of fire resistance is to allow evacuation to proceed safely.

Kirby et al. (2004) present a risk-based determination of standard fire periods. They indicate that the risk is the product of the frequency , probability, and consequences of an occurrence. The frequency and consequence may be taken as proportional to the building height and the probability obtained from a Monte Carlo simulation considering fractile values of fire load density, a range of compartment heights, ventilation height as a fraction of compartment height, a range of floor areas and a range of ventilation areas as a proportion of floor area.

The effect of sprinklers was considered by factoring the fire load density by 0,61. The results were standardized against 80% fractile fire load density for a building 18 m high, yielding a risk of 64,8. This allowed a complete set of standard fire exposure data to be derived, and the data are presented in Table 24 of BS 9999. It should noted that a comparison of Approved Document B and Table 24 indicates very similar values. Two points are, however, worth making: in the first place, Table 24 is more extensive than Approved Document B, and second, for certain low risk buildings it may be possible to reduce the standard fire period to 15 min for life safety purposes only.

Having outlined the basic philosophy behind structural fire safety engineering design, it is still necessary to consider the role that traditional test methods and the subsequent prescriptive approach to fire resistance design have to play.

Chapter 3

Prescriptive approach

Chapter 2 presented the rationale behind the concept of calculation methods applied to the design of structures to resist the effects of fire. These methods use an extension of the limit state approach that was originally formulated to deal with structural design at ambient conditions to give a structure as a whole a uniform factor of safety against collapse.

However, such calculation methods are not currently in a state that they can be applied to all structures. Also, such calculation methods can be too cumbersome to be used on simple structures where, for example, low periods of fire resistance are required. It should be noted that calculation methods may give solutions that are unacceptable for other design criteria, i.e., calculations for reinforced concrete may produce covers to the reinforcement that are below those required to satisfy durability requirements.

It is therefore necessary to consider methods that can be used where the calculation approach cannot be justified due to the nature of a problem or where no appropriate methods exist. These methods are essentially prescriptive in that the designer is told what parameter values to use rather than being able to calculate these values. It should be noted that the prescriptive approach historically preceded the calculation approach. The data used in the prescriptive approach are obtained by interpreting results from the standard fire test. For a history of the development of the standard fire test, reference should be made to Malhotra (1982a, 1994) or Babrauskas and Williamson (1978a, b).

3.1 STANDARD FIRE TEST

The basic principle of the standard fire test, that may more properly be known as the *standard furnace test*, is that a structural element is loaded to produce the same stresses that would be induced in that element when in place in the structure of which is considered a representative part. The element is then heated under load with the measured temperature régime in the furnace following a prescribed temperature–time relationship until failure of the element occurs. Traditionally, beams and slabs are heated from beneath, while columns are heated on all four sides and walls heated from one side only (see Figure 3.1).

The standard furnace test is regulated on an international basis by ISO 834 (1975) which has been subject to subsequent amendments. National bases tend to be in essential agreement with the international standard, although there may be slight variations in test details. In the UK, the current standard was BS 476 Parts 20-22 (1987), although it should be noted that much of the data contained within British Standard Codes of Practice and other relevant statutory standards were obtained from tests carried out to an earlier version of the standard, BS 476 Part 8 (1972). BS 476 has now been replaced by BS EN 1363. The remainder of this section utilizes the requirements of BS EN 1363 and BS EN 13501.

Traditionally, most building structure fires have been considered to occur with the bulk of the combustable material taken as cellulosic and the resultant standard furnace temperature–time curve established on this basis. For such fire tests, the temperature-time curve specified for the furnace is

$$\theta_g = 20 + 345 \log (8t + 1) \tag{3.1}$$

where θ_g is the furnace temperature (°C) and t is the time (min). Equation (3.1) is plotted in Figure 3.2. The standard curve gives temperatures of 842°C at 30 min, 945°C at 60 min, and 1049°C at 120 min. Whilst Equation (3.1) is mathematically concise, it is not ideal for calculating analytically explicit solutions to heat transfer equations (Section 5.1) when an element is exposed to the standard furnace curve on one or more boundaries. Thus several alternative expressions were derived including those of Williams-Leir (1973) and Fackler (1959).

Williams-Leir:

$$\theta_g = \theta_0 + 532(1 - e^{-0,01t}) - 186(1 - e^{-0,05t}) + 820(1 - e^{-0,2t}) \tag{3.2}$$

or

$$\theta_g = \theta_0 + 186 \, [2,86(1 - e^{-0,01t}) - ((1 - e^{-0,05t}) + 4,41 \, (1 - e^{-0,20t})] \tag{3.3}$$

where t is in minutes and θ_0 is the ambient temperature.

Fackler:

$$\theta_g = \theta_0 + 774(1 - e^{-0,49\sqrt{t}}) + 22,2\sqrt{t} \tag{3.4}$$

It should be further noted that whereas BS EN 1363 considers fire temperature to be related to a base temperature of 20°C and thus needs to set

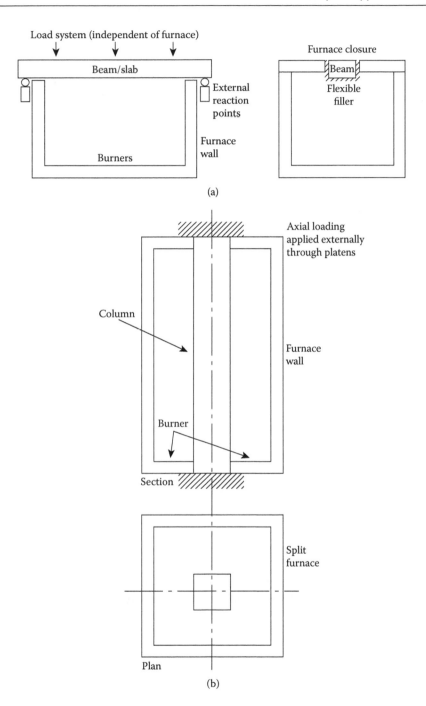

Figure 3.1 Standard furnace test layout: (a) beam and (b) column.

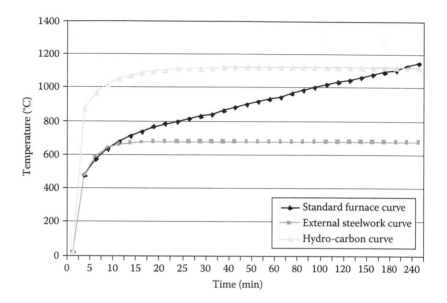

Figure 3.2 Standard temperature–time responses.

resultant limits on the variations in ambient conditions that are acceptable during a fire test, ISO 834 (1975) relates fire temperature to a measured ambient temperature θ_0:

$$\theta_g = \theta_0 + 345 \log (8t + 1) \tag{3.5}$$

A revision of ISO 834 was proposed to bring it into line with BS 476 by adopting a fixed base temperature of 20°C. That draft revision also gave an alternative exponential form of the temperature time curve as

$$\theta_g = 20 + 1325 \left(1 - 0,325e^{-0,2t} - 0204e^{-1,7t} - 0,471e^{-19t}\right) \tag{3.6}$$

This equation corresponds to the parametric curve for natural fire exposure given in EN 1992-1-2 with the parameter Γ defining the degree of exposure set equal to unity, i.e., $t^* = t$. The temperature–time relationships given by Equations (3.1) to (3.6) relate to members exposed to the full effect of a standard fire. This exposure is considered too severe for load-carrying members in a façade external to a structure, and the following temperature–time relationship has been proposed (EN 1991-1-2):

$$\theta_g = 20 + 660 \left(1 - 0,687e^{-0,32t} - 0,313e^{-3,8t}\right) \tag{3.7}$$

Equation (3.7) is also plotted in Figure 3.2.

With the advent of an increasing use of petrochemicals and plastics, it was recognized that the cellulosic curve alone was no longer satisfactory since the temperature rise in a petrochemical fire is much faster. Thus an additional curve was required. Where an element is exposed in an environment of which the predominant constituent is of petrochemical origin, a more appropriate temperature–time curve is given by:

$$\theta_g = 20 + 1100\,[1 - 0,325e^{-0,1667t} - 0,204e^{-1,417t} - 0,471e^{-15,833t}] \qquad (3.8)$$

The hydrocarbon curve of Equation (3.8) is designed for a very rapid temperature rise. In 10 min, the temperature reaches 1052°C, with the limiting temperature of 1120°C effectively reached at 40 min. Equation (3.8) is also plotted in Figure 3.2. Varley and Both (1999) suggested that the standard hydrocarbon curve is not severe enough to represent fires in tunnels and proposed either the RABT curve that rises to around 1200°C in 10 min and remains constant until 60 min before a linear cooling phase or the Rijkswaterstaat curve that rises to 1350°C at 60 min before cooling to 1200°C at 120 min.

The standard furnace test, whether conducted under the temperature–time régime imposed by the cellulosic or hydrocarbon curve continues until failure occurs from meeting any one of the following criteria (or limit states):

Insulation (I) — The average temperature on an unexposed face achieves a temperature of 140°C or a local value exceeds 180°C.

Integrity (E) — Cracks or openings occur in a separating element such that ignition can occur on the unexposed face.

Load-bearing capacity (R) — The element tested loses load-bearing capacity when it can no longer carry the applied loading. In practice, however, deflection limits are imposed, partly because large deflections occur at collapse from the formation of plastic hinges in beams or slabs or due to incipient buckling in walls or columns, and partly to avoid the specimen collapsing into the furnace and damaging the furnace and loading system.

For any members such limits should not be applied until the deflection reaches $L/30$. For (1) flexural members, limiting deflection is $L^2/400d$ (mm) or rate of deflection is $L^2/9000d$ (mm/min) where d is depth of the member and L is the span, both in mm, and for (2) vertically loaded members, limiting vertical contraction is $h/100$ (mm) or rate of contraction is $3h/1000$ (mm/min), where h is the initial height of the member (mm).

It must be stressed that the deformation or rate of deformation limits apply *only* to performance in a standard furnace test and *not* to elements within a structure.

The result from the fire test is quoted in time units of minutes when each limiting criterion (R, E, or I) appropriate to the element of the construction tested is reached. The final test grading is then expressed as the least time for any of the criteria rounded down to the nearest appropriate classification, i.e., 30, 60, 90, 120, 180 or 240 min. A fire test may be carried out for one of three reasons:

1. To determine the fire resistance grading for a given method of construction to enable the method to be accepted by regulatory bodies.
2. To assist in the development of new products or methods because initially it is more appropriate to accept results from tests rather than from calculations especially where calculations may need test results to justify the assumptions used therein.
3. To enable research on the influence of specific variables on a form of construction to provide a better understanding of the performance of structural elements or materials.

There can be substantial problems with such testing, especially where phenomena such as spalling occur (Purkiss et al. 1996; Van Herberhen and Van Damme, 1983). This was also borne out in a set of tests reported by Aldea et al. (1997) on reinforced concrete columns. The results are given in Table 3.1. The columns were 290 mm square and 2,10 m long. The actual concrete strengths are not quoted although the loading in all cases was 50% of the design load.

The highest strength columns did not contain polypropylene fibres. It is quite clear from these results that no correlation between steel area and fire endurance was noted for the C20 and C50 concrete columns. The high performance concrete column behaviour was dominated by spalling that prevented the columns reaching their full endurance period.

In recent years, the use of the standard furnace test to determine parametric variations has tended to become redundant with the advent of computer simulation. It is, however, this use of the fire test that provided much of the tabular data currently in design codes such as EN 1992-1-2 and

Table 3.1 Fire resistance column test results

Concrete	Reinforcement	
Grade	8H12 (1,08%)	4H25 (2,33%)
C20	3 hr 54 min	3 hr 13 min
C50	2 hr 32 min	3 hr 29 min
C90	1 hr 46 min[a]	1 hr 29 min[b]
—		

[a] Spalling started 8 min into test.
[b] Spalling started 12 min into test.
Source: Aldea, C.M. et al. (1997). In Phan, L.T. et al., Eds., *Proceedings of International Workshop on Fire Performance of High-Strength Concrete*, pp. 109–124. With permission.

similar documents such as the FIP/CEB Report (1978) for concrete construction. Although the fire test provided very useful amounts of data, the test has a series of inherent drawbacks stemming in part from the nature of the test and the test uses.

3.2 DRAWBACKS OF FIRE TEST

3.2.1 Expense

The fire test, especially one requiring a full 4-hr rating, is very expensive both in terms of specimen preparation and the cost of the actual test. It should also be noted that the data obtained from a particular test are applicable only to that test. If a test does not yield the required result (e.g., the desired fire resistance period was 60 min and the test showed 59 min fire resistance, its classified result would be 30 min), a new test with modifications would be necessary to establish the desired rating.

3.2.2 Specimen limitations

As test furnaces are restricted in size, it is generally impossible to test large elements of construction. As a result, representative specimens are tested. Restrictions (in the case of columns, a maximum height around 3 m and for beams and slabs, a maximum span of 4 m and maximum width of around 4 m) mean that it is difficult to test realistic multi-span beams or slender columns because any reduction in member size to achieve the required slenderness will give erroneous results. It is generally difficult to apply scaling to fire test results.

The availability or design of furnaces can also restrict the type of test that can be carried out. For example, the most critical situation in testing columns is likely to occur when a column is subjected to moments due to the applied loading and the thermal gradients induced by non-symmetric heating. However, most available column test furnaces can only heat columns on all four sides whilst the columns are under the influence of axial loads only.

3.2.3 Effects of restraint and continuity

Based on both the limitations of the loading arrangements in most furnaces and on specimen size noted above, it is generally only possible to test specimens with idealized end conditions, i.e., beams and slabs with simply supported ends and columns with some indeterminate degree of fixity arising from the loading platens of the test rig. Clearly, structural elements do not exist in isolation and, in practice, if a fire is restricted to part of a structure, some load redistribution will occur away from the fire affected zones to zones unaffected by the fire. This redistribution will generally enhance the fire performance of a member (see Figure 3.3).

This has been shown by tests that examined the effects of restraint (Selvaggio and Carlson, 1963; Ashton, 1966; Bahrends, 1966; Ehm and von Postel, 1966; Issen et al., 1970; Lawson, 1985; Cooke and Latham, 1987). For example, Ashton indicates that whilst restraint may not significantly alter load carrying capacity, deflections will be substantially reduced. It should be noted that in the tests reported by Ashton, the restraint applied was higher than that likely to apply in practice. Research undertaken in the US reported that a moderate degree of restraint would enhance the fire performance of a prestressed concrete slab; a high degree of restraint yielded fire performance similar to a situation where no restraint existed (Selvaggio and Carlson, 1963; Issen et al. 1970).

Restraint of columns will produce two effects. The first is to lessen potential losses in strength and elasticity from additional forces induced by resistance to the free thermal expansion. The second is that the failure mode of a column may change from a sudden buckling collapse to a mode in which progressively larger deformations occur when load capacity is limited by strain capacity (Dougill, 1966). Dougill continues by questioning the relevance of the result on a single column in a furnace test to its performance in a stiff structure. Dougill's analysis was adumbrated before the existence of transient strain was known. However, if the numerical stress–strain curves derived by Li and Purkiss (2005) are used instead of simple temperature-reduced stress-strain curves, Dougill's analysis remains valid.

3.2.4 Confidentiality of results

While confidentiality of any result is absolutely essential from a manufacturer's view, confidentiality also means that any data obtained from such tests are unavailable for the purposes of research and therefore cannot be used to acquire additional data for improved calibration of calculations on fire performance.

3.2.5 Loading

For routine fire testing, the specified loading is that causing the same stresses in the members to which they would be subjected under normal service or working loads, i.e., they are loaded to a value of η_{fi} times the ultimate load. The fire testing of members under service loading led to the observation that when concrete beams or slabs were tested as simply supported, they were unable to continue to carry the applied loadings when the reinforcing or prestressing steel reached a temperature around 500 to 550°C. A similar observation was made for steel beams. Thus, the concept of a critical failure temperature of steel, whether reinforcing or structural, was adumbrated.

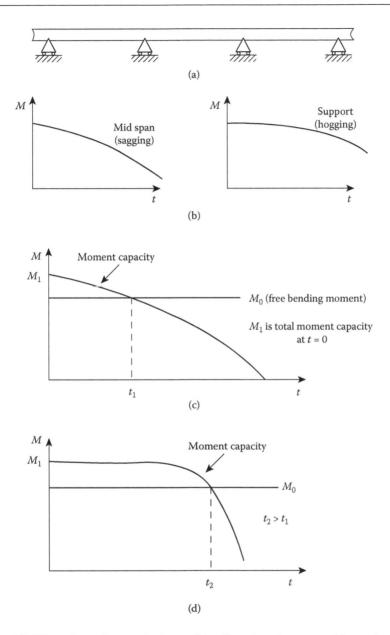

Figure 3.3 Effect of member continuity on fire-affected performance of flexural members: (a) basic structure; (b) decrease of moment capacity with time; (c) no hogging moment; and (d) hogging and sagging.

Early work using fire testing to provide data on the performances of reinforced and prestressed concrete structural elements under loading (other than service loading) generally concentrated on loading exceeding the service load to determine the effect of overload (Thomas and Webster, 1953; Ashton and Bate, 1960). Only recently have the effects of load levels lower than service loads been considered (Robinson and Latham, 1986; Robinson and Walker, 1987). The latter work finally led to the abandonment of the concept of a fixed critical temperature for structural steelwork (Section 8.2.1).

More recent tests reported loading less than the equivalent full service loading and possibly under a parametric fire curve rather than the standard furnace curve (Bailey and Lennon, 2008; Kelly and Purkiss, 2008; Bailey and Ellobody, 2009; Nadjai et al., 2011). The use of reduced loading and/or parametric fire curves makes the application of the results nonuniversal. Under parametric fire curves, authors may state a time equivalence to standard exposure, but such statements should be treated with a degree of respect due to uncertainties in the relations determining time equivalence.

3.2.6 Failure modes

The standard furnace test is capable only of studying the failure mode of a single structural element and cannot therefore be used to investigate failure modes of complete structures. It has been observed that failures of columns in concrete structures may result from shear caused by the expansion of the structure (Malhotra, 1978; Beitel and Iwankiw, 2005).

3.2.7 Reproducibility

There are two areas of concern here. The first is that replicate tests are rarely undertaken, partly since the relevant test standards require only a single test and partly due to expense. Dortreppe et al. (1995) report the results on a pair of identical reinforced concrete columns tested in the same furnace; one revealed a fire endurance of 84 min and the other 138 min. The authors indicated that a computer analysis predicted 85 min. It is not clear why the anomaly occurred, but the results indicate the possible dangers of relying on single tests to establish code data. The use of single tests is rarely condoned in other fields of structural testing the effect is to an extent mitigated by the use of a classification scheme for expressing the results.

The second area of concern is reproducibility of results from different test furnaces. To assess this effect, a series of tests on steel rolled hollow sections was carried out (Witteveen and Twilt, 1981/2). The results are presented in Figure 3.4 and indicate substantial differences in behaviour. It is likely that were such a test series carried out now, the differences between furnaces would be smaller.

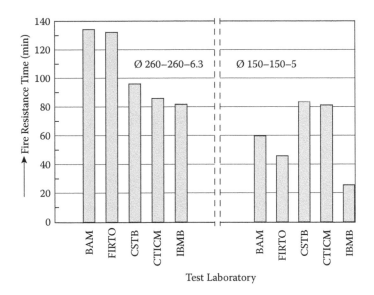

Figure 3.4 Effect of furnace characteristics on fire test results. (*Source:* Witteveen, J. and Twilt, L. (1981/2). *Fire Safety Journal,* 4, 259–270. With permission.).

Some of these differences are due to the loading arrangements, in that it is very difficult to provide a pure pin end or a pure fixed end at the loading platens. The actual condition is likely to be somewhere between the two and somewhat variable depending on the load applied and the construction of the loading system. Loading will vary from furnace to furnace even though most furnaces use hydraulic load applications.

Another difference arises because the test specifies only the temperature of the furnace at any time and not heat flux falling on the specimen. The rate of temperature rise of the specimen will depend on the incident heat flux that will be a function of the thermal characteristics of the furnace based on its thermal inertia (Harmathy, 1969). The rate of temperature rise also depends on the method of firing the furnace [positions of burners and type of fuel (oil or gas)] because these factors will affect the emissivity of the surface of the member. The results of measurements on three U.S. furnaces indicating differences between measured heat fluxes are presented in Figure 3.5 (Castle, 1974). All three furnaces were operated to the standard furnace curve and were within the tolerance limits of the relevant test codes.

With these differences it is perhaps not surprising that variations in test results occur among furnaces. To help mitigate these differences BS EN 1363-1 specifies that furnace linings shall consist of materials with densities less than 1000 kg/m³ with a minimum thickness of 50 mm comprising at least 70% of the internally exposed surface.

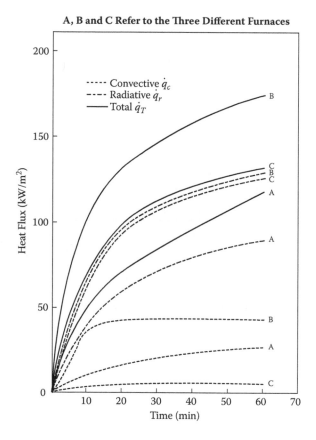

A, B and C Refer to the Three Different Furnaces

Figure 3.5 Variability of heat fluxes of different furnaces. (*Source:* Castle, G.K. (1974). *Journal of Fire and Flammability*, 5, 203–222. With permission.).

In spite of these drawbacks, the furnace test has provided a substantial data bank of results on which to base the prescriptive method of determining the fire resistances of structures. The furnace test still provides a convenient and necessary method of comparing the performances of different types of construction and also provides data where calculation methods are not suitable or possible.

3.3 PRESCRIPTIVE DETERMINATION OF FIRE RESISTANCE

The prescriptive approach to determining the fire resistance of a structure or, more correctly, the assemblage of the individual elements comprising the structure, can be defined in the flow diagram in Figure 3.6. The elements

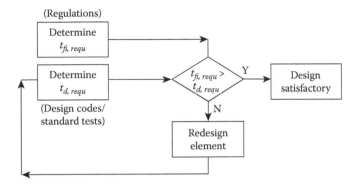

Figure 3.6 Prescriptive approach for determining fire resistance of structural elements.

are detailed to demonstrate the fire resistance required by the appropriate regulatory guidelines. It should be noted that this method is still permissible under most design codes including the Eurocodes.

The prescriptive method is essentially very quick but is not likely to be economical. Calculations may show that thinner members with lower covers (e.g., concrete) or little or no fire protection (e.g., structural steelwork) can provide requisite fire performance. In order to provide examples of the prescriptive method, it is convenient to consider each of the construction materials separately.

3.3.1 Concrete

The two main variables to consider are the specification of minimum overall dimensions of members and the minimum axis distance to the main reinforcement. The minimum overall dimensions are specified for walls and slabs to keep the temperature on the unexposed faces below the insulation limit of 140°C or to ensure that spalling will not be severe enough to cause the web of a beam or rib in a slab or column to lose enough concrete to prevent it from carrying its design loading. The minimum axis distance is specified to keep the temperature of the main reinforcement—the bottom flexural steel in a beam or slab or the vertical compression reinforcement in a column or wall—below a critical value generally considered around 500 to 550°C.

Around this temperature, the strength of reinforcement drops to a value equal to the stresses induced by service loading (the loading generally applied in fire testing). The values of member dimensions and axis distance depend on the type of aggregate in the concrete (siliceous or calcareous in normal weight concrete or light weight concrete) and on the fire resistance period.

For columns, the effects of load level and effective height on these dimensions are also considered. EN 1992-1-2 presents two tabular

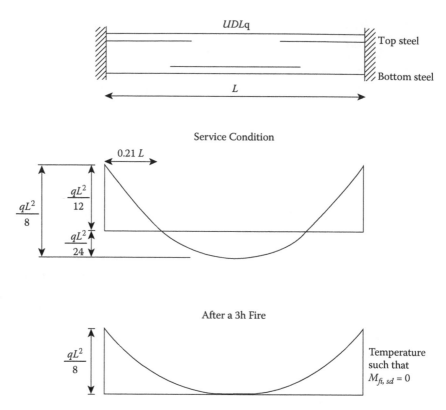

Figure 3.7 Redistribution of moments from midspan to support in a fire.

methods. The tabular data for Method A (Dortreppe and Franssen, undated) are to be used only when a value of $\alpha_{cc} = 1,0$ is adopted. Thus, in the UK where a value of $\alpha_{cc} = 0,85$ has been adopted, the formulae attached to Method A must be used. Method B is derived from the work of Dortreppe et al. (1995) and operates in terms of a mechanical reinforcement ratio ω_k. For beams and slabs, some allowance may be made for continuous members in comparison with simply supported members by a slight reduction in axis distances and overall depth values in recognition of the degree of redistribution of moments away from areas of sagging moment to areas of hogging moment during a fire (see Figure 3.7).

Additional requirements where spalling is considered critical may include the provision of supplementary reinforcement in the form of light mesh in the concrete or the use of polypropylene fibres where axis distances exceed certain values or the concrete is high strength or self-compacting.

As indicated in the previous paragraph, axis distances to reinforcement rather than covers are specified. This is actually more scientifically correct because computer-based heat transfer calculations revealed that the temperatures at the centres of reinforcing bars are identical to those at the same position in plain concrete (Ehm, 1967). Becker et al. (1974) indicated that this is correct only for a reinforcement area less than 4% of the gross section. Where the main reinforcement consists of more than one layer, the effective axis distance is used to determine the fire resistance requirement.

3.3.2 Structural steelwork

It is far more difficult to employ a prescriptive approach to steelwork. One reason is the diversity of protection systems available and another is the fact that a calculation approach for steelwork is much simpler and far less restrictive than that for concrete. Because of the diversity of systems available, prescriptive data can only apply to generic categories and not to specific products.

In addition, fire performance is generally determined on a single-sized member (406 × 178 × 60 UB for beams and 203 × 203 × 52 UC for columns in the UK) rather than a range of member sizes (Morris et al., 1988). Thus the results should be applied strictly only to members of that size. If applied to alternative sizes, the results will be conservative when applied to larger sections and unsafe for smaller section. Also, deflections (or surface strains) may be higher in an actual design over and above those revealed in a representative test. This could create problems with the stickability of fire protection material.

The situation has been partially remedied by manufacturers that provide data for their particular protection systems that allow the thickness of a protection system to be related to the section size and required fire resistance. In the UK, a convenient compendium of such data is *The Red Book* (2012) produced by the Loss Prevention Certification Board. It should be noted that current data in the compendium are based on steel 'failure' temperatures of 550°C for columns and 620°C for beams. Therefore these data make no allowance for the load intensity applied to a member.

The difference between the two failure temperatures is that board and spray systems are generally tested on columns that are noted to fail when a section attains a mean temperature of 550°C, when tested under full-service loading. Intumescent paint systems are generally tested on beams which, when loaded to full service, fail at a slightly higher temperature because of the thermal gradient induced by the heat sink effects of non-composite cover slabs on the top flanges of beams.

3.3.3 Masonry

Since masonry is generally used non-structurally and thus its vertical load-carrying capacity is not used, the essential requirement is for insulation to keep the temperature on the unexposed face below 140°C. This means that it is relatively simple to use a prescriptive approach. Tables of required wall thicknesses to achieve a required fire resistance are given in EN 1996-1-2.

3.3.4 Timber

Since timber is rarely used in situations requiring more than notional fire resistance, few data are available for a prescriptive approach, although some experimental data exist on the fire performance of stud partition walls (Meyer-Ottens, 1969). Where exposed timber is required to provide fire resistance, it is relatively simple to perform the necessary calculations since the core of the timber member inside the charred zone is little affected by temperature and normal ambient design methods may then be used.

Before considering the possible methods of determining the performances of structures or structural elements in a fire by calculation, it is useful first to consider the character of fires and the relationships of the characteristics to the standard fire test.

Chapter 4

Behaviour of natural fires

Unlike the temperature–time response in a furnace test that is imposed by the standard to which the test is carried out, the temperature–time response in a fire compartment is a function of compartment size, type of compartment, available combustible material, and air supply available for combustion. This situation is often designated a *natural* or *real fire* (the former term is preferred here, although EN 1991-1-2 uses *parametric fire*), and the heat it generates may be calculated from basic principles. However, attempts have been made to represent the solution to the natural compartment temperature–time response by empirical curves; one such is given in EN 1991-1-2.

The temperature–time curve generated from the standard furnace test is often called a pseudo fire as the temperature characteristic cannot be generated from basic principles. It is, however, possible to link the two types of responses either by the concepts of fire severity and equivalent fire period or through parametric curves.

4.1 DEVELOPMENT OF COMPARTMENT FIRES

The development of compartment fires can be broken down into three phases: pre-flashover (also known as growth period); post-flashover (fully developed fire); and the decay period (Figure 4.1).

4.1.1 Pre-flashover period

In the pre-flashover period, combustion is restricted to small areas of the compartment; therefore, only localized rises in temperature can occur. It should be noted that such rises may be substantial. The overall or average rise in temperature within a bounded fire compartment will be very small and indeed at this stage there may be no obvious signs of a fire.

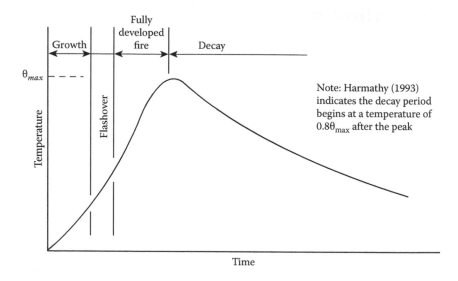

Figure 4.1 Phases of fully developed fire.

A large number of incipient fires never get beyond the stage of pre-flashover because of insufficient fire load or air supply (ventilation) to allow the fire to grow beyond pre-flashover. In many cases, human inter-vention causes flashover, for example, by opening a door or window and thereby suddenly increasing the air supply. The pre-flashover stage is often ignored in the calculations of the compartment temperature–time response since the overall effect on the compartment is small even though the pre-flashover period can be long compared to the subsequent stages of a fire.

Flashover occurs when fire ceases to be a local phenomenon within the compartment and spreads to all the available fuel within the compartment. Propagation of flames through any unburnt gases and vapours collected at ceiling level then ensues.

4.1.2 Post-flashover period

In this period, the rate of temperature rise throughout the compartment is high as the rate of heat release within the compartment reaches a peak. In compartment fires, maximum temperatures of over 1000°C are pos-sible. The rate of temperature rise continues until the rate of generation of volatiles from the fuel bed begins to decrease as the rate of fuel con-sumption decreases or when insufficient heat is available to generate such volatiles.

During the post-flashover or growth period, structural elements are exposed to the worst effects of a fire, and collapse or loss of integrity is likely. Once the rate of temperature rise reaches a peak, the fire continues into its decay phase.

4.1.3 Decay phase

As its name suggests, the temperature in the compartment starts to decrease as the rate of fuel combustion decreases. Due to thermal inertia, the temperature in the structure will continue to increase for a short while in the decay period; i.e., there will be a time lag before the structure starts to cool.

4.2 FACTORS AFFECTING GROWTH PHASE

Early work (Kawagoe, 1958) suggested that the burning rate of wood cribs that are conventionally used as test fuels in compartment fire tests was proportional to $A_v\sqrt{h}$ where A_v is the area of openings in the side of the compartment and h is the height of the ventilation opening. A fire of this nature is said to be *ventilation controlled* since the availability of the air supply governs the fire. The fire temperatures will not continue to increase indefinitely as the ventilation is increased, as a limiting value of ventilation beyond which the fire temperatures will not increase exists.

Beyond this point, the compartment temperatures are dependent entirely on the quantity of fuel available. After the limiting value of ventilation is reached, a fire is then said to be *fuel controlled*. Although the concept of a ventilation factor was originally deduced empirically, it is possible to derive from first principles the dependence of a ventilation-controlled fire on the $A_v\sqrt{h}$ factor based on the mass flow rate of air to the fire (Drysdale, 1998).

The other major factor is obviously the amount of available fuel. As noted above, tests are usually but not always conducted using wood cribs to simulate the combustible contents of a compartment. This is reasonable as the contents of most buildings are generally cellulosic and thus most materials can be expressed as equivalent calorific values of wood cribs. It should be noted, however, that the advent of plastics makes this practice less applicable. Care must be taken as plastics and other hydrocarbon-based materials emit heat far more quickly, although over a shorter period, and also emit substantial amounts of dense, possibly toxic, smoke (Latham et al., 1987). A comparison of the compartment temperature–time responses for a purely cellulosic compartment fire and one fuelled by a combination of part cellulosic material and part plastic material is given in Figure 4.2.

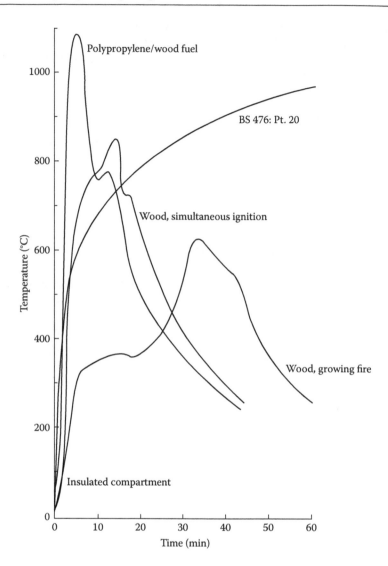

Figure 4.2 Comparison of temperature–time response in compartment due to cellu-losic and hydrocarbon combustibles. (*Source:* Latham, D.J. et al. (1987). *Fire Safety Journal*, 12, 139–152. With permission.)

4.3 CALCULATION OF COMPARTMENT TEMPERATURE–TIME RESPONSES

4.3.1 Basic formulation

Much of the work in this area was carried out in Sweden by Pettersson, Magnusson, and Thor and forms the basis of a Swedish Institute of Steel Construction Report (1976). This report along with an outline in Drysdale (1998) forms the basis of this section of text. The following assumptions were made to simplify the calculation model.

1. Combustion is complete and occurs entirely within the boundaries of the compartment.
2. There is no temperature gradient within the compartment.
3. The compartment walls can be characterized by a single set of heat transfer characteristics known as thermal inertia b and defined as $\sqrt{(\lambda \rho c)}$ where λ is the thermal conductivity, ρ is the density, and c is the specific heat.
4. All the fire load is ignited instantaneously.
5. The heat flow through the walls is assumed to be unidirectional.

The model is essentially generated by the solution of the compartment heat balance equation (see Figure 4.3):

$$\dot{h}_c = \dot{h}_L + \dot{h}_W + \dot{h}_R + \dot{h}_B \tag{4.1}$$

Each term of Equation (4.1) may be considered separately. Note that the heat stored in the gas $\dot{h}_B$ is negligible and may be ignored.

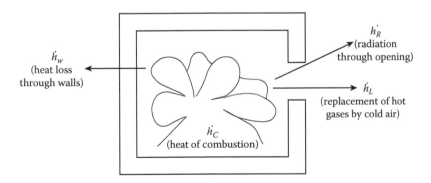

Figure 4.3 Heat balance terms for compartment fire.

4.3.1.1 Rate of heat release ($\dot{h}_C$)

It is assumed that the fire is fuel controlled; i.e., the rate of heat release is proportional to the ventilation factor, or

$$\dot{h}_c = 0,09 A_V \sqrt{h} \Delta h_C \tag{4.2}$$

where ΔH_C is the heat of combustion of wood (18,8 MJ/kg).

4.3.1.2 Rate of heat loss by radiation through openings ($\dot{h}_R$)

This is calculated using the Stefan-Boltzmann law assuming the external temperature is negligible compared with the gas temperature within the compartment

$$\dot{h}_R = A_V \varepsilon_f \sigma \theta_g^4 \tag{4.3}$$

where θ_g is the gas temperature, σ is the Stefan-Boltzmann constant and ε_f is the effective emissivity of the gasses.

4.3.1.3 Rate of heat loss from convection ($\dot{h}_L$)

$$\dot{h}_L = \dot{m}_F c_p (\theta_g - \theta_0) \tag{4.4}$$

where $\dot{m}_F$ is the rate of outflow of gases, cp is the specific heat of the gases, and θ_0 is the external temperature.

4.3.1.4 Rate of heat loss through compartment walls ($\dot{h}_W$)

This depends on both the gas temperature and internal surface temperature θ_i, the effective thermal conductivity of the compartment boundary λ_i, and the heat transfer coefficients (both convective and radiational) of the boundaries α.

The rate of heat loss is determined by discretizing the compartment walls into a series of n slices of thickness Δx and determining the constant heat flow characterized by the mid-depth temperature θ_i through the layers. When the temperatures θ_1 to θ_n have been determined, the heat flow may be determined from

$$\dot{h}_W = \frac{(A_t - A_V)(\theta_g - \theta_i)}{\dfrac{1}{\alpha_i} + \dfrac{\Delta x}{2\lambda_i}} \tag{4.5}$$

where $A_t - A_v$ is the net internal area of the compartment allowing for ventilation openings, α_i is the boundary heat transfer coefficient, λ_i is the thermal conductivity of a layer and θ_i is the mid-layer temperature. Equations (4.1) to (4.5) may be solved numerically to yield the compartment temperature θ_g.

4.3.1.5 Compartment temperature–time characteristics

From the above formulations for the rates of heat flow, Equation (4.1) may now be solved for the gas temperature θ_g and graphs or tables prepared for varying values of the fuel load per unit area of the compartment boundary q_t (MJ/m²) and a modified ventilation factor O defined as $A_v\sqrt{h}/A_t$ (m$^{0.5}$). Typical results from Pettersson et al. are given in Figure 4.4 for a compartment with walls constructed from brickwork or concrete.

For comparison, the standard furnace curve from BS 476: Part 20 (or ISO 834 or EN 13501) is also plotted. For large compartments, combustion will gradually spread throughout, but it has been shown that the peak temperatures measured at various points in the compartment occur at different times but have similar peak values (Kirby et al., 1994). Where a compartment has multiple vertical openings, horizontal openings in the roof, or a different construction from that of the standard or reference compartment, the results of analysis of compartment temperatures need modifying.

4.3.2 Modifications for allowing other compartment configurations

4.3.2.1 Vertical openings

In this case, the modified ventilation factor should be calculated using A_v as the total area of vertical openings and h_{eq} as the weighted average of the heights of the openings defined by

$$h_{eq} = \frac{\sum A_{v,i} h_i}{\sum A_{v,i}} \tag{4.6}$$

where $A_{v,i}$ and h_i are the area and height of the ith opening, and the summations are taken over all the vertical openings.

4.3.2.2 Horizontal openings

The effect of substantial horizontal openings is for all the combustion gases to be vented through the roof rather than via the windows. When this happens, the flow becomes unstable and cannot be predicted using a simple model. The effect of horizontal openings is to move the neutral layer having zero pressure with respect to atmospheric pressure from the centre of the

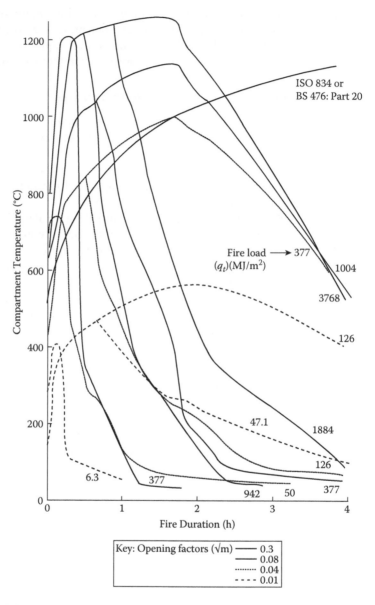

Figure 4.4 Temperature–time response curves for compartment fire with varying fuel loads and ventilation. (*Source:* Pettersson, O. et al. (1976). *Fire Engineering Design of Steel Structures*, Publication 50. Stockholm: Swedish Institute of Steel Construction. With permission.)

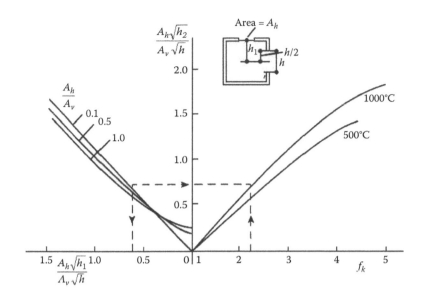

Figure 4.5 Nomogram for calculating f_k for horizontal openings where h_2 is an intermediate variable. (Source: Pettersson, O. et al. (1976). *Fire Engineering Design of Steel Structures*, Publication 50. Stockholm: Swedish Institute of Steel Construction. With permission.)

vertical opening toward the top of the opening. Horizontal openings are treated by applying a modification factor f_k to the opening factor calculated on the basis of vertical openings only.

The nomogram (from Pettersson et al., 1976) for calculating f_k is given in Figure 4.5. The nomogram requires values of $(A_h\sqrt{h_1})/(A_v\sqrt{h})$ and A_h/A_v where A_h is the area of the horizontal opening, h_1 is the vertical distance between the horizontal opening and the mid-height of the vertical opening, A_v is the area of the vertical opening, and h its height. To ensure that the horizontal opening is not dominant, i.e., the assumptions in the calculation of the compartment temperature–time response still hold, limits are placed on the values of $(A_h\sqrt{h_1})/(A_v\sqrt{h})$ for various compartment temperatures. Thus, for a compartment temperature of 1000°C, the limiting value of $(A_h\sqrt{h_1})/(A_v\sqrt{h})$ is 1,76 and for 500°C it is 1,37. If the curves are approximated by straight lines and therefore the distinction between fire compartment values is removed, Buchanan (2001) gives the following equation for calculating the equivalent total vertical openings:

$$\left[A_v\sqrt{h_v}\right]_{equiv} = A_V\sqrt{h_v} + 2,3A_h\sqrt{h_1} \tag{4.7}$$

and is only applicable for $0,3 \le (A_h\sqrt{h_1})/(A_v\sqrt{h_v}) \le 1,5$.

4.3.2.3 Compartment construction

To allow for the effects of differing compartment construction, the values of both fire load density and ventilation factor should be multiplied by a modification factor k_f. Values of k_f depend on both the materials of construction and the actual opening factor. Typical values are given in Table 4.1. To use the above model to calculate the compartment temperature–time response, the design fire load must be calculated.

4.3.3 Calculation of fire load

This can be achieved in one of two ways: by full calculation on the contents of the compartment or by generic data or empirical relationships.

4.3.3.1 Full calculation

In this method, the quantity and calorific value of each item in the compartment is identified, and then the total fire load in the compartment is established.

$$Q_{fi,k} = \sum M_{k,i} H_{u,i} \qquad (4.8)$$

Table 4.1 Transformation parameter k_f for various compartments

Compartment boundary	Opening factor $(A_v \sqrt{h_{eq}}/A_t)$					
	0,02	0,04	0,06	0,08	0,10	0,12
$\lambda = 0,81$ W/m°C, $\rho c = 1,67$ MJ/m³°C[a]	1,00	1,00	1,00	1,00	1,00	1,00
Concrete	0,85	0,85	0,85	0,85	0,85	0,85
Aerated concrete ($\rho = 500$ kg/m³)	3,00	3,00	3,00	3,00	3,00	2,50
50% concrete, 50% aerated concrete	1,35	1,35	1,35	1,50	1,55	1,65
50% aerated concrete, 33% concrete, brickwork remainder[b]	1,65	1,50	1,35	1,50	1,75	2,00
20% concrete, 80% sheet steel[c]	1,00	1,00	0,80	0,70	0,70	0,70
20% concrete, 80% plasterboard[d]	1,50	1,45	1,35	1,25	1,15	1,05
Sheet steel[e]	3,00	3,00	3,00	3,00	3,00	3,00

[a]　Taken as reference compartment with $\sqrt{\rho c \lambda} = 19,4$ Wh^{0,5}/m²C.

[b]　Actual remainder is 13 mm plasterboard (790 kg/m³), 100 mm diabase wool (50 kg/m²), and 200 mm brickwork (1800 kg/m³) from interior to exterior.

[c]　Values quoted are for fire load density of 60 MJ/m² (or less). For fire load of 500 MJ/m² (or greater), all coefficients are 0,50. For intermediate fire load densities, linear interpolation may be used.

[d]　Plasterboard comprises a skin of 2×13 mm thick boards (790 kg/m³) with air gaps of 100 mm.

[e]　Sheet steel is a sandwich panel enclosing 100 mm of diabase wool (50 kg/m³).

Source: Pettersson, O. et al. (1976). *Fire Engineering Design of Steel Structures*, Publication 50. Stockholm: Swedish Institute of Steel Construction. With permission.

Table 4.2 Calorific values of common combustible materials

Material	Calorific value (H_{ui}) MJ/kg
Acetylene	48,2
Acrylic	37 to 29
Alcohol	27 to 33
Celluloid	17 to 20
Cellulose	15 to 18
Coal	28 to 34
Foam rubber	34 to 40
Gasoline	43 to 44
Grain	16 to 18
Hydrogen	119,7
Methane	50,0
Paper, cardboard	13 to 21
Paraffin	40 to 42
Polyethylene	43 to 44
PTFE	5,00
PVC	16 to 17
Rubber tyres	31 to 33
Wood	17 to 20
Wool	21 to 26

Source: CIB W14 Report *(1983). Table 8.1, Page 25. With permission.

where $Q_{fi,k}$ is the fire load and $M_{k,i}$ and $H_{u,i}$ are the mass and calorific value of the ith piece of contents within the compartment, respectively. The value of the fire load per unit area of either the whole compartment or the floor area may then be determined as appropriate. Typical values of $H_{u,i}$ are given in Table 4.2. Clearly this method is extremely laborious for all but the simplest compartments and it is generally acceptable to use generic data.

4.3.3.2 Generic data

Based on a large amount of experience, it is now possible to provide typical data on the mean and standard deviation of a fire load for a given type of structure or occupancy. A substantial quantity of such data is given in the CIB W14 Workshop Report (1983). Typical values are presented in Table 4.3.

4.3.4 Parametric equation approach

One disadvantage of the data on compartment temperature–time response from the results of Pettersson et al. (1976) is that they were presented

Table 4.3 Generic fire load data

| Occupancy | Fire load (MJ/m²) | | |
	Average	Standard deviation	Coefficient of variation
Dwelling	140 to 150	20,1 to 24,7	14,5 to 16,5
Office	102 to 124	31,4 to 39,4	25,3 to 35,6
School	61,1 to 96,7	14,2 to 20,5	16,9 to 30,1
Hospital	116	36	31,0
Hotel	67	19,3	28,8

Note: Fire load density is calculated on total area of bounding fire compartment. Values quoted for dwellings, offices, and schools are ranges. It is not implied that, for example, the lowest standard deviation necessarily corresponds to the lowest mean.

Source: CIB W14 Report *(1983). Table 8.2a, Page 26. With permission.

in tabular or graphical form. This format is not ideal if such data are required for further calculations on the thermal responses of structural elements within a compartment. One possible solution to this difficulty is to determine empirical parametric equations relating the variables concerned. Although originally there may have been a theoretical background to these parametric equations, over the course of their development, they have been tweaked to give better correlation with measured compartment fires.

4.3.4.1 Formulation of Lie (1974)

This is a series of parametric equations derived from the heat balance equation and involving the opening factor and thermal boundaries of a compartment. The model consists of a nonlinear characteristic up to the maximum temperature reached during a fire, followed by a linear decay curve and then a constant residual temperature. The parameters needed are the opening factor O defined as $A_v\sqrt{h_{eq}}/A_t$ (m^{0,5}), the fuel load per unit area of the compartment $q_{t,d}$ (MJ/m²), and the fire duration t_d (min). The first two parameters are the same as those used in the work by Pettersson et al. The compartment temperature θ_g is given by

$$\theta_g = 20 + 250(10O)^{\frac{0,1}{O^{0,3}}} e^{\frac{-O^2 t}{60}} [3(1 - e^{-0,01t}) - (1 - e^{-0,05t}) + 4(1 - e^{-0,2t})] + C\sqrt{\frac{600}{O}}$$

(4.9)

where C is a constant that takes a value of zero for heavy-weight compartment boundary materials ($\rho \geq 1600$ kg/m³) and unity for lightweight materials ($\rho < 1600$ kg/m³). Equation (4.9) is valid only for a value of t less than that defined by Equation (4.10)

$$t \leq \frac{4,8}{O} + 60 \tag{4.10}$$

If t is greater than the value given by Equation (4.10), it should be set equal to the right side of Equation (4.10). The duration of the fire t_d in min is given by

$$t_d = \frac{q_{t,d}}{5,5O} \tag{4.11}$$

Note that both Lie and Pettersson et al. use a constant of proportionality of 0,09 rather than 0,1 in the equation to calculate fire duration. In the decay phase, i.e., when $t > t_d$, the gas temperature is given by

$$\theta_g = \theta_{max} - 600 \left(\frac{t}{t_d} - 1 \right) \geq 20 \tag{4.12}$$

where θ_{max} is the temperature reached at t_d. These parametric equations are valid for $0,01 \leq O \leq 0,15$. If O exceeds $0,15$, it should be set equal to $0,15$.

4.3.4.2 EN 1991-1-2 approach

The background theory to this approach was proposed by Wickström (1981/2, 1985a) who suggested that the compartment temperature–time relationship depends entirely on the ratio of the opening factor $A_v \sqrt{h_{eq}}/A_t$ to the thermal inertia $\sqrt{(\rho c \lambda)}$ of the compartment boundary and that the standard furnace curve could be attained by a ventilation factor of $0,04$ m^{0,5} and a thermal inertia of 1160 Ws/m² °C.

Thus the gas temperature θ_g can be related to a parametric time base t^* related to the real time t (hours) and the ventilation and compartment boundary by the following equation.

For the heating phase:

$$\theta_g = 20 + 1325 \left[1 - 0,324 e^{-0,2t^*} - 0,204 e^{-1,7t^*} - 0,472 e^{-19t^*} \right] \tag{4.13}$$

where t^* is defined by

$$t^* = t\Gamma \tag{4.14}$$

with Γ defined as

$$\Gamma = \frac{\left(\dfrac{O}{\sqrt{\rho c \lambda}}\right)^2}{\left(\dfrac{0{,}04}{1160}\right)^2} \tag{4.15}$$

where O is the opening factor and $\sqrt{(\rho c \lambda)}$ is the thermal inertia of the compartment boundary.

For the cooling phase:

For $t_d^* < 0,5$ hr:

$$\theta_g = \theta_{max} - 625(t^* - x t_{max}^*) \tag{4.16}$$

For $0,5 \le t_d^* \le 2,0$ hr:

$$\theta_g = \theta_{max} - 250(3 - t_{max}^*)(t^* - x t_{max}^*) \tag{4.17}$$

For $t_d^* > 2,0$ hr:

$$\theta_g = \theta_{max} - 250 (t^* - x t_{max}^*) \tag{4.18}$$

where θ_{max} is the maximum temperature reached during the heating phase and t_{max} $(= t^*_{max}/\Gamma)$ is given by

$$t_{max} = max\left[\frac{0{,}2 \times 10^{-3} q_{t,d}}{O\Gamma}, t_{lim}\right] \tag{4.19}$$

where t_{lim} depends upon the growth rate of the fire. Values for t_{lim} are given in Table 4.4. Also if $t_{max} = t_{lim}$, then t^* is defined by $t^* = t\Gamma_{lim}$ where Γ_{lim} is given by

$$\Gamma_{lim} = \frac{\left[\dfrac{0{,}1 \times 10^{-3} q_{t,d}}{b t_{lim}}\right]^2}{\left(\dfrac{0{,}04}{1160}\right)^2} \tag{4.20}$$

Table 4.4 Values of t_{lim}

Fire growth rate	Occupancy	t_{lim} *(min)*
Slow	Transport (public space)	25
Medium	Dwelling, hospital room, hotel room, office, school classroom	20
Fast	Library, shopping centre, cinema)	15

Source: Table E5 of EN 1991-1-2.

If the following limits apply ($O > 0,04$, $q_{t,d} < 75$ and $b < 1160$), the value of Γ_{lim} becomes $k\Gamma_{lim}$, where k is given by

$$k = 1 + \frac{O - 0,04}{0,04} \frac{q_{t,d} - 75}{75} \frac{1160 - b}{1160} \tag{4.21}$$

For fuel-controlled fires, t_{max} equals t_{lim}, but for ventilation-controlled fires, t_{max} is given by $0,2 \times 10^{-3} q_{t,d}/O$. If $t_{max} > t_{lim}$, $x = 1$ or if $t_{max} \leq t_{lim}$, $x = t_{lim}\Gamma/t^*_{max}$.

Note that EN 1991-1-2 places limits on the validity of Equations (4.15) to (4.19). The limits are $0,02 \leq O \leq 0,20$ (m^{0,5}), $50 \leq q_{t,d} \leq 1000$ (MJ/m^2), and $100 \leq \sqrt{(\rho c \lambda)} \leq 2200$ (J/m^{2}s^{0,5}K). Clause 3.1.2 (PD 6688-1-2) indicates that the parametric fire curve may be applied to floor areas greater than 500 m^2 and compartment heights greater than 4 m, and it also indicates that the results for tall compartments may be unduly onerous. Reliable elevated temperature data may be used to calculate the value of $\sqrt{(\rho c \lambda)}$. Clause 3.1.2 also sets the lower limit of the opening factor O as $0,01$ m^{0,5}.

One anomaly with both approaches is that the rate of increase of temperature in the growth period is independent of the fire load. It is not always necessary to be able to characterize the complete compartment temperature–time response. It may be sufficient to predict basic data such as maximum temperature or fire duration.

4.4 ESTIMATION OF FIRE CHARACTERISTICS

If the relative amounts of heat flow in Equation (4.1) are calculated for specific cases, it is possible to draw conclusions on the relative importance of its terms. Typical data on this are given in Table 4.5 (Heselden, 1968). Note that the largest portion of heat loss is through the effluent gases, and as the window area decreases, the proportion of heat transmitted through the walls of a standard brick or concrete compartment increases. Whilst the total heat output is entirely dependent on the fuel available, its rate of output will be ventilation controlled. Thus it has

Table 4.5 Relative heat losses in compartment fire

		Relative heat loss (% of Total)				
Opening	*Fire load*	*Effluent*	*Boundary gasses*	*Feedback to fuel*	*Radiation*	*Steel*
0,5	7,5	67	17	12	3	1
	15,0	67	14	11	6	2
	30,0	65	13	11	9	2
	60,0	61	11	11	13	4
0,25	7,5	56	24	16	2	2
	15,0	55	29	10	4	2
	30,0	52	20	11	11	6
	60,0	53	22	12	9	4
0,125	60,0	47	25	16	7	5

Notes: The figures in the "Opening" column indicate relative area of full height opening in the longer side of compartment. Fire load density expressed as kilograms of wood cribs per square metre of floor area. Although tests were undertaken to assess protection requirements for fire exposed steelwork, steelwork adsorbs only a relatively small proportion of the heat; the figures in the table can be held to refer to a normal compartment.

Source: Heselden, A.J.M. (1968). In *Proceedings of Symposium on Behaviour of Structural Steel in Fire*. London: Her Majesty's Stationary Office (now BRE), pp. 19–28. With permission.

been suggested that the rate of burning is proportional to the ventilation factor (Fujita, undated)

$$R = 0{,}1A_v\sqrt{h_{eq}} \tag{4.22}$$

where R is the rate of burning (kg/sec) and $A_v\sqrt{h_{eq}}$ is a ventilation factor ($m^{1,5}$). The outflow of heat through the gasses escaping through the ventilation sources will also be proportional to $A_v\sqrt{h_{eq}}$ and that through the walls to $A_t - A_v$ where A_t is the total internal area of the compartment. This led to the suggestion that the temperature reached in a compartment fire should be a function of these parameters. The analysis of a large number of fires by Thomas and Heselden (1972) plotted in Figure 4.6 indicates that a parameter η best describes the parametric effect, where η is

$$\eta = \frac{A_t - A_v}{A_v\sqrt{h_{eq}}} \tag{4.23}$$

For low values of η (high ventilation areas), the rate of heat loss is greatest but so is the heat loss through the windows, and thus the temperatures are

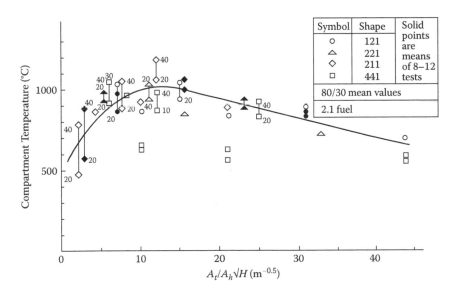

Figure 4.6 Relationship of maximum compartment temperatures and compartment geometry. (*Source:* Heselden, A.J.M. (1968). In *Proceedings of Symposium on Behaviour of Structural Steel in Fire*. London: Her Majesty's Stationery Office (now BRE), pp. 19–28.)

low. For high values of η (low ventilation areas), heat loss is lower but the rate of heat release is also low, giving rise to low temperatures. The curve in Figure 4.6 can be represented by the following equation (Law, 1978):

$$\theta_{f,\max} = 6000\frac{1-e^{-0,1\eta}}{\sqrt{\eta}} \tag{4.24}$$

The value of $\theta_{f,\max}$ given by Equation (4.24) is an upper limit that will not be achieved when the fire load is low. Thus the value of $\theta_{f,\max}$ must be modified to allow for this and enable the maximum fire temperature $\theta_{\max}$ to be calculated as

$$\theta_{\max} = \theta_{f,\max}(1-e^{-0,05\psi}) \tag{4.25}$$

where ψ is defined by

$$\psi = \frac{L_{fi,k}}{\sqrt{A_V(A_t - A_v)}} \tag{4.26}$$

where $L_{fi,k}$ is the total fire load expressed as an equivalent mass of wood having the same calorific value. PD 7974-3 suggests that Equations (4.24) and (4.25) are valid for $720 < b < 2500 \ J/m^2s^{0.5}$, although it may produce conservative values about 25% too high. The rate of burning R defines the rate of heat release and can also be used to define the fire duration t_d (in seconds) given by

$$t_d = \frac{L_{fi,k}}{R} \tag{4.27}$$

with values of R taken from Equation (4.22). Thomas and Heselden recognized that the rate of burning calculated from the equation was not entirely correct and gave only an approximate value; they found that the rate of burning depended on both the internal area of the compartment and its geometry. A more accurate estimate (Law, 1983) of the rate of burning is given by

$$R = 0{,}18A_v \sqrt{h_{eq}} \sqrt{\frac{W}{D}} \ (1 - e^{-0.036\eta}) \tag{4.28}$$

where D is the depth and W is the width of the compartment. Note that Equation (4.28) only holds for ventilation-controlled fires. For fuel-controlled fires, the rate of burning depends on the quantity and type of fuel.

Attempts have been made over the years to relate compartment fires to the standard furnace test, as related in Chapter 3. Since all fire testing is performed using a standard temperature–time régime and such tests have provided a substantial amount of data about the performance of structural elements and design data, it would be useful to enable the data obtained from such tests to be used in cases where a natural fire characteristic would be a more reasonable model for the response of a compartment. Several attempts have been made to provide this link. They are based on fire severity or time equivalence.

4.5 FIRE SEVERITY AND TIME EQUIVALENCE

4.5.1 Fire severity

The initial ideas in this field came from Ingberg (1928) who, after a series of compartment tests with known fire loads, suggested that fire severity could be calculated by considering equivalence of the areas under the furnace curve and the compartment curve above a base of either 150 or 300°C (Figure 4.7). In this way, Ingberg derived a correlation between fire load measured in his tests as load per unit floor area and the standard fire resistance periods.

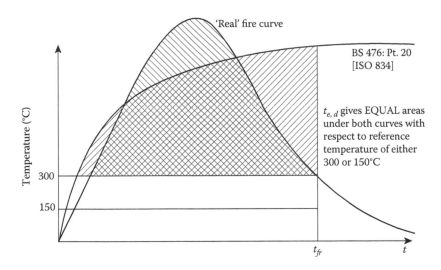

Figure 4.7 Equivalence of fire severity based on areas beneath standard and compartment temperature–time curves.

Although this method provided a basis for fire grading as in the Report of the Committee on the Fire Grading of Buildings (Ministry of Works, 1946), it has little theoretical justification and appears not to consider the effect of ventilation that would also affect equivalence since compartment temperatures are affected by both air and fuel supplies (Robertson and Gross, 1970). This flaw led to the rejection of Ingberg's ideas, and thus alternative approaches needed to be found. They led to the idea, of time equivalence based on equal temperature rise within an element. It must be reiterated that time equivalence only gives a measure of the total heat input from a fire and does not distinguish short high intensity fires and long low intensity fires (Law, 1997).

4.5.2 Time equivalence

Two methods are available to calculate time equivalence. The first is to consider equivalence based on the same temperature rise in the element, and the second is to consider equivalence of heat input. Law (1997) gives much of the background to the determination of time equivalence using temperature rise.

4.5.2.1 Temperature base

Law (1973) made the first attempt to correlate the effects of a natural fire and a furnace test on a temperature base. Law calculated the time taken by an insulated steel member to reach a specified temperature of 550°C

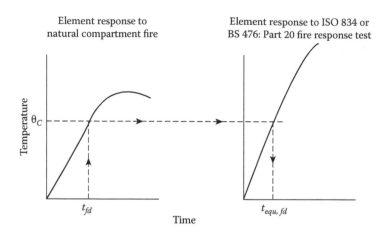

Element response to
natural compartment fire

Element response to ISO 834 or
BS 476: Part 20 fire response test

Figure 4.8 Equivalence of fire severity based on temperatures reached in elements within compartment.

by exposure to a natural fire, modelled by assuming its peak temperature maintained throughout the fire and the equivalent time taken to reach the same temperature when exposed to the standard furnace test (Figure 4.8).

In the authors' view, time equivalence should be used only where the structural element (or structure) behaviour can be characterized by a single temperature. Thus, it is acceptable for protected steelwork, unprotected steelwork (with possible modification), and concrete in flexure where fire performance for sagging depends on the reinforcing temperature and on average concrete temperature for hogging. For concrete columns when the maximum (average) temperatures in the reinforcement and in the concrete do not occur at the same time, it is unclear which maximum temperature governs the strength criterion. Time equivalence cannot be used for timber as the controlling phenomenon is char depth, not temperature rise. Law (1973) found that the required fire resistance was a function of the fire load, internal area of the compartment, and ventilation area

$$t_{e,d} = \frac{KL_{fi,k}}{\sqrt{A_v A_t}} \qquad (4.29)$$

where $t_{e,d}$ is the required equivalent fire resistance (min), $L_{fi,k}$ is the total fire load (kg of wood equivalent), A_v is the area of vertical openings (m²), and A_t is the total internal area (m²). The parameter K has a value near unity for SI units. Equation (4.29) holds only for insulated steel and should be applied with care to uninsulated steel. Equation (4.29) has gone through three modifications to improve its accuracy. The first was presented at the 1983 CIB W14 Workshop, and took the form

$$t_{e,d} = \frac{0,067 q_{t,d}}{\sqrt{\dfrac{A_v \sqrt{h_{eq}}}{A_t}}} \qquad (4.30)$$

where $q_{t,d}$ is the fire load per unit area of compartment boundary (MJ/m²). A second modification considering the effect of compartment thermal boundaries was published by the CIB in 1985 (Thomas, 1986), and took the form

$$t_{e,d} = c w q_{f,d} \qquad (4.31)$$

where $q_{f,d}$ is the fire load per unit floor area, c is a parameter depending on the construction materials of the compartment walls, and w is related to the compartment geometry by the following equation

$$w = \sqrt{\frac{A_f}{A_v}} \sqrt{\frac{A_f}{A_t \sqrt{h_{eq}}}} \qquad (4.32)$$

where A_f is the floor area (m²). It should be noted for an area of vertical openings greater than 10% of the floor area, w can be taken conservatively as 1,5. The parameter c may be taken conservatively as 0,1 or directly as a specific heat value given in Table 4.6. A third approach is cited in the informative Annex F to EN 1991-1-2 where the equivalent fire duration $t_{e,d}$ is given by

$$t_{e,d} = k_c q_{f,d} k_b w_f \qquad (4.33)$$

where $q_{f,d}$ is the fire load density related to the floor area, w_f is a ventilation factor given by Equation (4.32), k_b is a factor reliant on the thermal properties of the boundary listed in Table 4.7, but may be taken conservatively as 0,07. k_c is a correction factor taken as 1,0 for protected steel and concrete and 13,70 for unprotected steel. Annex B of PD 6688-1-2 effectively sets the value of k_c as 1,0. Note also that PD 6688-1-2 restricts the use of time equivalence for unprotected steelwork to less than 30 min.

Table 4.6 Parameter c for calculating equivalent fire duration

Compartment thermal absorbtivity (b) = √ρcλ) (Wh^{0,5}/m²K)	c (m²min/MJ)
b < 12	0,09
12 ≤ b ≤ 42	0,07
b > 42	0,05

Source: Thomas, P.H., Ed. (1986). Design guide: structural fire safety. In Fire Safety Journal 10, 77–137. With permission.

Table 4.7 Values of k_b from EN 1992-1-2 and PD 7974-3

	k_b (min m²/MJ)	
Thermal absorbtivity (b) (J/m²s⁰·⁵K)	*EN 1992-1-2*	*PD 7974-3*
b > 2500	0,04	0,055
720 ≤ b ≤ 2500	0,055	0,07
b < 720	0,07	0,05

Source: Table F.2 of EN 1991-1-2 and Table 4 of PD 7974-3.

Annex E of EN 1991-1-2 relates the design fire load $q_{f,d}$ to the generic fire load $q_{f,k}$ for a particular occupancy by the following equation

$$q_{f,d} = m\delta_{q1}\delta_{q2}\delta_n q_{f,k} \tag{4.34}$$

where δ_{q1} is a partial safety factor depending on the danger of fire activation as a function of compartment floor area, δ_{q2} is a partial safety factor dependent upon the danger of fire activation based on type of occupancy, and δ_n is a factor allowing for the presence of active fire measures if not considered in the fire model. EN 1991-1-2 recommends a factor of 1,0 for δ_n. The values of δ_{q1} and δ_{q2} are specified in Table E.1. The value of m should be taken as 0,8 for cellulosic fire loads. PD 6688-1-2 indicates that the method in Annex E should not be used and that reference should be made to PD 7974-1. The ventilation factor w_f subject to a lower limit of 0,5 is given by

$$w_f = \left(\frac{6,0}{H}\right)^{0,3}\left[0,62 + \frac{90(0,4-\alpha_v)^4}{1+b_v\alpha_h}\right] \tag{4.35}$$

where α_v is the ratio between the area of vertical openings and the floor area and should lie between 0,025 and 0,25, α_h is the ratio between the area of the horizontal openings and the floor area, H is the compartment height, and b_v is given by

$$b_v = 12,5(1+10\alpha_v - \alpha_v^2) \geq 10,0 \tag{4.36}$$

For small compartments with no roof openings, floor areas less than 100 m², and no horizontal openings, the ventilation factor w_f may be taken as

$$w_f = \frac{A_f}{A_t}O^{-0,5} \tag{4.37}$$

where O is the opening factor.

4.5.2.2 Normalized heat load base

The theory behind this approach was developed by Harmathy and Mehaffey (1985) and is presented below. The normalized heat load $\bar{h}''$ in a furnace test to the standard furnace curve can be related to the duration of the test $t_{e,d}$ by

$$t_{e,d} = 0{,}11 + 0{,}16 \times 10^{-4}\,\bar{h}'' + 0{,}13 \times 10^{-9}(\bar{h}'')^2 \tag{4.38}$$

The heat flow from a compartment normalized with respect to the thermal boundaries of the compartment $\bar{h}'$ is given by

$$\bar{h}' = \frac{(11{,}0\delta + 1{,}6)L_{f,d}A_f \times 10^6}{A_t\sqrt{\lambda\rho c} + 935\sqrt{\Phi L_{f,d}A_f}} \tag{4.39}$$

where A_f is the floor area, A_t is the total area of compartment boundaries, $\sqrt{(\lambda\rho c)}$ is the surface averaged thermal inertia of the compartment boundary, Φ is a ventilation factor related to the rate of mass inflow of air into the compartment, $L_{f,d}$ is the fire load (kg of wood equivalent) per unit area of floor, and δ defines the amount of fuel energy released through the openings and is given by

$$\delta = 0{,}79\sqrt{\frac{H^3}{\Phi}} \tag{4.40}$$

where H is the compartment height. The value of Φ may be taken as its minimum value $\Phi_{\min}$ given by Equation (4.41) as this produces a conservative answer,

$$\Phi_{\min} = \rho_{air}A_v\sqrt{gh_{eq}} \tag{4.41}$$

where h_{eq} and A_v are the height and area of the window opening, respectively, ρ_{air} is the density of air and g is the acceleration due to gravity ($9{,}81$ m/s^2). The normalized heat load to the furnace $\bar{h}''$ can be related to the normalized heat load in a compartment fire $\bar{h}'$ by the following equation

$$\bar{h}'' = \bar{h}'\exp\left(\beta\sqrt{\left(\frac{\sigma_{h''}}{\bar{h}''}\right)^2 + \left(\frac{\sigma_{h'}}{\bar{h}'}\right)^2}\right) \tag{4.42}$$

where β is the statistical acceptance or rejection limit on the variables and is 1,64 for a 5% limit. The ratio of the variance of the normalized heat load to the normalized heat load in the compartment is given by

$$\frac{\sigma_{h'}}{\bar{h}'} = \frac{\sigma_{\bar{L}}}{\bar{L}_{f,d}}\frac{A_t\sqrt{\lambda\rho c} + 467{,}5\sqrt{\Phi_{\min}\bar{L}_{f,d}A_f}}{A_t\sqrt{\lambda\rho c} + 935\sqrt{\Phi_{\min}\bar{L}_{f,d}A_f}} \tag{4.43}$$

where $\sigma_{\bar{L}}/\bar{L}_{f,d}$ is the coefficient of variation in the combustible fire load in the compartment. The coefficient of variation for the normalized heat input to the furnace is given by

$$\frac{\sigma_{b''}}{\overline{b}''} = 0,9 \frac{\sigma_{t_{e,d}}}{\overline{t}_{e,d}} \tag{4.44}$$

where $\sigma_{t_{e,d}}/\bar{t}_{e,d}$ is the coefficient of variation in the test results from a furnace and can typically be taken in the order of 0,1. Thus when values of the fire load and compartment geometry are determined, the equivalent furnace test value can be found. The alternative methods of determining the behaviour characteristics of a fire compartment are compared in the following example.

Example 4.1: Determination of behaviour characteristics of fire compartment

Consider a compartment 14 m × 7 m × 3 m high with six windows (each 1,8 m wide × 1,5 m high) with a fire load related to floor area of 60 kg/m² of wood equivalent.

Compartment construction: dense concrete ($\sqrt{(\lambda\rho c)} = 32$ Wh^{0,5}/m²K)

Total area of windows:

$A_v = 6 \times 1,5 \times 1,8 = 16,2$ m²
$A_f = 14 \times 7 = 98$ m²
$A_t = 2 \times 98 + 2 \times 14 \times 3 + 2 \times 7 \times 3 = 322$ m²

Calorific value of fire load = 18 MJ/kg
Fire load per unit surface area $q_{t,d}$:

$q_{t,d} = 60 \times 18 \times A_f/A_t = 60 \times 18 \times 98/322$
     = 329 MJ/m² (equivalent to 1080 MJ/m² based on floor area)

1. Equivalent Fire Durations

(a) *From Equation (4.29):*

$$t_{e,d} = \frac{KL_{fi,k}}{\sqrt{A_v A_t}} = \frac{1,0 \times 60 \times 98}{\sqrt{16,2 \times 322}} = 81 \,\text{min}$$

(b) *From Equation (4.30):*

$$t_{e,d} = \frac{0,067 q_{t,d}}{\sqrt{\dfrac{A_v \sqrt{b_{eq}}}{A_t}}} = \frac{0,067 \times 329}{\sqrt{\dfrac{16,2 \times \sqrt{1,5}}{322}}} = 89 \,\text{min}$$

(c) From CIB method:

A_f = 98 m^2 or 10% A_f = 9,8 m^2 and A_v = 16,2 m^2 so A_v is greater than 10% A_f. Therefore the approximate method (ii) may be used.

(i) Exact Method:

Determine value of w from Equation (4.32):

$$w = \sqrt{\frac{A_f}{A_v}}\sqrt{\frac{A_f}{A_t\sqrt{h_{eq}}}} = \sqrt{\frac{98}{16,2}}\sqrt{\frac{98}{322\sqrt{1,5}}} = 1,23$$

Determine boundary conditions: $\sqrt{\rho c \lambda}$ = 32 Wh^{0,5}/m^{2}K. Hence from Table 4.5, c = 0,07. From Equation (4.31),

$$t_{e,d} = cwq_{f,d} = 0,07 \times 1,23 \times (60 \times 18) = 93 \, min$$

(ii) Approximate Method:

w = 1,5 and c = 0,1, hence $t_{e,d}$ = 162 min.

(d) EN 1991-1-2 Annex F:

Take k_c = 1,0. From the data in the text, $\sqrt{(\rho c \lambda)}$ = 32 × 60 = 1920 J/m^{2}s^{0,5}K, k_b = 0,055. As there are no horizontal openings, the value of b is not required.

$$\alpha_v = A_w/A_f = 16,2/98 = 0,165$$

From Equation (4.35) with b = 0

$$w_f = \left(\frac{6,0}{H}\right)^{0,3}\left[0,62 + \frac{90(0,4 - \alpha_v)^4}{1 + b_v\alpha_h}\right]$$

$$= \left(\frac{6,0}{3}\right)^{0,3}\left[0,62 + \frac{90(0,4 - 0,165)^4}{1}\right] = 1,101$$

The value of w_f satisfies the limiting condition of 0,5. Using the approximate Equation (4.37) for w_f gives

$$w_f = \frac{A_f}{A_t}O^{-0,5} = \frac{A_f}{A_t}\left(\frac{A_v\sqrt{h_{eq}}}{A_t}\right)^{-0,5} = \frac{98}{322}\left(\frac{16,2\sqrt{1,5}}{322}\right)^{-0,5} = 1,23$$

$$q_k = 60 \times 18 = 1080 \, MJ/m^2$$

From Equation (4.33)

$$t_{e,d} = q_{f,d}k_bw_f = 0,055 \times 1,101 \times 1080 = 65 \, min$$

The conservative value of w_f gives $t_{e,d}$ as 73 min. Using the k_b factor value of 0,07 recommended in PD 7974-3 gives values for $t_{e,d}$ of 83 and 93 min, respectively.

(e) Harmathy and Mehaffey:

Boundary conditions:

$$\sqrt{(\rho c\lambda)} = 32 \times \sqrt{3600} = 1920 \ Js^{0,5}/m^2 degK$$
$$L_{f,d} = 60 \ kg/m^2$$
$$L_{f,d}A_f = 5880 \ kg$$

From Equation (4.41) calculate Φ_{min}:

$$\Phi_{min} = \rho_{air} \ A_v \ \sqrt{g \ h_{eq}} = 1,21 \times 16,2 \ \sqrt{9,81 \times 1,5} = 75,2 \ kg/s$$

From Equation (4.40), calculate δ:

$$\delta = 0,79 \left(\frac{H^3}{\Phi_{min}} \right)^{0,5} = 0,79 \left(\frac{3^3}{75,2} \right)^{0,5} = 0,473$$

To ease subsequent calculations:

$$\sqrt{\Phi L_{f,d}A_f} = \sqrt{75,2 \times 60 \times 98} = 665$$
$$A_t\sqrt{(\rho ck)} = 322 \times 1920 = 618240$$

Calculate $\bar{h}'$ from Equation (4.40):

$$\bar{h}' = \frac{(11,0 \ \delta + 1,6) \ (L_{f,d} \ A_f) \times 10^6}{A_t \ \sqrt{\lambda \rho c} + 935 \ \sqrt{\Phi \ L_{f,d} \ A_f}} = \frac{(11,0 \times 0,473 + 1,6) \ 5880 \times 10^6}{618240 + 935 \times 665}$$

$$= 32260 \ s^{0,5}K$$

The normalized standard deviation for the fire load will be taken as 0,3. This is reasonable if the data for office loading in Table 4.3 are examined. From Equation (4.43), calculate $\sigma_{h'}/\bar{h}'$:

$$\frac{\sigma_{h'}}{\bar{h}'} = \left(\frac{\sigma_L}{\bar{L}_{f,d}} \right) \left(\frac{A_t \ \sqrt{\lambda \rho c} + 467,5 \ \sqrt{\Phi_{min} \ \bar{L}_{f,d} \ A_f}}{A_t \ \sqrt{\lambda \rho c} + 935 \ \sqrt{\Phi_{min} \ \bar{L}_{f,d} \ A_f}} \right)$$

$$= 0,3 \frac{618240 + 467,5 \times 665}{618240 + 935 \times 665} = 0,225$$

As recommended by Harmathy and Mehaffey, $\sigma_{t_{e,d}}/\bar{t}_{e,d}$ is taken as 0,1. Thus from Equation (4.44),

$$\frac{\sigma_{h''}}{\bar{h}''} = 0,9 \frac{\sigma_{t_{e,d}}}{\bar{t}_{e,d}} = 0,9 \times 0,1 = 0,09$$

Calculate $\bar{b''}/\bar{b'}$ from Equation (4.42):

$$\frac{\bar{b''}}{\bar{b'}} = \exp\left(\beta \sqrt{\left(\frac{\sigma_{b''}}{\bar{b''}} \right)^2 + \left(\frac{\sigma_{b'}}{\bar{b'}} \right)^2} \right) = \exp\left(1,64 \sqrt{0,225^2 + 0,09^2} \right) = 1,488$$

or

$\bar{b''} = 1,488 \times 32260 = 48003 \ \mathrm{s}^{0.5}\mathrm{K}$

Calculate $t_{e,d}$ from Equation (4.38)

$$t_{ed} = 0.11 + 0,16 \times 10^{-4}\bar{b''} + 0,13 \times 10^{-9}(\bar{b''})$$

$$= 0,11 + 0,16 \times 10^{-4} \times 48003 + 0,13 \times 10^{-9} \times 480 = 1,18 \ \mathrm{hrs} = 71\mathrm{mins}$$

A comparison of the values of $t_{e,d}$ is presented in Table 4.8. With the exception of the conservative approach adopted by the second of the two CIB approaches, the answers are reasonably consistent although the method in EN 1991-1-2 gives a lower (not conservative) value than earlier methods.

2. Maximum Temperatures

(a) Calculate η and $\theta_{f,max}$ from Equations (4.23) and (4.24):

$$\eta = \frac{A_t - A_V}{A_v \sqrt{h_{eq}}} = \frac{322 - 16,2}{16,2\sqrt{1,5}} = 15,4$$

$$\theta_{f,max} = 6000 \frac{1 - e^{-0,1\eta}}{\sqrt{\eta}} = 6000\frac{1 - e^{-1,54}}{\sqrt{15,4}} = 1201°\mathrm{C}$$

Table 4.8 Comparison of calculated equivalent fire durations from example 4.1

Method		Equivalent fire duration (minutes)
Equation (4.29)		81
Equation (4.30)		89
CIB Workshop:	Exact	93
	Approximate	162
EN 1991-1-2		65 (73)
PD 7974-3		83 (93)

Note: The values in parentheses for EN 1991-1-2 and PD 7974-3 were determined using conservative values of relevant parameters.

To correct $\theta_{f,max}$ for the type of fire, calculate ψ from Equation (4.26):

$$\psi = \frac{L_{fi,k}}{\sqrt{A_V(A_t - A_v)}} = \frac{60 \times 98}{\sqrt{16,2(322 - 16,2)}} = 83,5$$

Calculate θ_{max} from Equation (4.25):

$$\theta_{max} = \theta_{f,max}(1 - e^{-0,05\psi}) = 1201(1 - e^{-0,05 \times 83,5}) = 1183°C$$

(b) Theory due to Lie using Equations (4.9) to (4.12):
Opening factor, O:

$O = A_v \sqrt{h}/A_t = 16,2 \times \sqrt{1,5}/310 = 0,0616$

Fire load (in kg/m^2) of compartment $= 60 \times 98/322 = 18,26$ kg/m^2

Fire duration t_d using Equation (4.11):

$t_d = L_{fi,k}/(5,5O) = 18,26/(5,5 \times 0,0616) = 53,9$ min

Maximum allowable value of t using Equation (4.10):

$t_{max} = 4,8/O + 60 = 4,8/0,0616 + 60 = 137,92$ min $= 2,30$ hr

Thus Equation (4.9) will hold up to the total fire duration, and is evaluated in Table 4.9 and plotted in Figure 4.9.

The maximum temperature θ_{max} attained in the fire is 970°C. In the decay phase, the fire reaches ambient at 2,31 hr (139 min), and the temperature profile is linear between the maximum and the point at which ambient is reached.

(c) EN 1991-1-2 using Equations (4.13) to (4.19):
Calculate Γ from Equation (4.15):

$O = 0,0616$ m^{0,5} and $\sqrt{(\rho c \lambda)} = 1920$ Ws^{0,5}/m^2°C

$\Gamma = (O/\sqrt{(\rho c\lambda)})^2/(0,04/1160)^2 = 0,866$

Calculate t_{max} from Equation (4.19):

$t_{max} = 0,20 \times 10^{-3} q_{td}/(O\Gamma) = 1,223$ hr $= 74$ min

The values of the increase in gas temperature θ_g over ambient may now be calculated up to the design time t_d. These values are given in Table 4.9 and plotted in Figure 4.9. The maximum gas temperature θ_{max} (above 0°C) at a real time of 74 min is 953°C. As the fire load density is typical of an office, from Table 4.4, t_{lim} is 20 min (0,333 hr). This is less than t_{max}. Therefore $x = 1,0$ in the decay phase.

Since the parametric fire duration is between 0,5 and 2,0 hr, Equation (4.16) is used to calculate the decay phase. The fire decays to a temperature of 20°C at 2,96 hr parametric time or 3,42 hr real time.

Table 4.9 Comparison of parametric curves
(Lie and EN 1991-1-2)

Time (minutes)	Lie (°C)	EN 1991-1-2 (°C)
0	20	20
5	568	683
10	768	716
15	844	747
20	877	774
25	895	799
30	910	822
35	925	842
40	939	861
45	954	878
50	970	894
55	951	909
60	895	922
65	840	934
70	784	946
75	729	944
80	673	909
85	618	874
90	562	839
95	506	804
100	451	769
105	395	734
110	340	699
115	284	664
120	229	630
125	173	595
130	118	560
135	62	525
140	20	490
145	20	455
150	20	420
155	20	385
160	20	350
165	20	315
170	20	280
175	20	245
180	20	210

(continued)

Table 4.9 Comparison of parametric curves
(Lie and EN 1991-1-2) (Continued)

Time (Minutes)	Lie (°C)	EN 1991-1-2 (°C)
185	20	176
190	20	141
195	20	106
200	20	71
205	20	36
210	20	20
215	20	20

Figure 4.9 demonstrates that the maximum temperatures are similar, but Lie predicts a shorter period to maximum temperature, a faster cooling rate, and a shorter total overall duration of the complete fire of around two thirds that predicted by EN 1991-1-2.

3. Fire Duration

Calculate the rate of burning from Equations (4.27) and (4.28).

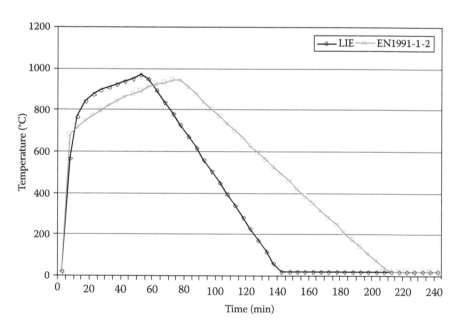

Figure 4.9 Comparison between parametric curves due to Lie and EN 1991-1-2.

Table 4.10 Comparison of maximum fire temperatures

Method	Maximum temperature (°C)
Lie	970
EN 1991-1-2	953
Equation (4.23)	1183

Table 4.11 Comparison of fire durations

Method	Fire duration (min)
Lie	54
EN 1991-1-2	74
Equation (4.27)	50
Equation (4.28)	46

(a) Approximate Equation (4.27):

From Equation (4.22):

$$R = 0,1 \times 16,2 \times \sqrt{1,5} = 1,98 \text{ kg/sec}$$
$$t_d = L_{fi,k}/R = 60 \times 98/1,98 = 2970 \text{ sec} = 49,5 \text{ min}$$

(b) More exact equation [Equation (4.26)]:

$$R = 0,18 \times 16,2 \times \sqrt{1,5} \times \sqrt{(14/7)} \times (1 - e^{-0,036 \times 15,4}) = 2,15 \text{ kg/sec}$$
$$t_d = L_{fi,k}/R = 60 \times 98/2,15 = 2734 \text{ sec} = 45,6 \text{ min}$$

Table 4.10 lists the maximum temperatures reached in both parametric curves and the value predicted by Equation (4.21) where it is observed that both Lie and EN 1991-1-2 predict similar maximum temperatures and Equation (4.23) gives a value some 20% higher. Table 4.11 shows the fire duration predicted by both parametric curves and the value predicted by Equations (4.25) and (4.26) where it is observed Lie Equations (4.25) and (4.26) predict similar values of around 50 min and EN 1991-1-2 a value around 50% longer.

4.6 LOCALIZED FIRES

Two types of localized fires require a brief discussion.

4.6.1 Plume fires

A plume fire can occur when a small intense fire at floor level generates flames spread along any ceiling or floor soffit. Where the possible fire scenario permits localized fires, any structural elements engulfed by such fires should be checked. Plume fires also permit calculation of smoke release (Annex C, EN 1991-1-2, or PD 7974-2).

4.6.2 5 MW design fire

A design fire has a heat output of 5 MW over an area of 3 m × 3 m and is used to design smoke extract in sprinklered premises. It is not used in structural assessment. The background to the 5 MW design fire is given in Law (1995).

4.7 ZONE MODELLING AND COMPUTATIONAL FLUID DYNAMICS (CFD)

This section is intended only to provide brief introductions to both topics.

4.7.1 Zone modelling

In zone modelling, a number of constituent equations are assembled to describe to the required level of accuracy phenomena such as the behaviour of a thermal plume rising above the local fire source, the ceiling layer of hot combustion gasses and the outward flow through any ventilation area. In general zone, models comprise an upper zone filled with the effects of combustion and a lower zone surrounding the upper zone.

Zone modelling is only appropriate if this type of behaviour can be justified for the compartment in question. Although zone modelling can be handled manually, computer modelling allows parametric assessment and will provide a time-based solution. Zone modelling is more appropriate for smaller simple compartments rather than large or tall compartments where one-dimensional flow may not be appropriate.

4.7.2 Computational fluid dynamics (CFD)

These factors allow the solution of the fundamental equations of conservation of mass, momentum, energy, and chemical species in a three-dimensional spatial grid on a time base to yield a complete set of data within a compartment boundary. However, CFD models require decisions on:

- Treatment of turbulent mixing of combustion gasses, often dealt with by using a time-averaged approach
- The algorithm to be used in calculating the numerical solution at interior node points
- Approach to determining boundary conditions
- Treatment of combustion and thermal radiation

Any model must give a convergent solution to the variables under consideration. Not until a convergent solution is obtained can the validity of the results be assessed. The convergence of the results may well be dependent upon the grid chosen and a sensitivity analysis should possibly be undertaken. It is recommended to use simple models to determine the bounds of acceptability. It should be noted that CFD modelling is a specialist discipline.

For both these methods, reference needs to made to specialist literature, e.g., Cox (1995).

Having established the behaviour of a compartment fire and possible relationships for design purposes between the standard furnace curve and natural fires, it is now pertinent to consider the effect of the fire temperature–time response on structures or structural members within a compartment or on the boundary of a compartment.

Chapter 5

Properties of materials at elevated temperatures

Data about the behaviours of materials at elevated temperatures are needed to allow both the Fourier equation of heat transfer and the structural simulation to be solved. It is convenient, therefore, to divide this chapter into two distinct sections corresponding to the two stages of the analysis.

The major portion of this chapter will concentrate on steel, both structural and reinforcing, and concrete. A substantial amount of data about both materials has been published in two RILEM reports (Anderberg, 1983; Schneider, 1986a). The reports are compendia of the existing data compiled on an *ad hoc* basis using various test methods that are not currently standardized. In addition, data are given on timber, masonry and aluminium, although the need for data about timber is less because current design methods do not generally need temperature-dependent property data. For masonry, no detailed design methods in current use require such data.

5.1 THERMAL DATA

The Fourier equation of heat transfer is given by

$$\nabla(a(\Delta\theta)) = \dot{\theta} \tag{5.1}$$

where θ is the space-dependent temperature and a is the temperature-dependent thermal diffusivity. It should be noted that the thermal diffusivity is related to the density ρ, thermal conductivity λ and specific heat c_v:

$$a = \frac{\lambda}{\rho c_v} \tag{5.2}$$

Data for calculating the thermal response are normally required only for concrete, steel, and aluminium. Although calculations for the thermal response can be carried out for masonry, it is not generally necessary. Strength calculations for timber are carried out on the cores, which is considered either to be temperature-unaffected or to have an allowance made for the temperature effect by a factor applied to the allowable stresses based on member thickness. Thus, the thermal data are not generally required, although in some cases where temperature rise in a core is likely to be significant, a knowledge of the thermal diffusivity is needed.

5.1.1 Steel

The values of the properties concerned are sensibly independent of either the use of steel (structural or reinforcing) and of the strength or grade of the steel.

5.1.1.1 Density

The density of steel may be taken as its ambient value of 7850 kg/m³ over the normally experienced temperature range.

5.1.1.2 Specific heat

Malhotra (1982a) suggested that the specific heat of steel c_a (J/kg°C) may be taken as

$$c_a = 475 + 6,01 \times 10^{-4} \theta_a^2 + 9,64 \times 10^{-2} \theta_a^2 \tag{5.3}$$

Equation (5.3) and test data from Pettersson et al. (1976) and Stirland (1980) reproduced from Malhotra (1982a) are plotted in Figure 5.1. Owing to the discontinuity in the specific heat of steel at around 750°C, Equation (5.3) only holds up to this value. EN 1994-1-2 gives equations that hold up to a temperature of 1200°C:

For $20°C \le \theta_a \le 600°C$:

$$c_a = 425 + 0,773\theta_a - 1,69 \times 10^{-3} \theta_a^2 + 2,22 \times 10^{-6} \theta_a^3 \tag{5.4}$$

For $600°C \le \theta_a \le 735°C$:

$$c_a = 666 - \frac{1302}{\theta_a - 738} \tag{5.5}$$

For $735°C \le \theta_a \le 900°C$:

$$c_a = 545 + \frac{17820}{\theta_a - 731} \tag{5.6}$$

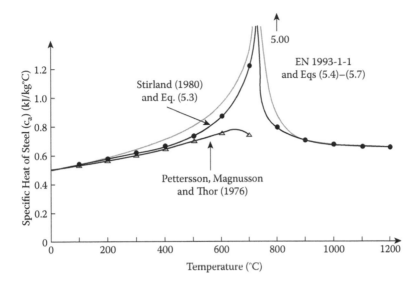

Figure 5.1 Variation of specific heat of steel with temperature. (Source: Malhotra, H.L. (1982a) *Design of Fire-Resisting Structures*. Glasgow: Surrey University Press, With permission.)

For $900 \leq \theta_a \leq 1200°C$:

$$c_a = 650 \tag{5.7}$$

Equations (5.4) to (5.7) are also plotted in Figure 5.1. Note that a constant value for c_a of 600 J/kg°C may be taken in simple calculation models.

5.1.1.3 Thermal conductivity

Typical values for the thermal conductivity λ_a of steel (W/m°C) are given in Figure 5.2 (Pettersson et al., 1976; Malhotra, 1982a). The values of thermal conductivity are slightly dependant on steel strength. The reason for this is not known, but, in any case, it is not very significant. EN 1993-1-2 gives the following equations for λ_a (W/m°C):

For $20°C \leq \theta_a \leq 800°C$:

$$\lambda_a = 54 - 33,3 \times 10^{-3} \theta_a \tag{5.8}$$

For $\theta_a \geq 800°C$:

$$\lambda_a = 27,3 \tag{5.9}$$

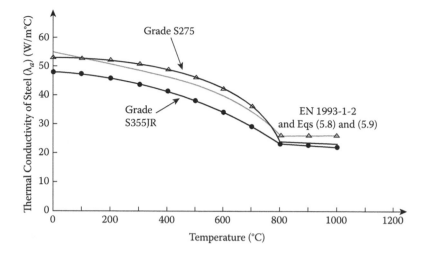

Figure 5.2 Variation of thermal conductivity of steel with temperature. (Sources: Pettersson, O. et al. (1976). *Fire Engineering Design of Steel Structures.* Stockholm: Swedish Institute of Steel Construction; Malhotra, H.L. (1982a) *Design of Fire-Resisting Structures.* Glasgow: Surrey University Press, With permission.)

Equations (5.8) and (5.9) are also plotted in Figure 5.2. Note that it is permissible for approximate calculations to take the thermal conductivity of steel as 45 W/m°C.

5.1.1.4 Thermal diffusivity

Using the data given in Sections 5.1.1.2 and 5.1.1.3 and the standard density values, the thermal diffusivity of steel (m²/hr) shows a sensibly linear relationship with temperature up to 750°C according to the following equation (Malhotra, 1982a):

$$a_a = 0,87 - 0,84 \times 10^{-3}\theta_a \tag{5.10}$$

5.1.2 Concrete

With concrete, the situation is much more complex in that values of the required thermal parameters are dependent on the mix proportions, the type of aggregate, the original moisture content of the concrete, and the age of the concrete. The data presented in this section can thus only be taken as representative of typical concretes.

5.1.2.1 Density

Even though a weight loss caused by the evaporation of both free and bound water occurs when concrete is heated, this loss is not generally enough to cause substantial changes in density and thus a calculation may be considered accurate enough to take ambient values. However, EN 1992-1-2 suggests that change in density with temperature to be used in thermal calculations may be taken as:

For $20°C \le \theta_c \le 115°C$:

$$\rho(\theta_c) = \rho(20°C) \tag{5.11}$$

For $115°C \le \theta_c \le 200°C$:

$$\rho(\theta_c) = \rho(20°C)\left(1 - 0{,}02\frac{\theta_c - 115}{85}\right) \tag{5.12}$$

For $200°C \le \theta_c \le 400°C$:

$$\rho(\theta_c) = \rho(20°C)\left(0{,}98 - 0{,}03\frac{\theta_c - 200}{200}\right) \tag{5.13}$$

For $400°C \le \theta_c \le 1200°C$:

$$\rho(\theta_c) = \rho(20°C)\left(0{,}95 - 0{,}07\frac{\theta_c - 400}{800}\right) \tag{5.14}$$

where $\rho(20°C)$ is the ambient density. For structural calculations, the density of concrete must be taken as its ambient value over the whole temperature range.

5.1.2.2 Specific heat

Figure 5.3 presents values of specific heat for a variety of concretes (Schneider, 1986a). Note that the type of aggregate has a substantial effect on the values. EN 1992-1-2 gives the following equations for the specific heat of dry normal weight concrete (siliceous or calcareous aggregates) in J/kg°C:

For $20°C \le \theta_c \le 100°C$:

$$c_c(\theta_c) = 900 \tag{5.15}$$

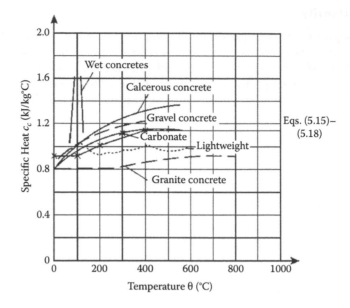

Figure 5.3 Variation of specific heat of concrete with temperature. (*Source:* Schneider, U. (1986a) *Properties of Materials at High Temperatures-Concrete*, 2nd ed., RILEM Report. Kassel: Gesamthochschule. With permission.)

For $100°C \le \theta_c \le 200°C$:

$$c_c(\theta_c) = 900 + (\theta_c - 100) \tag{5.16}$$

For $200°C \le \theta_c \le 400°C$:

$$c_c(\theta_c) = 1000 + \frac{\theta_c - 200}{2} \tag{5.17}$$

For $400°C \le \theta_c \le 1200°C$:

$$c_c(\theta_c) = 1100 \tag{5.18}$$

where the moisture content is not explicitly evaluated in the thermal analysis, a peak $c_{c,peak}$ is added to Equation (5.16) at 100° to 115°C before decaying linearly to 200°C. The values of $c_{c,peak}$ are given in Table 5.1. Equations (5.15) to (5.18) are also plotted in Figure 5.3. For lightweight concrete, a constant value of 840 J/kg°C may be taken (EN 1994-1-2).

Table 5.1 Values of $c_{c,peak}$

Moisture content (%)	$c_{c,peak}$ (J/kg°C)
0	900
1,5	1470
3,0	2020
10,0	5600

5.1.2.3 Thermal conductivity

Figure 5.4 presents values of thermal conductivity for various concretes (Schneider, 1986a). It will be observed that the normal weight aggregate concretes fall into a band with the values for lightweight concrete being substantially lower. EN 1992-1-2 gives the following equations as limits

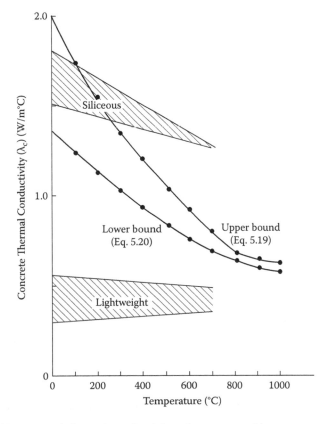

Figure 5.4 Variation of thermal conductivity of concrete with temperature. (*Source:* Schneider, U. (1986a) *Properties of Materials at High Temperatures-Concrete,* 2nd ed., RILEM Report. Kassel: Gesamthochschule. With permission.)

between which the values of thermal conductivity (W/m°C) of siliceous aggregate normal weight concretes lie:

$$\lambda_c = 2,0 - 0,2451\left(\frac{\theta_c}{100}\right) + 0,0107\left(\frac{\theta_c}{100}\right)^2 \qquad (5.19)$$

and

$$\lambda_c = 1,36 - 0,136\left(\frac{\theta_c}{100}\right) + 0,0057\left(\frac{\theta_c}{100}\right)^2 \qquad (5.20)$$

The specification of which curve should be used is normally given in the country's National Annex. For example, in the UK, the lower limit defined in Equation (5.20) is recommended. However, the discrepancy between the analytical curves given above and the results in Schneider is not explained. EN 1994-1-2 indicates it is permissible to take a constant value of 1,6 W/m°C. For lightweight concrete, EN 1994-1-2 gives the following relationship:

For $20°C \le \theta_c \le 800°C$:

$$\lambda_c = 1,0 - \left(\frac{\theta_c}{1600}\right) \qquad (5.21)$$

For $\theta_c > 800°C$:

$$\lambda_c = 0,5 \qquad (5.22)$$

5.1.2.4 Thermal diffusivity

The results plotted in Figure 5.5 indicate, as expected, that two distinct bands of results for normal weight and lightweight concrete exist (Schneider, 1986a).

Using the values of $\lambda_c = 1,60$ W/m°C and $c_c = 1000$ J/kg°C recommended in EN 1992-1-2 for simple calculation methods together with a density $\rho_c = 2400$ kg/m³, it is suggested that an approximate value of the thermal diffusivity a_c can be determined using Equation (5.2) to give $a_c = \lambda_c/\rho c_c = 1,6/(1000 \times 2400) = 0,67 \times 10^{-6}$ m²/s. This value may be high when compared to the $0,417 \times 10^{-6}$ m²/s suggested by Wickström (1985a), and $0,35 \times 10^{-6}$ m²/s and $0,52 \times 10^{-6}$ m²/s suggested by Hertz (1981b) for granite and quartzite concretes, respectively.

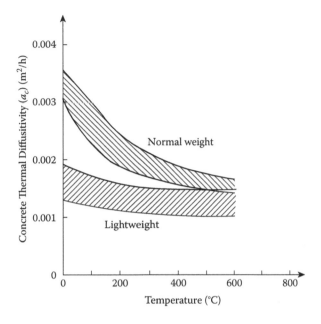

Figure 5.5 Variation of thermal diffusivity of concrete with temperature. (*Source:* Schneider, U. (1986a) *Properties of Materials at High Temperatures-Concrete,* 2nd ed., RILEM Report. Kassel: Gesamthochschule. With permission.)

5.1.3 Masonry

The most significant variable characterizing the high temperature performance of masonry is the density rather than the type of brick (clay or calcium silicate) as the density is a measure of the porosity of the brick.

5.1.3.1 Density

Again, the density should be taken as the ambient value.

5.1.3.2 Specific heat

As shown in Figure 5.6, the specific heat is sensibly independent of the density of the brick (Malhotra, 1982a). Harmathy (1993) gives the following expression for the specific heat of masonry (kJ/kg°C):

$$c_{pm} = 0,851 + 0,512 \times 10^{-3} \theta_m - \frac{8,676 \times 10^3}{\left(\theta_m + 273\right)^2} \tag{5.23}$$

where θ_m in degrees Celsius is the temperature of the masonry.

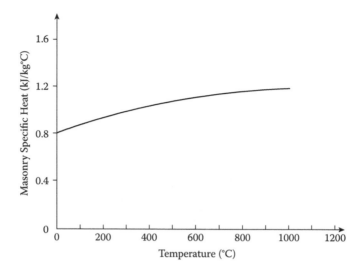

Figure 5.6 Variation of specific heat of masonry with temperature. (*Source:* Malhotra, H.L. (1982a) *Design of Fire-Resisting Structures.* Glasgow: Surrey University Press, With permission.)

5.1.3.3 Thermal conductivity

As shown in Figure 5.7, the thermal conductivity of masonry is dependent on the density of the brick, with high density bricks having higher values of thermal conductivity (Malhotra, 1982b). Welch (2000) indicates that the effect of moisture on the effective thermal conductivity λ' is given by

$$\lambda' = \lambda_0 \left(1 + M\right)^{0,25} \tag{5.24}$$

where λ_0 is the dry thermal conductivity (W/m°C) and M the moisture content (%).

5.1.4 Timber

As mentioned in the introduction to this section, the only thermal property generally needed to determine temperatures within an uncharred core is either the thermal diffusivity (mm²/s) given in Schaffer (1965) as

$$a_w = 0,2421 - 0,1884S \tag{5.25}$$

where S is the specific gravity of the timber, or the thermal conductivity (W/m°C)

$$\lambda_w = \left(2,41 + 0,048M\right)S + 0,983 \tag{5.26}$$

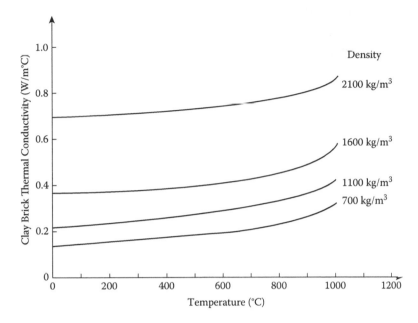

Figure 5.7 Variation of thermal conductivity of masonry with temperature. (*Source:* Malhotra, H.L. (1982b) *Matériaux et Constructions,* 15, 161–170. With permission.)

where M is the moisture content in percent by weight. It appears that both the thermal diffusivity and thermal conductivity are independent of temperature, and the specific heat of oven dry wood in White and Schaffer (1978) is temperature dependent and given in kJ/kg°C as

$$c_{pw} = 1,114 + 0,00486\theta_w \tag{5.27}$$

where θ_w is the temperature of the wood.

5.1.5 Aluminium

Due to the lower softening and melting points of aluminium compared with steel, materials data are required over a more limited temperature range, i.e., up to 300°C. This is true for aluminium in both its pure state and when alloyed.

5.1.5.1 Density

The density used in calculations may be taken as that pertaining at ambient conditions (i.e., 2700 kg/m³).

5.1.5.2 Specific heat

Touloukian and Ho (1973) and Conserva et al. (1992) suggest that over a temperature range of 0 to 300°C, the specific heat may be taken as constant with a value between 900 and 1000 J/kg°C with the slight scatter in the values arising from the effects of the trace elements used in the various alloys. A summary of test results on the specific heat can be found in Kammer (2002). The specific heat resulting from these tests varies little with the type of alloy and depends on the temperature.

According to data in EN 1999-1-2, the specific heat of aluminium varies from 913 J/kg°C at room temperature (20°C) to 1108 J/kg°C at elevated temperature (500°C)—2,1 and 1,7 times higher than that of steel, respectively. EN 1999-1-2 gives the following expression for specific heat c_{al} (J/kg°C) for an aluminium temperature θ_{al} between 0 and 500°C:

$$c_{al} = 0,41\theta_{al} + 903 \tag{5.28}$$

5.1.5.3 Thermal conductivity

The data obtained from tests on the thermal conductivity of aluminium are limited in the literature, especially at elevated temperatures. Tests reported by Holman (2010), Brandes (1993), and Kammer (2002) show that the thermal conductivity is different for different alloys. The common feature, however, is that the thermal conductivity of aluminium is high compared to steel.

Touloukian and Ho (1973) and Conserva et al. (1992) suggest that over a temperature range 0 to 300°C, the thermal conductivity of aluminium may be taken as constant with a value of 180 to 240 W/m°C. The scatter in the values quoted is again due to the effects of the trace elements used in the various alloys and the temperature dependence. EN 1999-1-2 gives two equations for the thermal conductivity of aluminium λ_{al} (W/m°C) dependent upon the alloy for an aluminium temperature θ_{al} between 0 and 500°C:

For alloys in 3000 and 6000 series:

$$\lambda_{al} = 0,07\theta_{al} + 190 \tag{5.29}$$

For alloys in 5000 and 7000 series:

$$\lambda_{al} = 0,1\theta_{al} + 140 \tag{5.30}$$

5.1.5.4 Emissivity

The emissivity resulting from tests varies from 0,03 to 0,11 for new plain aluminium and 0,05 to 0,31 for heavily oxidised aluminium (Holman,

2010; Kammer, 2002). These values apply for unprotected members not engulfed in flames and depend on the alloy and the thickness of the corrosion layer related to the age and environment of the structure. In EN 1999-1-2, the coefficient of emissivity is specified as 0,3 for clean uncovered surfaces and 0,7 for painted and covered (e.g., sooted) surfaces.

5.2 MATERIALS DATA

In order to determine the structural response in a fire, it is necessary to formulate constitutive laws for the mechanical behaviours of the relevant materials at elevated temperatures. A complete formulation is required only where a full analysis is undertaken in order to calculate deformations and displacements. Where it is only necessary to calculate load capacity, a more limited data set can be utilized. Indeed, much early work on evaluating material behaviour was directed to determining specific properties such as tensile strength of steel or compressive strength of concrete at elevated temperatures. Only much later was the need for constitutive models appreciated.

Standard test procedures were laid down by organizations such as British Standards for ambient conditions. However, there are no such standards in force for testing at elevated temperatures, although proposals for such standards are under consideration by Comité Européen de Normalisation (CEN). It should be recognized that as creep (or relaxation) is much higher at elevated temperatures, the rate of loading (stress) or strain used in elevated temperature testing has a far more significant role than at ambient conditions. The rate of heating used to condition the test specimen will also affect the final test results.

5.2.1 Testing régimes

Most early experimental investigations utilized steady-state testing régimes whereby a specimen was heated at a uniform rate of temperature rise in a furnace and allowed to condition by soaking for a predetermined period at the test temperature so that it could attain constant temperature through the cross-section before being loaded to determine the required property. Strength evaluation was generally carried out using a constant rate of stress loading whether for tensile tests on steel or compressive tests on concrete.

A disadvantage of using constant rate of stress loading is that the complete stress–strain curve for concrete cannot be obtained. If, however, the test is carried out at a constant rate of deformation, the resultant loads can be measured, and the complete stress–strain curve can be obtained, provided the test rig used is stiff enough. Classical creep tests are performed by loading a heated specimen at constant load and measuring the resultant strains over a suitable period.

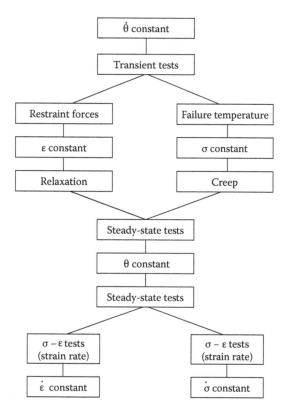

Figure 5.8 Testing régimes to determine mechanical behaviours of materials at elevated temperatures. (*Source:* Malhotra, H.L. (1982b) *Matériaux et Constructions*, 15, 161–170. With permission.)

Observations on specimens heated under constant stress indicated a behaviour pattern that could not be explained purely from the results of steady-state tests. It thus became necessary to consider transient testing in which the temperature was allowed to change. The complete range of possible testing régimes is shown in Figure 5.8 (Malhotra, 1982b). Similar to the thermal data, it is convenient to consider each material separately. After the presentation of typical experimental data, analytical models derived from the results will be considered in Section 5.3.

5.2.2 Steel

The primary thrust on determining the effect of temperature on the properties of steel was strength behaviour characterized by the yield or proof stress, strength, and then a complete stress–strain curve.

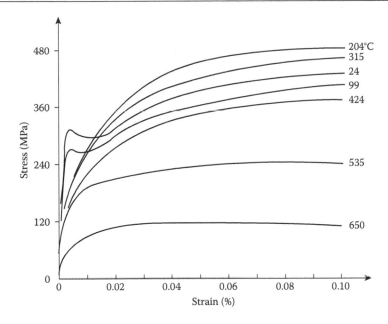

Figure 5.9 Stress–strain curves for structural steel at elevated temperatures. (*Source:* Harmathy, T.Z. and Stanzak, W.W. (1970) In *Symposium on Fire Test Performance, American Society of Testing and Materials*, STP 464, pp. 186–208. With permission.)

5.2.2.1 Strength characteristics

A typical set of stress–strain curves for U.S. Grade A36 steel (yield strength 300 MPa) is shown in Figure 5.9. Note that the strength loss at elevated temperatures is substantial even though at relatively low temperatures there is a slight strength gain (Harmathy and Stanzak, 1970). It should also be noted that at ambient temperatures and only slightly above ambient, a distinct yield plateau is observed. At much higher temperatures, no yield plateau occurs, and the curve resembles that for high yield steel. Reinforcing and pre-stressing steels follow very similar patterns (Figure 5.10; Harmathy and Stanzak, 1970; Anderberg, 1978a).

Due to the large strains exhibited at elevated temperatures in fire-affected members, it is more usual to quote the 1,0 or 2,0%, or in exceptional cases the 5,0%, proof stress rather than the conventional ambient value of 0,2%. Where the variation of proof stress after normalizing the results for ambient strengths of either reinforcing steels or pre-stressing steels is considered, the resultant strength loss is approximately above 350°C which is sensibly independent of the steel type (Figure 5.11a and b; Holmes et al. 1982).

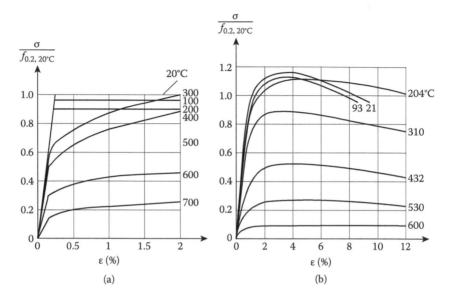

Figure 5.10 Stress–strain curves for (a) reinforcing and (b) pre-stressing steels at elevated temperatures. (*Source:* Anderberg, Y. (1978a) *Armeringsståls Mekaniska Egenskaper vid Höga Temperaturer.* Bulletin 61, University of Lund, Sweden; Harmathy, T.Z. and Stanzak, W.W. (1970) In *Symposium on Fire Test Performance, American Society of Testing and Materials,* STP 464, pp. 186–208. With permission.)

5.2.2.2 Unrestrained thermal expansion

The free thermal expansion of steel is relatively independent of the type of steel (Anderberg, 1983). EN 1992-1-2 gives the following expressions for structural and reinforcing steels:

For $20°C \leq \theta_s \leq 750°C$:

$$\varepsilon_s(\theta_s) = -2{,}416 \times 10^{-4} + 1{,}2 \times 10^{-5}\theta_s + 0{,}4 \times 10^{-8}\theta_s^2 \qquad (5.31)$$

For $750°C \leq \theta_s \leq 860°C$:

$$\varepsilon_s(\theta_s) = 11 \times 10^{-3} \qquad (5.32)$$

For $860°C \leq \theta_s \leq 1200°C$:

$$\varepsilon_s(\theta_s) = -6{,}2 \times 10^{-4} + 10^{-5}\theta_s + 0{,}4 \times 10^{-8}\theta_s^2 \qquad (5.33)$$

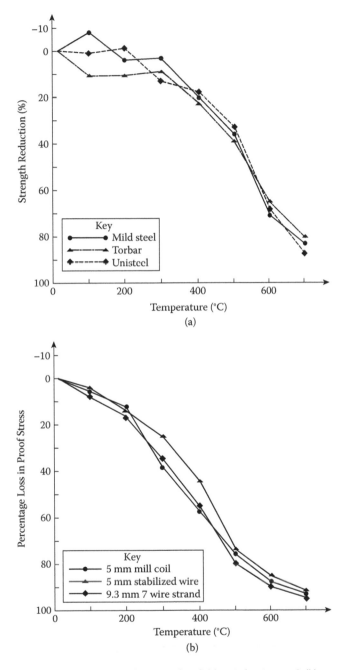

Figure 5.11 Variation of normalized strength of (a) reinforcing and (b) pre-stressing steels with temperature. (*Source:* Holmes, M., Anchor, R.D., Cooke, G.M.E. et al. (1982) *Structural Engineer*, 60, 7–13. With permission.)

The expression for pre-stressing steel, $20°C \leq \theta_s \leq 1200°C$ is

$$\varepsilon_s(\theta_s) = -2,016 \times 10^{-4} + 10^{-5}\theta_s + 0,4 \times 10^{-8}\theta_s^2 \qquad (5.34)$$

Note that EN 1993-1-2 uses $\Delta l/l$ as the symbol for thermal strain rather than $\varepsilon_s(\theta_s)$. EN 1994-1-2 indicates that a simple calculation for thermal strain is

$$\varepsilon_S(\theta_s) = 14 \times 10^{-6}(\theta_s - 20) \qquad (5.35)$$

Figure 5.12 shows a comparison of thermal expansions of steels obtained using Equations (5.31) to (5.35). Note that after 850°C, the three expansion curves are very close.

5.2.2.3 Isothermal creep

For steel, isothermal creep (i.e., creep measured at constant stress and constant temperature) tends to become significant only above ~450°C when reinforcement and structural steels approach their limiting carrying capacities. Typical data on the isothermal creeps of steels are given in Figure 5.13 (Anderberg, 1988).

It is a general practice to analyse creep data using the Dorn temperature-compensated time approach with secondary creep related

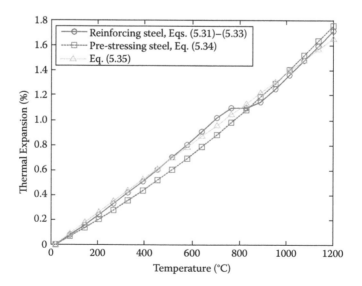

Figure 5.12 Thermal expansions of steels calculated using different expressions.

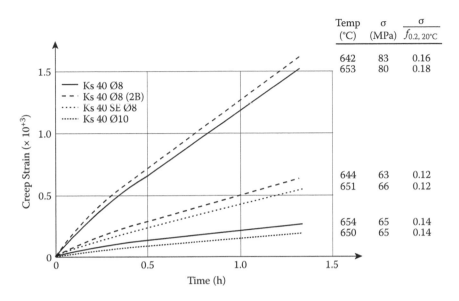

Figure 5.13 Isothermal creep strains for reinforcing steels at elevated temperature. (*Source:* Anderberg, Y. (1983) *Properties of Materials at High Temperatures-Steel.* RILEM Report, University of Lund, Sweden. With permission.)

to the Zener-Hollomon parameter (Section 5.3.1.2). Based on the nature of the test, it is very difficult to obtain repeatable or consistent values of the parameters used to analyze steel creep data, and thus any values must be treated with caution. To alleviate this problem, the practice of using strength data derived from anisothermal creep tests evolved because creep is included implicitly in those data.

5.2.2.4 Anisothermal creep data

A specimen is preloaded to a given stress, heated to failure at a known temperature rate, and the resultant strains are measured. Typical data for British structural steels of Grade S275 (originally 43A) and Grade S355JR (originally 50B) are given in Figure 5.14 (Kirby, 1986; Kirby and Preston, 1988). It should be noted that the results from anisothermal creep tests are very sensitive to the exact composition of a steel. As a result, other steels that show similar trends to those in Figure 5.14 will exhibit differing values (Anderberg, 1983).

5.2.3 Concrete

It is impossible in a text such as this to provide complete data on concrete behaviour owing to the very wide variations in concrete due to mix

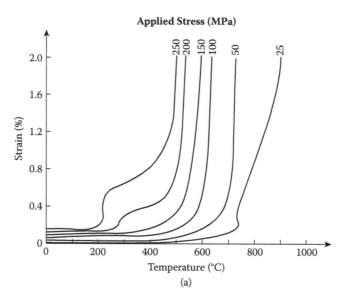

(a)

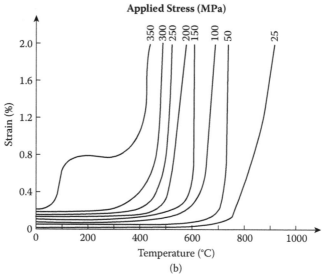

(b)

Figure 5.14 Anisothermal creep strain data for structural steels: (a) grade S275 and
(b) grade S355JR. (Source: Kirby, B.R. and Preston, R.R. (1988) Fire Safety
Journal, 13, 27–37. With permission.)

proportions, aggregate type, age, and other factors. Thus only representative results will be presented.

5.2.3.1 Stress–strain data

Early researchers tended to be interested in measuring specific properties such as compressive strength or elastic modulus rather than obtaining a complete stress–strain profile. Although most work tended to be performed on specimens that were heated without applied loads, it was soon established that heating under an applied stress (or preload) revealed substantially smaller strength reductions (for example, Malhotra, 1956; Abrams, 1968). Data from Malhotra and Abrams are plotted in Figure 5.15.

It was also noted that where different researchers used various test methods such as measuring elastic modulus using cylinders in compression (Maréchal, 1970; Schneider, 1976), dynamic modulus (Philleo, 1958), and cylinders in torsion (Cruz, 1962), the absolute values of the results were different, but the trends were similar (Figure 5.16). The first researcher to establish a complete stress–strain curve for concrete was Furamura (1966), whose results showed that, besides reductions of the compressive stress and elastic modulus, the slope of the descending branch of the curve also reduced (Figure 5.17). Baldwin and North (1973) demonstrated that if Furamura's results were normalized using peak (or maximum) compressive

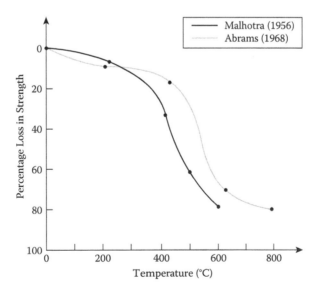

Figure 5.15 Variation of concrete strength with temperature. (*Sources:* Malhotra, H.L. 1956. *Magazine of Concrete Research,* 8, 85–94; Abrams, M.S. (1968) In *Temperature and Concrete,* Special Publication SP-25. Detroit: American Concrete Institute, pp. 33–58. With permission.)

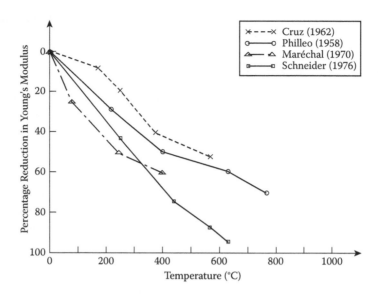

Figure 5.16 Variation in Young's modulus for concrete with temperature. (*Sources:* Philleo, R. (1958) *Proceedings of American Concrete Institute,* 54, 857–864; Cruz, C.R. (1962) *Journal of Portland Cement Association Research and Development Laboratories,* 5, 24–32; Maréchal, J.C. (1970) *Annales de L'Institute Technique du Bâtiment et des Travaux Publics,* 23, 123–146; Schneider, U. (1976) *Fire and Materials,* 1, 103–115. With permission.)

stress and the strain value at peak stress (often called the peak strain), the curves reduced to a single curve that could be curve fitted by

$$\frac{\sigma_C}{\sigma_{0,C}} = \frac{\varepsilon_C}{\varepsilon_{0,C}} \exp\left(1 - \frac{\varepsilon_c}{\varepsilon_{0,c}}\right) \qquad (5.36)$$

where σ_c and ε_c are the stress and strain, respectively, $\sigma_{0,c}$ is the maximum or peak value of stress, and $\varepsilon_{0,c}$ is the strain corresponding to the peak stress value. Note that although Equation (5.36) is specific to Furamura's data, the principle of normalization holds on any set of stress–strain curves for a given concrete. Detailed derivation of Equation (5.36) can be found in Furamura et al. (1987). A more general equation that can be fitted to any concrete stress–strain curve is from Popovics (1973),

$$\frac{\sigma_c}{\sigma_{0,c}} = \frac{\varepsilon_c}{\varepsilon_{0,c}} \frac{n}{n - 1 + \left(\frac{\varepsilon_c}{\varepsilon_{0,c}}\right)^n} \qquad (5.37)$$

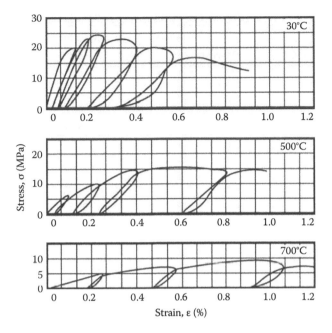

Figure 5.17 Stress–strain curves for concrete with no pre-load at elevated temperatures. (*Source:* Furamura, F. (1966) *Transactions of Architectural Institute of
Japan*, Abstract 7004, p. 686. With permission.)

To fit Equation (5.37) to any test data, only a single parameter n is needed.
This is determined from Equation (5.38)

$$\frac{1}{n} = 1 - \frac{\sigma_{0,c}}{\varepsilon_{0,c}E_c} \tag{5.38}$$

Popovics suggested that n was dependent only on the concrete strength.
However, the analysis of the experimental stress–strain curves obtained
during tests at elevated temperatures indicated that n is also likely to
be dependent on the size and volume fraction of an aggregate since the
aggregate will also affect the nonlinearity of the stress–strain curve (see
Table 5.2). The parameter n can be interpreted as a measure of the degree
of nonlinearity in the stress–strain curve that is affected by the sizes and
volume fractions of aggregates (Hughes and Chapman, 1966).

When a specimen is preloaded during the heating cycle and the stress–
strain characteristic obtained, the characteristics are affected much less by
temperature as seen in Figure 5.18 (Purkiss and Bali, 1988). This is almost
certainly due to the effect of the preload (or stress) that keeps the cracks

Table 5.2 Variations of stress–strain curve parameter *n* with concrete mix

Reference	Mix details	w/c ratio	Preload[a]	n
Furamura (1966)	Quartzite; a/s/c[b]: 2,9/2,8/1	0,70	0	1,58
Anderberg and Thelandersson (1972)	Quartzite 20 mm, a/s/c[b]: 1,92/2,88/1	0,60	0	2,00
Purkiss (1972)	Quartzite 10 mm, a/s/c[b]: 1,2/1/1	0,454	0	6,90
Schneider (1976)	Quartzite OPC 240 kg/m³	0,8	0	3,57
			0,10	2,44
			0,30	2,22
Bali (1984)	Quartzite 10 mm, a/s/c[b]: 3,5/2,5/1	0,65	0	7,25
			0,20	2,31
			0,60	2,68

[a] Preload is stress applied during heating to concrete strength.
[b] a/s/c = aggregate/sand/cement ratio.

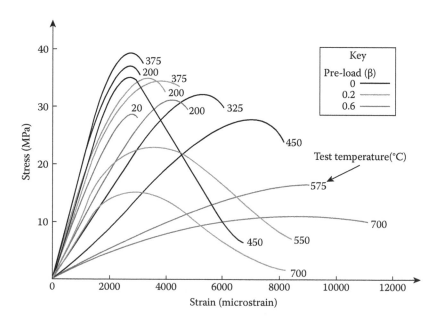

Figure 5.18 Stress–strain curves for concrete with pre-load at elevated temperatures. (*Source:* Purkiss, J.A. and Bali, A. (1988) In *Proceedings of 10th Ibausil.* Weimar: Hochschule für Architektur und Bauwesen, Section 2/1, pp. 234–239. With permission.)

that would otherwise have formed from thermal incompatibility of the aggregate and the matrix closed or at least reduced. This postulate, at least for residual crack density measurements, is confirmed by Guise (1997).

5.2.3.2 Creep

The creep of concrete at elevated temperatures is far greater than that at ambient conditions. Since creep can be considered an Arrhenius-type phenomenon, the creep rate is proportional to $\exp(-U/\Theta)$, where U is the activation energy and Θ is the absolute temperature. Over the time period considered in most creep tests at elevated temperatures for 5 hr or less, the variation of creep strains with time can be represented by a power law. Normally only primary and secondary creeps are observed, although at very high temperatures and stresses increasing creep rates can be observed, thereby indicating the possibility that incipient creep rupture could occur. Typical creep data for concrete are given in Figure 5.19 (Anderberg and Thelandersson, 1976).

5.2.3.3 Free thermal expansion

Free thermal expansion is predominantly affected by aggregate type and is not linear with respect to temperature as shown in Figure 5.20 (Schneider, 1986a). This nonlinear behaviour is in part due to chemical changes in the aggregate (e.g., the breakdown of limestone around 650°C) or physical changes in the aggregate (e.g., the α-β quartz phase transformation around 570°C in siliceous aggregates). The nonlinear behavior is also caused in part by thermal incompatibilities between the aggregate and the matrix. The presence of free moisture will affect the results below 150°C since the water driven off may cause net shrinkage. EN 1992-1-2 gives the following equations for the free thermal strain of normal weight concrete (siliceous aggregate):

For $20°C \leq \theta_c \leq 700°C$:

$$\varepsilon_{th,c} = -1{,}8 \times 10^{-4} + 9 \times 10^{-6}\theta_c + 2{,}3 \times 10^{-11}\theta_c^3 \tag{5.39}$$

For $700°C \leq \theta_c \leq 1200°C$:

$$\varepsilon_{th,c} = 14 \times 10^{-3} \tag{5.40}$$

The equations for calcareous aggregate concrete are
For $20°C \leq \theta_c \leq 805°C$:

$$\varepsilon_{th,c} = -1{,}2 \times 10^{-4} + 6 \times 10^{-6}\theta_c + 1{,}4 \times 10^{-11}\theta_c^3 \tag{5.41}$$

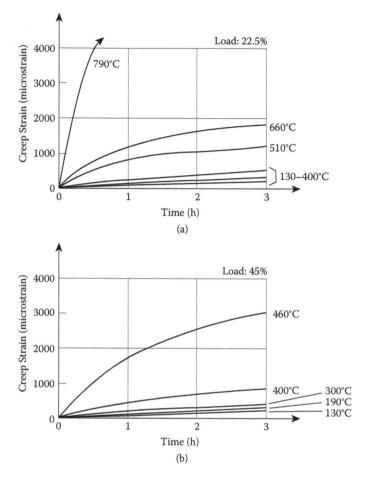

Figure 5.19 Isothermal creep data for concrete at elevated temperatures. (*Source:* Anderberg, Y. and Thelandersson, S. (1976) *Stress and Deformation Characteristics of Concrete 2: Experimental Investigation and Material Behaviour Model,* Bulletin 54. University of Lund, Sweden. With permission.)

For $805°C \leq \theta_c \leq 1200°C$:

$$\varepsilon_{th,c} = 12 \times 10^{-3} \qquad (5.42)$$

For approximate calculations, the coefficients of thermal strains may be taken as:

For siliceous aggregate concrete:

$$\varepsilon_{th}(\theta_c) = 18 \times 10^{-6}(\theta_c - 20) \qquad (5.43)$$

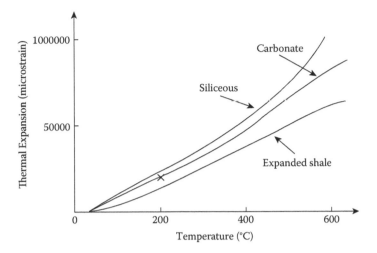

Figure 5.20 Thermal expansion of concrete. (*Source:* Schneider, U. (1986a) *Properties of Materials at High Temperatures-Concrete*, 2nd ed., RILEM Report. Kassel: Gesamthochschule. With permission.)

For calcareous aggregate concrete:

$$\varepsilon_{th}(\theta_c) = 12 \times 10^{-6}(\theta_c - 20) \tag{5.44}$$

EN 1994-1-2 gives the following expression for lightweight concrete:

$$\varepsilon_{th}(\theta_c) = 8 \times 10^{-6}(\theta_c - 20) \tag{5.45}$$

5.2.3.4 Transient tests

If a concrete specimen is heated at a constant rate under constant applied stress and the strains are measured, these strains represent a function of the applied stress as shown in Figure 5.21 (Anderberg and Thelandersson, 1976). The strains are initially recorded as tensile and become compressive as the temperature continues to increase, and finally the strain rate is very high approaching failure. At very high stress levels, the strains may be compressive over a whole temperature range. The magnitude of the measured strains is a function of the heating rate, the concrete mix (including aggregate type, although most tests have been conducted on siliceous aggregate) and the stress level.

If an attempt is made to calculate the total strains from free thermal expansion strains, instantaneous elastic strains, and strains derived from classical creep tests, an additional term known as transient strain

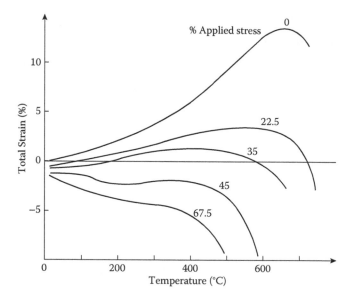

Figure 5.21 Variation of total strain with temperature for concrete heated under load. (*Source:* Anderberg, Y. and Thelandersson, S. (1976) *Stress and Deformation Characteristics of Concrete 2: Experimental Investigation and Material Behaviour Model,* Bulletin 54. University of Lund, Sweden. With permission.)

must be incorporated in the calculation to yield the requisite strain balance. This transient strain is essentially due to stress-modified and thermally-induced incompatibilities between the aggregate and the cement–mortar matrix.

Note that these transient strains are only exhibited on the first heating cycle, but not the first cooling cycle. Any subsequent heating and cooling cycles do not exhibit such strains. Furthermore, these transient strains can be determined only from measurements of total strain, free thermal strain, and elastic strains.

The transient strains were first identified by Anderberg and Thelandersson (1976). A substantial amount of data on transient strains has also been reported following work at Imperial College in London (Khoury et al., 1985a and b, 1986; Khoury, 1992) that analyzed the effects of applied stress and aggregate type along with heating rate. Although Khoury et al. produced master transient strain curves, they did not attempt to predict the amount of transient strain after the elastic strain on first loading is taken into account. The prediction of transient strains is covered in Section 5.3.2.

5.2.3.5 Tensile strength of concrete at elevated temperatures

There are only limited data available on tensile strength whether based on direct or splitting strength. Felicetti and Gambarova (2003) indicate that for direct tension tests, the tensile strength drops roughly linearly to around 0,25 times the ambient strength at 600°C. They also indicate that the ratio between splitting tensile strength and direct tensile strength for any concrete is around 1,2 to 1,6 in the temperature range of 20 to 750°C, although there is no consistent pattern to the exact values.

5.2.3.6 Bond strength

As would be expected, the bond strength between concrete and steel, whether reinforcing or pre-stressing, decreases with increasing temperature. The magnitude of the loss is a function of the reinforcement (smooth or deformed) and type of concrete. The exact results obtained will also be dependent on the test method used, as there is no standardized test procedure. Some typical results are given in Figure 5.22 (Schneider, 1986a).

Bond strength is rarely critical in reinforced concrete as the reinforcement on the bottom face of a beam or slab will carry only a small proportion of the applied load, and the reinforcement at the support will only be slightly affected by temperature and thus able to carry full bond stresses. Bond is likely to be more critical in pre-stressed concrete, although few failures have directly occurred due to loss of bond.

5.2.3.7 High strength concrete (HSC) and self-compacting concrete (SCC)

HSC has a cylinder strength greater than 60 MPa. HSC and SCC can be considered together as SCC currently is produced with strengths similar to those of HSC.

High performance concrete — Phan and Carino (1998) provide an excellent overview of the variations of strength properties with temperature. Their summary of loss in strength for both normal weight and lightweight concretes is given in Figure 5.23a and b. Loss in elastic modulus appears in Figure5.23c. The performance of HSCs is erratic in that some perform similarly to normal strength concretes (NSCs) and others perform substantially worse because of a strength loss around 40% at 200 to 300°C before a slight regain of strength around 400 to 450°C. Otherwise, there is little difference between HSC and normal concrete.

The degradation in elastic modulus is similar to that of normal concrete. Purkiss (2000) suggested that the effect of age was probably a factor in the behaviour of HSC at elevated temperatures due probably to moisture

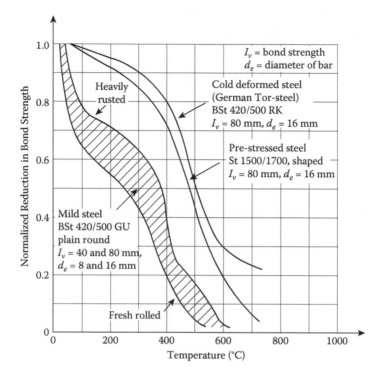

Figure 5.22 Variation of bond strength between reinforcement and concrete at elevated temperatures. (*Source:* Schneider, U. (1986a) *Properties of Materials at High Temperatures-Concrete*, 2nd ed., RILEM Report. Kassel: Gesamthochschule. With permission.)

content. Castillo and Durani (1990) indicate that preloading specimens before heating exacerbates the strength loss below 400°C although at around 500°C a strength increase around 20% is noted (Figure 5.24). They too indicate little difference in elastic modulus behaviours of HSC and NSC.

Gillen (1997) indicates little difference between the behaviours of normal aggregate HSC and lightweight aggregate HSC. Fu et al. (2005) provide data on the complete stress–strain curve for various HSCs under both stressed and unstressed conditions. Their results appear slightly anomalous in that the concretes heated under stress appear to perform similarly or worse with regard to strength than those heated unstressed. The effect on the elastic modulus of heating under load appears erratic, with the normalized modulus from the stressed specimens dropping off more slowly. Schneider (1986b, 1988) indicates that for NSC, the elastic modulus increases when the concrete is heated under load. Hassen and Colina (2006) provide some transient strain data for high performance concretes.

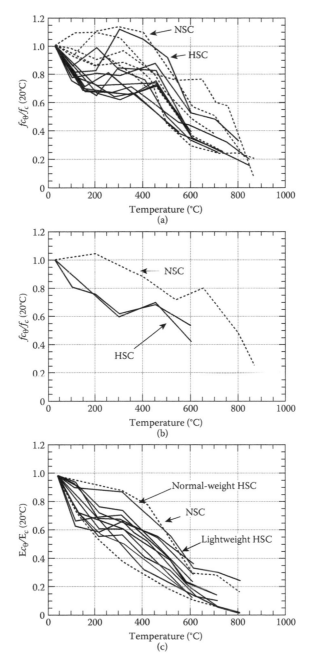

Figure 5.23 Compressive strength–temperature relationships for (a) NWA concrete, (b) LWA concrete, and (c) modulus of elasticity–temperature relationships for NWA and LWA concrete from unstressed test results.

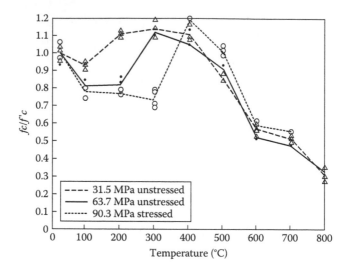

Figure 5.24 Variation of compressive strength with increased temperature. (*Source:* Castillo, C. and Durani, A.J. (1990) *ACI Materials Journal*, 87, 47–53. With permission.)

Self-compacting concrete — Persson (2003) provides much of the available data on SCCs of strengths between 15 and 60 MPa, summarized by the following formulae:

For compressive strength:

$$\frac{\sigma_{c,\theta}}{\sigma_{c,20}} = -0,0000012\theta_c^2 - 0,000131\theta_c + 0,99 \tag{5.46}$$

For elastic modulus:

$$\frac{E_{c,\theta}}{E_{c,20}} = 0,0000013\theta_c^2 - 0,00221\theta_c + 1,04 \tag{5.47}$$

For strain at peak stress:

$$\frac{\varepsilon_{c,\theta}}{\varepsilon_{c,20}} = -0,0000022\theta_c^2 - 0,00279\theta_c + 0,95 \tag{5.48}$$

Tests on dynamic and static moduli (Persson, 2003) give the following relationships between temperature and concrete strength and the relevant modulus.
For dynamic modulus $E_{dyn,\theta}$ (GPa):

$$E_{dyn,\theta} = (0,837 - 0,00079\theta_c)\sigma_{c,20} \tag{5.49}$$

For static modulus $E_{stat,\theta}$ (GPa):

$$E_{stat,\theta} = \left(0,00211 + 37,4 \times 10^{-6}\theta_c - 91,6 \times 10^{-9}\theta_c^2\right)\sigma_{c,20}^2 \qquad (5.50)$$

subject to 5 MPa $\leq \sigma_{c,20} \leq$ 60 MPa. Persson (2003) also indicates that when polypropylene fibres are introduced, the strength drops by 2,3% per kg/m³ for fibre dosages of 0 to 4 kg/m³. This is roughly in agreement with the 5% for a dosage of 3,0 kg/m³ quoted by Clayton and Lennon (1999) for high performance concrete.

5.2.3.8 Fibre concretes

Steel fibre concrete — Lie and Kodur (1995, 1996) give the following set of equations for thermal conductivity (W/m°C) for steel fibre-reinforced siliceous aggregate concrete:

For $0°C \leq \theta_c \leq 200°C$:

$$\lambda_c = 3,22 - 0,007\theta_c \qquad (5.51)$$

For $200°C \leq \theta_c \leq 400°C$:

$$\lambda = 2,24 - 0,0021\theta_c \qquad (5.52)$$

For $400°C \leq \theta_c \leq 1000°C$:

$$\lambda = 1,40 \qquad (5.53)$$

For fibre-reinforced carbonate concrete, the values of thermal conductivity become:

For $0°C \leq \theta_c \leq 500°C$:

$$\lambda_c = 2,00 - 0,001775\theta_c \qquad (5.54)$$

For $500°C \leq \theta_c \leq 1000°C$:

$$\lambda_c = 1,402 - 0,00579\theta_c \qquad (5.55)$$

The values for steel fibre-reinforced concrete are only slightly higher than those for nonfibre concrete. The data provided by Lie and Kodur (1995, 1996) for specific heat indicate that except for small peaks at around 100 and 420°C, the specific heat for siliceous aggregate can be taken as 1000 J/kg°C over a temperature range of 0 to 1000°C. This is also true for carbonate aggregate up to 600°C after which a severe peak of 8000 J/kg°C occurs at 700°C, followed by a return to the earlier value. Lie and Kodur (1995, 1996) provide the following data on coefficients of thermal expansion, α_c. For siliceous aggregate:

For $0°C \leq \theta_c \leq 530°C$:

$$\alpha_c = -0{,}00115 + 0{,}0000016\theta_c \tag{5.56}$$

For $530°C \leq \theta_c \leq 600°C$:

$$\alpha_c = -0{,}0364 + 0{,}0000083\theta_c \tag{5.57}$$

For $600°C \leq \theta_c \leq 1000°C$:

$$\alpha_c = 0{,}0135 \tag{5.58}$$

calcareous aggregate:

For $0°C \leq \theta_c \leq 750°C$:

$$\alpha_c = -0{,}00115 + 0{,}000001\theta_c \tag{5.59}$$

For $750°C \leq \theta_c \leq 1000°C$:

$$\alpha = -0{,}05187 + 0{,}0000077\theta_c \tag{5.60}$$

Purkiss (1987) notes that the fibres reduce the thermal expansion around 20% for temperatures up to 500°C, but after this there is little difference. This effect is sensibly independent of fibre volume fraction. Lie and Kodur (1995, 1996) also give a formulation for stress–strain curves. Independent of the aggregate, the stress–strain curves are given by:

For $\varepsilon_c \leq \varepsilon_{0,\theta}$:

$$\frac{\sigma_{c,\theta}}{\sigma_{c,0}} = 1 - \left(\frac{\varepsilon_{0,\theta} - \varepsilon_c}{\varepsilon_{0,\theta}}\right)^2 \tag{5.61}$$

For $\varepsilon_c > \varepsilon_{0,\theta}$:

$$\frac{\sigma_{c,\theta}}{\sigma_{c,0}} = 1 - \left(\frac{\varepsilon_{0,\theta} - \varepsilon_c}{3\varepsilon_{0,\theta}}\right)^2 \tag{5.62}$$

where the strain at maximum stress $\varepsilon_{0,\theta}$ is given by

$$\varepsilon_{0,\theta} = 0{,}003 + \left(7{,}0\theta_c + 0{,}05\theta_c^2\right) \times 10^{-6} \tag{5.63}$$

and the strength reduction factors $\sigma_{c,0}/\sigma_{c,0,20}$ are given by

For $0°C \leq \theta_c \leq 150°C$:

$$\frac{\sigma_{c,0,\theta}}{\sigma_{c,0}} = 1 + 0,000769(\theta_c - 20) \tag{5.64}$$

For $150°C \leq \theta_c \leq 400°C$:

$$\frac{\sigma_{c,0,\theta}}{\sigma_{c,0}} = 1,1 \tag{5.65}$$

For $\theta_c > 400°C$:

$$\frac{\sigma_{c,0,\theta}}{\sigma_{c,0}} = 2,011 - 2,353 \frac{\theta_c - 20}{1000} \tag{5.66}$$

Fairyadh and El-Ausi (1989) present data on the splitting tensile strength at elevated temperatures of both glass and steel fibre-reinforced concrete. They indicate that the splitting tensile strength of concrete with no fibres reduces to zero at 700°C. When normalized, the results for 1,0% fibres show less degradation than those for 0,7 and 0,5% that are sensibly similar. All the steel fibre concretes retain around 25% of their strength at 800°C. The glass fibre results are anomalous in that the mix with 0,5% fibres showed less degradation than those with 1,0 or 0,7%. The strength retention at 800°C was around 14%.

Refactory concretes and slurry-infiltrated fibre concrete (SIFCON) — Robbins and Austen (1992) carried out tests on fibre-reinforced refactory concretes at elevated temperatures and reported a loss in strength around 60% at 850°C. The influence of fibre type is more marked at higher fibre contents, although the fibre content would appear not of primary importance when the results are normalized.

SIFCON effectively is a slurry concrete with maximized fibre content around 10%. Purkiss et al. (2001) reported the data on both thermal diffusivity and thermal expansion. They indicate that at 20 to 100°C the thermal diffusivity varies linearly with temperature, with a value of 2 mm²/sec at 20°C to 0,5 mm²/sec for SIFCON, and 0,7 mm²/sec for the matrix and remains constant thereafter. This behaviour is similar to that of normal fibre-reinforced concrete with fibre content to 3% (Purkiss, 1987). The thermal expansion of SIFCON is sensibly constant over a temperature range of 20 to 800°C at a value of 22 to 23 µstrain/°C. The matrix expands by 2000 µstrain at 100°C and remains at this value thereafter.

5.2.3.9 Multiaxial behaviour

Many concrete structures such as shear walls, tunnels, reactor vessels, dams, and offshore structures undergo multiaxial stress states. The analysis of such structures requires a multiaxial stress–strain constitutive relationship and a multiaxial strength criterion. Extensive research conducted since the 1970s describes the behaviour of concrete under multiaxial stress states (Popovics, 1970; Chen, 1982; ACI, 1984). Most of them, however, cover concrete only at ambient temperature.

Experiments on concrete behaviour under biaxial compressive loading at different temperatures were reported by Ehm and Schneider (1985). The results show that both stiffness and biaxial compressive strength decrease with increased temperature. The strength envelopes in biaxial compression at various temperatures established by Ehm and Schneider (1985) are shown in Figure 5.25. Note that the equibiaxial compressive strength decreases with temperature, but, at a given temperature, the decrease in biaxial compressive strength is smaller than the decrease of uniaxial compressive strength.

The confinement effect is more pronounced in heated concrete because elevated temperatures cause the degradation of the microstructure and an increase in porosity. As a result, the strength envelope shape changes toward a more elongated shape at elevated temperatures.

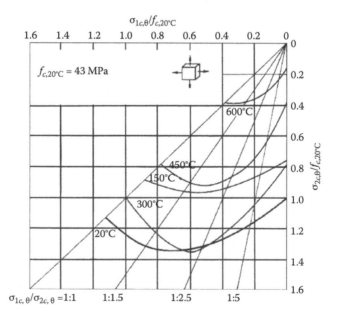

Figure 5.25 Bi-compressive strength envelopes at various different temperatures. (Source: Ehm, C. and Schneider, U. (1985) Cement and Concrete Research, 15, 27–34. With permission.)

Failure tests of high performance concrete under biaxial and triaxial tension–compression at high temperatures ranging from 200 to 600°C were reported by He and Song (2010). The results show that the strengths of high performance concrete under multiaxial tension–compression for all stress ratios are less than the corresponding uniaxial tensile and compressive strengths at the same temperature. Tension failure was found when the stress ratio between tension and compression was below 0,1.

Currently, no theoretical strength model is available for concrete at elevated temperatures. This is partly due to a lack of understanding of the failure mechanisms of concrete at high temperatures and partly because of very limited experimental data. In this circumstance, one may use existing failure criteria for concrete at ambient temperature and consider reduced uniaxial compressive and tensile strengths to account for the effects of temperatures.

Several failure criteria such as the Mohr-Coulomb, Drucker-Prager, and Bresler-Pister criteria developed for general brittle materials have been used for concrete. Figure 5.26 shows the failure surfaces for plane stress obtained using all these criteria (Wang et al., 2013). Note that both Mohr-Coulomb and Drucker-Prager strength criteria use only two parameters to represent strengths in uniaxial tension and compression. The Bresler-Pister strength criterion uses three parameters representing strengths in uniaxial tension, compression, and equal biaxial compression. The latter allows the shape of a failure surface to be adjustable. Existing experimental data at elevated temperatures (Ehm and Schneider, 1985; He and Song, 2010) seem to support the failure surfaces of shapes similar to the Bresler-Pister surface, but with a tension–tension quadrant similar to that of Mohr-Coulomb.

Some advances have been made in developing concrete constitutive models at high temperatures. Thelandersson (1982) attempted to formulate a multiaxial constitutive model for concrete at elevated temperatures using volumetric thermal strain as a scalar damage parameter. He was able to obtain reasonable correlation between the model and uniaxial test results. He also noted that many more data were required to improve and validate the model.

Khennane and Baker (1992) developed a thermoplasticity model based on the plasticity theory for modelling of concrete under biaxial stress states. Their model does not include the effects of degradation of elastic properties. Another example of a concrete model based on thermoplasticity theory was proposed by Heinfling (1998) and takes into account the increasing temperature sensitivity of compressive strength to hydrostatic pressure.

Concrete models at high temperatures based on the damage theory were developed by Gawin et al. (2004) and Baker and de Borst (2005). The models use a thermal damage parameter to capture the degradation of Young's modulus with temperature.

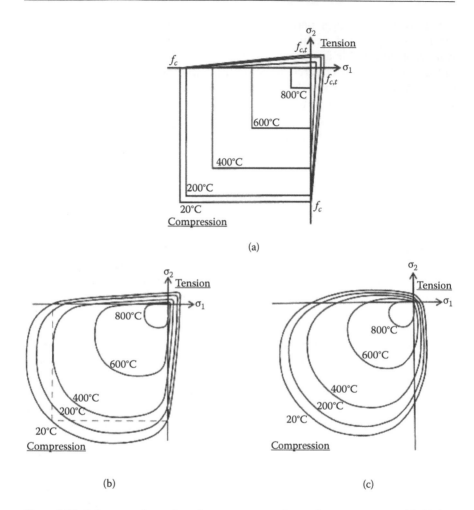

Figure 5.26 Failure surfaces for plane stress at elevated temperatures. (a) Mohr-Coulomb, (b) Drucker-Prager and (c) Bresler-Pister strength criteria. (*Source:* Wang, Y., Burgess, I., Wald, F. et al. (2013) *Performance-Based Fire Engineering of Structures.* Boca Raton: FL: CRC Press.)

The application of a plastic damage model to concrete at high temperatures by Nechnech et al. (2002) incorporated the effect of transient creep using Anderberg's formula. The notion of thermal damage is also introduced to account for the temperature dependence of the elastic modulus. Plasticity is formulated in the effective stress space and the isotropic state of damage is modelled by means of two damage scalars. However, the model was only developed in two dimensions and thus has limited applications.

Luccioni et al. (2003) proposed a similar plastic damage model for concrete at high temperatures that used a thermal damage parameter to measure the deterioration produced by high temperatures to incorporate the temperature effects. de Sa and Benboudjema (2011) investigated the influence of different modelling assumptions such as transient creep and isotropic or orthotropic damage by directly incorporating temperature-dependent Young's modulus into the expression of the stiffness tensor instead of using a thermal damage parameter.

More recently, Gernay (2012) developed a multiaxial constitutive model by using combined damage and plasticity theories for concrete in a fire situation including transient creep and cooling phases. The development of permanent strains in the material is treated in the plastic part of the model whereas the degradation of elastic properties is handled in the damage part of the model. To allow for capturing properly the isotropic state of damage in concrete, including stiffness recovery when the stress changes from tension to compression through closure of tensile cracks, a fourth-order damage tensor is used to map the effective stress into the nominal stress.

The analysis of concrete structures at elevated temperatures under multiaxial stress states is very complicated. It usually involves the use of nonlinear finite element methods, for which interested readers can read the relevant textbooks, for example, Belytschko et al. (2000), Zienkiewicz and Taylor (2000).

5.2.4 Timber

The situation with timber is different when compared to steel or concrete. When timber is subjected to a fire, the outer layer of the timber member chars and loses all strength while retaining a role as an insulating layer that prevents excessive temperature rise in the core. The central core is slightly affected by temperature with some small loss of strength and elasticity. Only three properties are thus required to determine the fire performance of timber: the rate of charring and both the strength and elasticity loss in the central core. The design values of these properties are provided in EN 1995-1-2. The guidance of the use of these properties is covered in the National Annex (NA). In the UK, this is specified in NA to BS EN 1995-1-2:2004.

5.2.4.1 Rate of charring

As noted in Section 3.2.6, the result from a standard furnace test is very much affected by the furnace characteristics, notably heat flux falling on a specimen. This is even more relevant for measurements of the charring rate of timber, which is very much affected by heat flux rather than the absolute rate of temperature rise. For example, Hadvig (1981) reports results in which the charring on the bottom face of a beam is up to 20% higher than that on the side faces. However, when timber members were exposed to the standard furnace temperature–time curve, the charring rate on a given face of the member was sensibly constant up to 90 min. This rate is dependent on the timber type (or density).

After 90 min, tests on timber exposed to the standard furnace curve seem to indicate a substantial rise in the rate of charring leading to a rapid loss of section. This effect is likely to be exacerbated by the fact that in most tests, the size of timber member used is such that little of the central core remains after 90 min. Values of charring rates from a wide range of tests are given in Table 5.3. Note the consistency in the data which, for design

Table 5.3 Experimentally derived timber char rates

Reference	Type of specimen	Timber type	Char rate (mm/min)
Wardle (1966)	Beam	Spruce	0,5 to 0,6
		Douglas Fir	0,6
		Baltic Fir (Laminated)	0,6
	Column	Fir	0,55
		Fir (Glulam)	0,66
Schaffer (1967)	Panel	Douglas Fir	
		Southern Pine	0,68
		White Oak	
Rogowski (1969)	Column	Hemlock	0,55 (par)
			0,67 (perp)
		Fir	0,64 (par)
			0,78 (perp)
		Redwood	0,71 (par)
			0,74 (perp)
		Cedar	0,71 (par)
			0,85 (perp)
Tenning (1969)	Beam	Glulam	0,62
		Laminated Pine	0,5 to 0,66
		Oak	0,4
		Teak	0,35
Ödeen (1969)	Beam	Fir	0,6 to 0,62
		Oak	0,4
		Teak	0,37
Fredlund (1988)	Slab	Spruce	0,265
		Pine	0,339
		Chipboard	0,167

Note: Two values for results from Rogowski (1969) are quoted as the tests were carried out on laminated timber columns. The values quoted as par are parallel to the laminations and perp are perpendicular to the laminations. None of the other tests on laminated sections differentiated the rates in the two directions.

purposes, allows the classification of various timbers into a relatively small number of categories.

Tests on panels can produce higher charring rates than those on beams or columns. The charring rates are, however, dependent to a limited extent on both moisture content and density. Schaffer (1967) gave equations relating the charring rate β_0 (mm/min), the moisture content M (percent by weight) and the dry specific gravity S values for three different timber types as follows:

For Douglas fir:

$$\frac{1}{\beta_0} = 0,79\left[(28,76+0,578M)S+4,187\right] \tag{5.67}$$

For southern pine:

$$\frac{1}{\beta_0} = 0,79\left[(5,832+0,120M)S+12,286\right] \tag{5.68}$$

For white oak:

$$\frac{1}{\beta_0} = 0,79\left[(20,036+0,403M)S+7,519\right] \tag{5.69}$$

5.2.4.2 Strength and elasticity loss

Few data exist on strength losses in timber subjected to fire, partly because it is known that temperatures at charred boundaries drop rapidly to near ambient in the core. Existing data indicate that both strength and elasticity losses are low (Figure 5.27; Gerhards, 1982). It also seems that the reduction per unit rise in temperature is sensibly independent of the timber type. Sano (1961) reported the following strength results (MPa) for compression parallel to the grain on two timbers tested at temperatures between −60 and +60°C:

For spruce:

$$\sigma_{par} = 49,3-0,424\theta_w \tag{5.70}$$

For ash:

$$\sigma_{par} = 57,4-0,392\theta_w \tag{5.71}$$

where θ_w is the temperature in the wood.

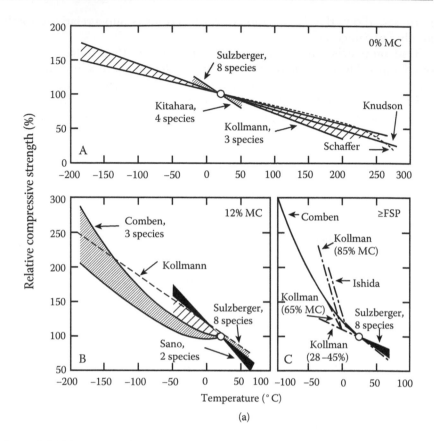

Figure 5.27 Variation of (a) strength and (b) Young's modulus for timber at elevated temperature. (*Source:* Gerhards, C.C. (1982) *Wood and Fiber*, 14, 4–36. With permission.) (*continued*)

5.2.5 Masonry

Few data are reported on the strength of masonry at elevated temperatures although some data are available on residual properties (Section 13.3.1.5). The residual property data suggest that strength degradation in mortar is likely to be the controlling factor in strength performance of masonry, although this has not yet been demonstrated by tests.

Limited data cover the thermal expansion of lightweight masonry units utilizing scoria aggregate. Such data are given for three types of masonry units in Figure 5.28, Note that in certain cases, the coefficient of thermal expansion is negative (Gnanakrishnan and Lawther, 1989).

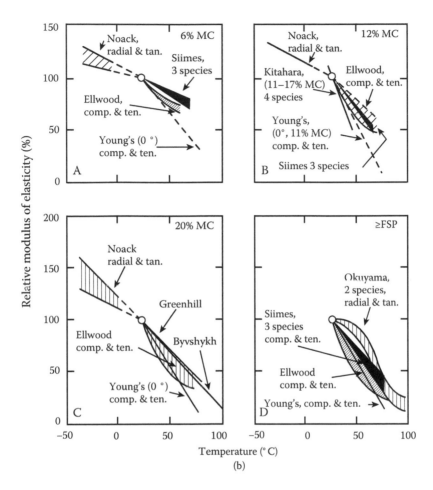

Figure 5.27 (continued). Variation of (a) strength and (b) Young's modulus for timber at elevated temperature. (Source: Gerhards, C.C. (1982) Wood and Fiber, 14, 4–36. With permission.)

5.2.6 Aluminium

5.2.6.1 Strength characteristics

Aluminium has no clearly defined yield point in its stress–strain curve. It is common to use the proof strength that produces 0,2% permanent tensile strain in a tensile test specimen to approximate its yield strength. The values of the 0,2% proof strength in the standard were obtained from steady-state tensile tests, i.e., tensile tests conducted at a certain strain rate and constant

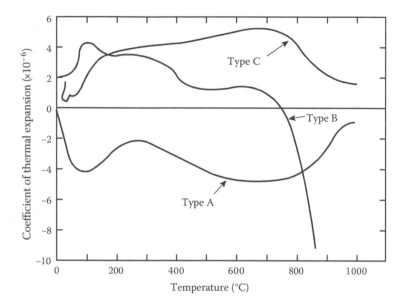

Figure 5.28 Thermal expansion of masonry. (*From:* Gnanakrishnan, N. and Lawther, R. (1989) in Proceedings of International Symposium on Fire Engineering for Building Structures and Safety, Melbourne: Institution of Engineers, Australia, pp. 93–99. With permission.)

temperature after a certain period of exposure to the temperature. Owing to the relatively low temperatures reached in aluminium before collapse ensues, there appears little need for creep data as insignificant creep will have occurred and it will be sufficiently accurate to consider basic stress–strain behaviour with some account taken of creep in the actual results.

Elasticity modulus data $E_{al,\theta}$ for aluminium alloys at elevated temperatures were determined using tensile tests (Kaufman, 1999) and bending tests (Maljaars et al., 2008). The results indicate that $E_{al,\theta}$ depends on the alloy series, but is almost independent of the alloy type within a series. The following two simple equations were derived from the data of Maljaars et al. (2010) for a temperature θ_{al} range from 20 to 350°C:

For 5xxx series alloys:

$$E_{al,\theta} = 72000 - 10\theta_{al} - 0{,}21\theta_{al}^2 \qquad (5.72)$$

For 6xxx series alloys:

$$E_{al,\theta} = 69000 - 10\theta_{al} - 0{,}21\theta_{al}^2 \qquad (5.73)$$

where the units used in Equations (5.72) and (5.73) are degrees Celsius for θ_{al} and MPa for $E_{al,\theta}$, respectively. Note that the modulus of elasticity at

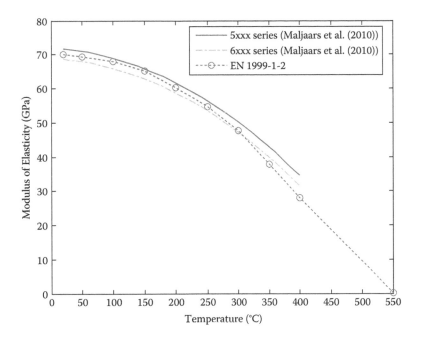

Figure 5.29 Reduction of modulus of elasticity of aluminium alloys at elevated temperatures.

elevated temperatures provided in EN 1999-1-2 is independent of the alloy series. Figure 5.29 shows a comparison of Equations (5.72) and (5.73) and the data recommended in EN 1999-1-2.

Kaufman (1999) reported the results of steady-state tensile tests of 158 alloys and tempers at various elevated temperatures. The results showed that the reduction of the 0,2% proof strength is between 80 and 20% for temperatures between 175 and 350°C. Figure 5.30 shows the reduction of the 0,2% proof strength as a function of temperature for 5xxx and 6xxx series alloys. It is evident that the reduction of the 0,2% proof strength is similar for alloys in the same series, although slight scatter exists in 5xxx alloys.

Similar data are provided in EN 1999-1-2 for both 5xxx and 6xxx series alloys. For simplicity, lower limits of the 0,2% proof strength at various temperatures are also provided in EN 1999-1-2 disregarding the type of alloys. Unlike steel, aluminium has a remarkable curved stress–strain curve that may be described using the Ramberg-Osgood equation as follows,

$$\varepsilon_{al,\theta} = \frac{\sigma_{al,\theta}}{E_{al,\theta}} + 0,002 \left(\frac{\sigma_{al,\theta}}{f_{al,\theta}} \right)^n \tag{5.74}$$

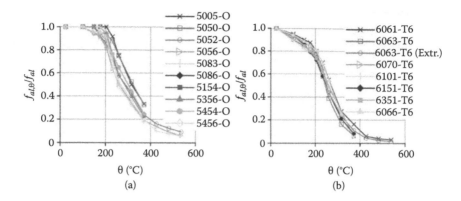

Figure 5.30 Reduction of the 0,2% proof strength of aluminium alloys at elevated temperatures. (a) 5xxx series alloys and (b) 6xxx series alloys. (*Source:* Maljaars, J., Twilt, L., Fellinger, J.H.H. et al. (2010) *HERON*, 55, 85–122. With permission.)

where $\varepsilon_{al,\theta}$ is the strain, $\sigma_{al,\theta}$ is the stress, $f_{al,\theta}$ is the 0,2% proof strength, and n is the temperature-dependent constant suggested as follows (Maljaars et al., 2010):

For alloy 5083-O/H111 for $175°C \leq \theta_{al} \leq 350°C$:

$$n = 8,8 - 0.016\theta_{al} \tag{5.75}$$

For alloy 6060-T66 for $175°C \leq \theta_{al} \leq 350°C$:

$$n = 19 - 0.040\theta_{al} \tag{5.76}$$

Equations (5.74) to (5.76) together with the definitions of reductions of the modulus of elasticity $E_{al,\theta}$ and the 0,2% proof strength $f_{al,\theta}$ described above provide the stress–strain curves of aluminium alloys at elevated temperatures.

It should be noted that, depending upon the alloy, a limiting temperature for a strength loss of 30% that represents an equivalent partial safety factor around 1,4 is between 150 and 300°C. This means that any design method must consider the exact nature of the alloy. Unlike steel, a single limiting temperature independent of strength grade cannot be established for aluminium. It should also be noted that on cooling, some alloys tend to exhibit brittle rather than ductile failure (Zacharia and Aidun, 1988).

5.2.6.2 Thermal expansion

Conserva et al. (1992) suggest that over a temperature range of 0 to 300°C, the unrestrained thermal strain is sensibly linear and therefore a constant coefficient of thermal expansion of between 24 and 26 μstrain/°C may be adopted. Davis (1993) gives average coefficients of linear thermal expansion for temperatures between 20, 50, 100, 200, and 300°C. Kammer (2002) provides similar results for both pure aluminium and aluminium alloys for temperatures up to 500°C.

These results show that the coefficient of linear thermal expansion α_{al} is almost independent of the alloy (Gale and Totemeier, 2003). The relative elongation $\varepsilon_{th} = \Delta l/l$ given in EN 1999-1-2 is based on the linear thermal expansion coefficient obtained by curve fitting for α_{al} between 20 and 500°C, and is expressed as follows:

$$\frac{\Delta l}{l} = 0,1 \times 10^{-7} \theta_{al}^2 + 22,5 \times 10^{-6} \theta_{al} - 4,5 \times 10^{-4} \qquad (5.77)$$

or slightly less accurately

$$\frac{\Delta l}{l} = 2,5 \times 10^{-5} \left(\theta_{al} - 20\right) \qquad (5.78)$$

where Δl is the temperature-induced elongation and l is the length at 20°C. Figure 5.31 shows the test data of Davis (1993) and Kammer (2002) and

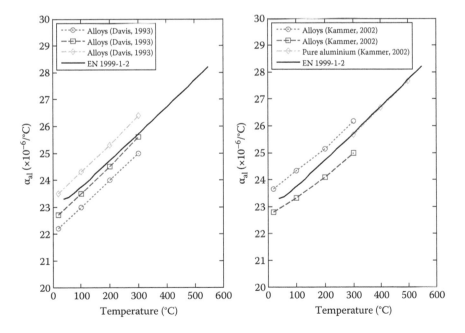

Figure 5.31 Thermal expansions of aluminium.

the values specified in EN 1999-1-2 for the coefficient of linear thermal expansion of aluminium.

5.3 CONSTITUTIVE STRESS–STRAIN LAWS

To perform computer analyses of steel and concrete elements or structures exposed to fire, it is necessary to be able to formulate constitutive stress–strain models for both materials. The models may be established using fundamental principles or by curve fitting established data with an empirical approach. Whilst the former is clearly preferable, it is often necessary to resort to the latter because of a paucity of data needed to establish a fundamental model or practical aspects in that the model is applicable only to a single identified material, e.g., a single steel strength (or grade). It is essential to be aware of potential limitations of any model.

Following Anderberg and Thelandersson (1976) the total strain ε_{tot} can be decomposed into three components: (1) free thermal strain ε_{th}, (2) stress-related strain ε_σ, and (3) transient creep strain ε_{tr}, such that

$$\varepsilon_{tot} = \varepsilon_{th} + \varepsilon_\sigma + \varepsilon_{tr} \tag{5.79}$$

Additional subscripts (s and c) will be used to indicate steel and concrete, respectively, as it is necessary to cover each material separately. The free thermal strain terms in Equation (5.79) have already been covered in Sections 5.2.2.2 and 5.2.3.3, and the remaining terms may now be considered.

5.3.1 Steel

5.3.1.1 Elastic strain

The mathematical model used to determine instantaneous elastic strain is dependent upon the characterization used for the stress–strain curve. Common models are discussed below.

Linear elastic, perfectly plastic — In this case (Figure 5.32a), the behaviour is taken as elastic up to yield, then plastic post-yield, i.e.,

For $0 \leq \varepsilon_{\sigma,s} \leq \varepsilon_{y,s,\theta}$:

$$\sigma_{s,\theta} = E_{s,\theta}\varepsilon_{\sigma,s} \tag{5.80}$$

For $\varepsilon_{\sigma,s} \geq \varepsilon_{y,s,\theta}$:

$$\sigma_{s,\theta} = \sigma_{y,s,\theta} \tag{5.81}$$

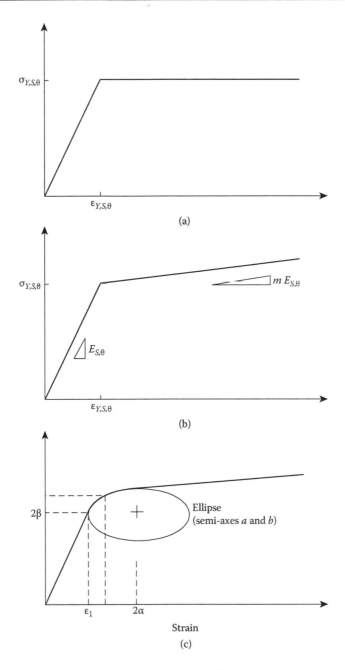

Figure 5.32 Idealization of stress–strain behaviour for steel at elevated temperatures: (a) linear elastic, perfectly plastic model, (b) bilinear strain hardening model, and (c) Dounas and Golrang model.

where $\varepsilon_{\sigma,s}$ is the stress-related strain, $\varepsilon_{y,s,\theta}$ is the temperature-dependent yield strain, $\sigma_{s,\theta}$ is the temperature-dependent stress, $\sigma_{y,s,\theta}$ is the temperature-dependent yield stress, and $E_{s,\theta}$ is the temperature-dependent modulus of elasticity. Note that the yield strain, yield stress, and modulus of elasticity are not independent since

$$E_{s,\theta} = \frac{\sigma_{y,s,\theta}}{\varepsilon_{y,s,\theta}} \tag{5.82}$$

This model has the advantage of simplicity but it may not be very accurate close to the yield point and in the post-yield phase. The latter inaccuracy can be ameliorated by adopting a linear elastic strain hardening model.

Linear elastic, strain hardening — In the elastic range, the model is identical to the above, but introduces post yield strain hardening (Figure 5.32b) with a post-yield gradient of $m_\theta E_{s,\theta}$, where m_θ is the strain hardening parameter and typically can be taken in the 0,1 to 0,15 range. Thus the governing equation for post-yield behaviour is modified to

$$\sigma_{s,\theta} = \sigma_{y,s,\theta} + m_\theta E_{s,\theta}\left(\varepsilon_{s,\theta} - \varepsilon_{y,s,\theta}\right) \tag{5.83}$$

The use of a strain hardening model, while providing a better model for post-yield behaviour, does not totally overcome the problem of inaccuracies near the yield point. Dounas and Golrang (1982) proposed that a combination of straight lines for pre- and post-yield and a quarter ellipse at the yield point could be used (Figure 5.32c). This complexity, however, is probably unnecessary when a simpler single expression model of Ramberg and Osgood (1943) exists.

Ramberg-Osgood — This model was originally proposed for the behaviour of aluminium which has no definite yield plateau. It is thus also applicable to steel. The Ramberg-Osgood equation in its simplest form is

$$\varepsilon_{s,\theta} = \frac{\sigma_{s,\theta}}{E_{s,\theta}} + 0.01\left(\frac{\sigma_{s,\theta}}{\sigma_{y,s,\theta}}\right)^n \tag{5.84}$$

where n is the parameter defining the fit of the equation to experimental data. These parameters will be temperature dependent. By using data fit, the following expressions for $E_{s,\theta}$, $\sigma_{y,s,\theta}$ and n are suggested by Burgess et al. (1990): For $20°C \le \theta_c \le 100°C$:

$$E_{s,\theta} = E_s$$

$$\sigma_{y,s,\theta} = \sigma_{y,s}\left(5,36 \times 10^{-6}\theta_s^2 - 1,04 \times 10^{-3}\theta_s + 1,019\right) \tag{5.85}$$

$$n = 237 - 1,58\theta_s$$

For $100°C \leq \theta_c \leq 400°C$:

$$E_{s,\theta} = E_s \left(1,078 - 0,778 \times 10^{-3} \theta_s\right)$$
$$\sigma_{y,s,\theta} = 0,968\sigma_{y,s} \qquad (5.86)$$
$$n = 15,3 \times 10^{-7} \times (400 - \theta_s)^{3,1} + 6$$

For $400°C \leq \theta_c \leq 700°C$:

$$E_{s,\theta} = E_s (1,641 - 2,185 \times 10^{-3} \theta_s)$$
$$\sigma_{y,s,\theta} = \sigma_{y,s} \left(1,971 - 2,506 \times 10^{-3} \theta_s\right) \qquad (5.87)$$
$$n = 6$$

For $700°C \leq \theta_c \leq 800°C$:

$$E_{s,\theta} = E_s \left(0,169 - 8,33 \times 10^{-5} \theta_s\right)$$
$$\sigma_{y,s,\theta} = \sigma_{y,s} (1,224 - 1,44 \times 10^{-3} \theta_s) \qquad (5.88)$$
$$n = 0,4\theta_s - 22$$

where E_s and $\sigma_{y,s}$ are the Young's modulus and yield strength of steel at ambient temperature, respectively. The Ramberg-Osgood stress–strain curve does not have peak stress, as shown in Figure 5.33. Therefore it must

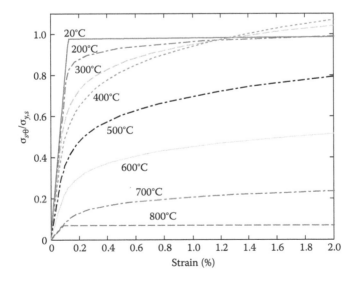

Figure 5.33 Ramberg-Osgood stress–strain curves at different temperatures ($\sigma_{y,s} = 275$ MPa, $E_s = 210$ GPa).

be used with care when very high strain levels are anticipated. Another disadvantage of the model is that it gives the strain in terms of the stress, and an explicit equation cannot be derived for the stress in terms of the strain.

5.3.1.2 Creep

As mentioned in Section 5.2.2.3, steel creep is generally analyzed in terms of the Dorn (1954) temperature-compensated time approach in which real time t is transformed into a temperature-compensated time Θ using Equation (5.89)

$$\Theta = \int_0^t \exp\left(\frac{-\Delta H}{R(\theta_a + 273)}\right) dt \tag{5.89}$$

where $\Delta H/R$ is the activation energy and θ_a is the temperature in the steel (°C). When creep data are plotted to a base of temperature-compensated time, two creep periods can be identified (Figure 5.34):

Primary creep — Over this portion, the creep is nonlinear and can be taken as parabolic (Harmathy, 1967), with the creep strain $\varepsilon_{cr,s}$ given by

$$\varepsilon_{cr,s} = \frac{\varepsilon_{cr,s,0}}{\ln 2} \cosh^{-1}\left(2^{\frac{Z\Theta}{\varepsilon_{cr,s,o}}}\right) \tag{5.90}$$

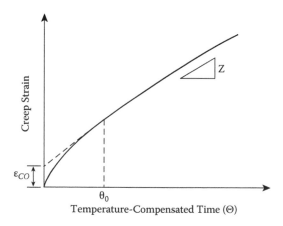

Figure 5.34 Idealization of isothermal creep data for steel.

where $\varepsilon_{cr,s,0}$ is the intercept of the secondary creep line with strain axis and Z is the Zener-Hollomon parameter. Equation (5.90) is valid up to a temperature-compensated time of Θ_0 given by

$$\Theta_0 = \frac{\varepsilon_{cr,s,0}}{Z}$$

(5.91)

Secondary creep — This is generally taken as a straight line with a slope of Z. Alternative creep equations have been proposed by Plem (1975) and Anderberg (1978a). Note that values of $\Delta H/R$ are very sensitive to the metallurgical characteristics of a steel as are values of Z and $\varepsilon_{cr,s,0}$. These two parameters are functions of the stress in the steel and can be calculated from

$$\varepsilon_{cr,s,0} = A\sigma_s^B$$

(5.92)

For $\sigma_s \le \text{SIG1}$:

$$Z = C\sigma_s^D$$

(5.93)

For $\sigma_s > \text{SIG1}$:

$$Z = He^{F\sigma_s}$$

(5.94)

Typical values of $\Delta H/R$, SIG1, A, B, C, D, H, and F are given in Table 5.4. Note that great care should be taken in using any set of values from this table to set up a creep model for steel, as the parameters appear very sensitive to the type of steel. If applied to a steel not listed in the table, results may be erroneous.

It is possible to set up empirical equations to yield the strain induced in steel, excluding the unrestrained thermal strain at constant temperature and under constant stress by analyzing existing data (Fields and Fields, 1989). It is, however, necessary to be aware of the procedure used to determine the empirical constants and any limitations adopted, since log–log plots of data are used often. This method compresses the data and may give the appearance of a better fit than actually exists.

5.3.1.3 Design curves

To standardize the parameters required for design, EN 1992-1-2, EN 1993-1-2 and EN 1994-1-2 give a stress–strain curves for both reinforcing and structural steels (Figure 5.35). The relationships between the various

Table 5.4 Steel creep parameters

Steel	$f_{0,s,20}$ (MPa)	$\Delta H/R$ (K)	A	B	SIGI (MPa)	D	C (min^{-1})	H (min^{-1})	F
1312	254	55800	$5,56 \times 10^{-6}$	1,722	108	7,804	$6,083 \times 10^{9}$	$1,383 \times 10^{23}$	0,0578
1312	263	53900	$2,66 \times 10^{-6}$	2,248	108	7,644	$8,95 \times 10^{8}$	$5,10 \times 10^{21}$	0,0601
1411	340	66000	$3,52 \times 10^{-7}$	2,08	118	8,402	$6,767 \times 10^{12}$	$4,417 \times 10^{27}$	0,0603
A36-66	304	38900	$4,07 \times 10^{-6}$	1,75	103	4,70	$6,217 \times 10^{6}$	$2,10 \times 10^{14}$	0,0434
2172	331	50000	$2,085 \times 10^{-8}$	2,30	108	5,38	$1,33 \times 10^{10}$	$1,083 \times 10^{19}$	0,0446
G40-12	333	36100	$1,766 \times 10^{-7}$	1,00	103	3,35	$4,733 \times 10^{7}$	$6,17 \times 10^{12}$	0,319
A421-65	1470	30600	$9,262 \times 10^{-5}$	0,67	172	3,00	$3,253 \times 10^{6}$	$1,368 \times 10^{12}$	0,0145
Ks40φ10	483	45000	$2,85 \times 10^{-8}$	1,037	84	4,70	$1,16 \times 10^{9}$	$4,3 \times 10^{16}$	0,0443
Ks40φ 8	456	40000	$3,39 \times 10^{-7}$	0,531	90	4,72	$7,6 \times 10^{5}$	$1,25 \times 10^{13}$	0,0512
Ks40φ8	504	47000	$1,99 \times 10^{-5}$	1,28	120	7,26	$4,05 \times 10^{4}$	$5,00 \times 10^{17}$	0,0384
Ks40SEφ8	558	40000	$3,86 \times 10^{-8}$	1,117	96	3,83	$5,8 \times 10^{7}$	$4,113 \times 10^{13}$	0,0414
Ks60φ8	710	40000	$2,06 \times 10^{-6}$	0,439	90	2,93	$8,517 \times 10^{8}$	$2,65 \times 10^{14}$	0,0313
Ps50φ5	500	40000	$1,10 \times 10^{-6}$	0,557	100	4,47	$9,783 \times 10^{6}$	$2,133 \times 10^{15}$	0,0368
Ps50φ8	749	41000	$1,28 \times 10^{-7}$	0,844	133	3,94	$1,367 \times 10^{8}$	$1,02 \times 10^{15}$	0,0265

Source: Anderberg, Y. (1983) Properties of Materials at High Temperatures–Steel. RILEM Report, University of Lund, Sweden. With permission.

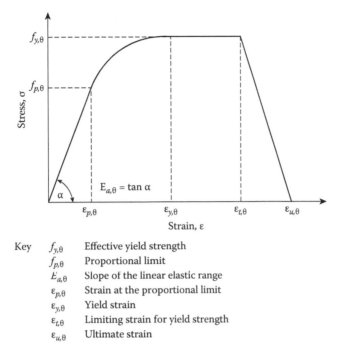

Key $f_{y,\theta}$ Effective yield strength
      $f_{p,\theta}$ Proportional limit
      $E_{a,\theta}$ Slope of the linear elastic range
      $\varepsilon_{p,\theta}$ Strain at the proportional limit
      $\varepsilon_{y,\theta}$ Yield strain
      $\varepsilon_{t,\theta}$ Limiting strain for yield strength
      $\varepsilon_{u,\theta}$ Ultimate strain

Figure 5.35 **Stress–strain relationship for carbon steel at elevated temperatures from EN 1993-1-2. (*Source:* EN 1993-1-2. © British Standards Institute.)**

parameters of the mathematical model in Figure 5.35 are given in Table 5.5 and the variations of strength or elastic modulus in Table 5.6. These stress–strain parameters include an allowance for creep. Hertz (2004) gives an equation that may be used for both hot and residual strength properties for both reinforcing and pre-stressing steels,

$$\xi(\theta) = k + \frac{1-k}{1 + \dfrac{\theta}{\theta_1} + \left(\dfrac{\theta}{\theta_2}\right)^2 + \left(\dfrac{\theta}{\theta_8}\right)^8 + \left(\dfrac{\theta}{\theta_{64}}\right)^{64}} \qquad (5.95)$$

where θ_1, θ_2, θ_8, and θ_{64} are experimentally determined parameters and k is the ratio between the minimum and maximum values of the property considered. For hot strengths, $k = 0$ but for residual strengths, k exceeds 0. Hertz (2006) provides strength data on quenched steel reinforcement.

Table 5.5 Relationships of various parameters of mathematical model in Figure 5.35

Strain range	Stress (σ)	Tangent modulus
$\varepsilon < \varepsilon_{p,\theta}$	$\varepsilon E_{a,\theta}$	$E_{a,\theta}$
$\varepsilon_{p,\theta} < \varepsilon < \varepsilon_{y,\theta}$	$f_{p,\theta} - c + \dfrac{b}{a}[a^2 - (\varepsilon_{y,\theta} - \varepsilon)^2]^{0,5}$	$\dfrac{b(\varepsilon_{y,\theta} - \varepsilon)}{a[a^2 - (\varepsilon_{y,\theta} - \varepsilon)^2]^{0,5}}$
$\varepsilon_{y,\theta} < \varepsilon < \varepsilon_{t,\theta}$	$f_{y,\theta}$	0
$\varepsilon_{t,\theta} < \varepsilon < \varepsilon_{u,\theta}$	$f_{y,\theta}\left[1 - \dfrac{\varepsilon - \varepsilon_{t,\theta}}{\varepsilon_{u,\theta} - \varepsilon_{t,\theta}}\right]$	-
$\varepsilon = \varepsilon_{u,\theta}$	0	-

where

$$a^2 = (\varepsilon_{y,\theta} - \varepsilon_{p,\theta})\left(\varepsilon_{y,\theta} - \varepsilon_{p,\theta} + \frac{c}{E_{a,\theta}}\right)$$

$$b^2 = c(\varepsilon_{y,\theta} - \varepsilon_{p,\theta})E_{a,\theta} + c^2$$

$$c = \frac{(f_{y,\theta} - f_{p,\theta})^2}{(\varepsilon_{y,\theta} - \varepsilon_{p,\theta})E_{a,\theta} - 2(f_{y,\theta} - f_{p,\theta})}$$

with $\varepsilon_{y,\theta} = 0,02$; $\varepsilon_{t,\theta} = 0,15$ and $\varepsilon_{u,\theta} = 0,20$, and $\varepsilon_{p,\theta} = \dfrac{f_{p,\theta}}{E_{a,\theta}}$

Table 5.6 Structural strength data for carbon steel

Steel temperature (θ_a) (°C)	Reduction factors		
	Yield strength $(k_{y,\theta})$	Proportional limit $(k_{p,\theta})$	Elastic modulus $(k_{E,\theta})$
20	1,000	1,000	1,000
100	1,000	1,000	1,000
200	1,000	0,807	0,900
300	1,000	0,613	0,800
400	1,000	0,420	0,700
500	0,780	0,360	0,600
600	0,470	0,180	0,310
700	0,230	0,075	0,130
800	0,110	0,050	0,090
900	0,060	0,0375	0,0675
1000	0,040	0,0250	0,0450
1100	0,020	0,0125	0,0250
1200	0,000	0,000	0,000

Source: Table 3.1 of EN 1993-1-2.

Hertz (2004) also gives strength reduction functions $\xi_{s,02}(\theta_s)$ for the 0,2% proof strength:

For $0°C \leq \theta_s \leq 600°C$:

$$\xi_{s,02}(\theta_s) = 1 + \frac{\theta_s}{767 \ln\left(\dfrac{\theta_s}{1750}\right)} \tag{5.96}$$

For $600°C \leq \theta_s \leq 1000°C$:

$$\xi_{s,02}(\theta_s) = 0,108 \frac{1000 - \theta_s}{\theta_s - 440} \tag{5.97}$$

5.3.2 Concrete

There is still much debate as to the most reliable approach to formulating the stress–strain section of a constitutive model. The original research in this field was carried out by Anderberg and Thelandersson (1976), but the basis of their analysis has been questioned by Schneider (1982; 1986b; 1988). An alternative approach based on Anderberg and Thelandersson for calculating transient strain was proposed by Deiderichs (1987). A further model using the concept of plastic hardening for determining the elastic response was developed by Khennane and Baker (1993).

It is the authors' firm view that the transient strain component, however calculated, cannot be ignored or neglected. This was clearly demonstrated by Mustapha (1994) and Purkiss and Mustapha (1996) with the analysis of reinforced concrete columns heated on three sides with the load applied axially and eccentrically. The results from the computer analysis along with the experimental data of Haksever and Anderberg (1981/2) are presented in Figure 5.36. The annotations of the material's models are M1 (slightly modified Anderberg and Thelandersson with variable slope to the descending branch of the stress–strain curve); M2 (Schneider); and M3 (no transient strain). Where the compression zone of the concrete is relatively cool (Column C2), all three models predict very similar trends except that the fire endurance period, if no transient strain is included, it is overestimated by a factor around 2 to 2,5. For columns C1 and C3, where the compression zone of the column is under the full heating effects of the furnace, the effect of ignoring transient strain is unacceptable. In column C1, the predicted horizontal deflection is of the wrong sign and over-predicts the fire endurance by a factor of 2.

For column C3, the three models predict similar trends in horizontal deflection up to about 75 min before the predictions with transient strain

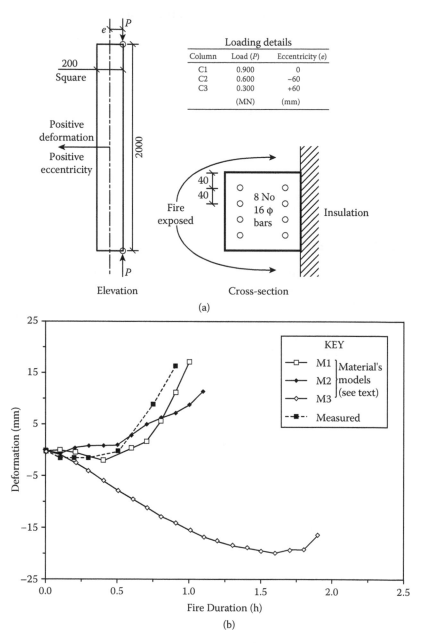

Figure 5.36 Comparison of experimental and calculated behaviours of columns in a fire for varying materials models: (a) general details, (b) column C1, (c) column C2, and (d) column C3. (*Source:* Purkiss, J.A. and Mustapha, K.N. (1995) In *Concrete 95: Toward Better Concrete Structures*, pp. 263–272. With permission.) (*continued*)

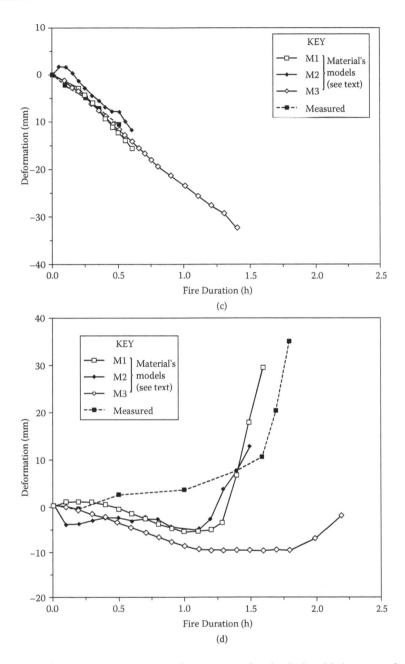

Figure 5.36 (continued). Comparison of experimental and calculated behaviours of columns in a fire for varying materials models: (a) general details, (b) column C1, (c) column C2, and (d) column C3. (*Source:* Purkiss, J.A. and Mustapha, K.N. (1995) In *Concrete 95: Toward Better Concrete Structures,* pp. 263–272. With permission.)

follow the test results, whilst prediction with no transient strain continues to produce deflections of the same sign until very close to failure.

It should be observed that the quantitative correlation between experimental results and prediction is the least good for column C3. This is almost certainly because the moments due to the thermal gradient and the eccentric load are of opposite signs. At the start of the test or simulation, the moment due to the eccentric load will control behaviour before it is overtaken later in the test by the opposite sense thermal gradient. In a simulation, this could produce temporary instability.

Rather than discuss the formulation of the individual terms of each constitutive model, it is more convenient to consider the individual models as entities.

5.3.2.1 Anderberg and Thelandersson

Anderberg and Thelandersson (1976) proposed that the stress-related component of the total strain $\varepsilon_{tot,c} - \varepsilon_{th,c}$ can be decomposed into three components:

$$\varepsilon_{tot,c} - \varepsilon_{th,c} = \varepsilon_{\sigma,c} + \varepsilon_{cr,c} + \varepsilon_{tr,c} \qquad (5.98)$$

where $\varepsilon_{tot,c}$ is total concrete strain, $\varepsilon_{th,c}$ is the free thermal strain, $\varepsilon_{\sigma,c}$ is the instantaneous stress-related strain, $\varepsilon_{cr,c}$ is the classical creep strain, and $\varepsilon_{tr,c}$ is the transient strain. The instantaneous stress-related strain was calculated assuming a parabolic stress–strain profile for strains beyond the peak and then a linear descending portion with a fixed (i.e., temperature independent) slope E_c^*. Thus the complete stress–strain relationship (Figure 5.37) is given by:

For $0 \leq \varepsilon_{\sigma,c} \leq \varepsilon_{1,c}$:

$$\frac{\sigma_c}{\sigma_{0,c}} = \frac{\varepsilon_{\sigma,c}}{\varepsilon_{0,c}}\left(2 - \frac{\varepsilon_{\sigma,c}}{\varepsilon_{0,c}}\right) \qquad (5.99)$$

For $\varepsilon_{\sigma,c} \geq \varepsilon_{1,c}$:

$$\sigma_c = \varepsilon_{\sigma,c}E_c^* + \sigma_{0,c}\left(1 - \frac{E_c^*}{E_c}\right)^2 \qquad (5.100)$$

where

$$\frac{\varepsilon_{1,c}}{\varepsilon_{0,c}} = 1 - \frac{E_c^*}{E_c} \qquad (5.101)$$

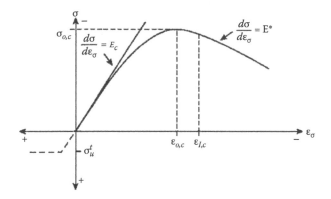

Figure 5.37 Stress–strain curves for concrete with constant descending (unloading) branch. (*Source:* Anderberg, Y. and Thelandersson, S. (1976) *Stress and Deformation Characteristics of Concrete 2: Experimental Investigation and Material Behaviour Model,* Bulletin 54. University of Lund, Sweden. With permission.)

The assumption of a temperature-independent slope for the descending branch is questionable when the experimental data on stress–strain behaviour are examined. The equation for analyzing creep data was obtained by curve fitting the creep results rather than working from basic fundamental principles. The equation finally obtained is

$$\varepsilon_{cr,c} = 0,00053 \frac{\sigma_c}{\sigma_{0,c,\theta}} \left(\frac{t}{180} \right)^{0,5} e^{0,00304(\theta_c - 20)} \tag{5.102}$$

where σ_c is the applied stress, $\sigma_{0,c,\theta}$ is the strength at temperature θ_c (°C), and t is the time in minutes.

To calculate the total creep strains occurring during a transient test, the accumulated creep strains were calculated using fictitious times and stresses at the previous time step. The transient strains can be calculated once the instantaneous stress-related strains and accumulated creep strains are known.

It remained to attempt to relate the resultant transient strains to some known parameter, and it was discovered that at temperatures below 550°C for the siliceous aggregate concrete used in the experimental investigation, the transient strain could be related to the free thermal strain, thus giving the following relationship

$$\varepsilon_{tr,c} = -k_2 \frac{\sigma_c}{\sigma_{0,c,20}} \varepsilon_{th,c} \tag{5.103}$$

where $\sigma_c/\sigma_{0,c,20}$ is the ratio of the concrete stress to the ambient strength and k_2 is an experimentally determined parameter. The minus sign is needed because the transient strain and thermal strain are of opposite signs. Anderberg and Thelandersson report a value of 2,35 for k_2, whereas other analyses reported in the same publication for different data give values of 1,8 and 2,0. Purkiss and Bali (1988) reported that they saw no significant correlation between the transient and thermal strains. Above 550°C, the picture is less clear, but it appears that from Anderberg and Thelandersson, the temperature-dependent rate of transient strain is constant:

$$\frac{\partial \varepsilon_{tr,c}}{\partial \theta} = 0,0001 \frac{\sigma_c}{\sigma_{0,c,20}} \tag{5.104}$$

5.3.2.2 Deiderichs

Deiderichs (1987) adopts a similar analysis, except that the classical creep strain is ignored and the instantaneous elastic strain calculated using the ambient modulus of elasticity. Thus the transient strain $\varepsilon_{tr,c}$ is given by:

$$\varepsilon_{tr,c} = \frac{\varepsilon_{c,\theta} - \frac{\sigma_c}{E_{c,20}} - \varepsilon_{th,c}}{\frac{\sigma_c}{\sigma_{0,20,c}}} \tag{5.105}$$

The Deiderichs model calculates the transient strain by a simple expression, but gave no guide to the determination of the values of transient strain. Subsequently Li and Purkiss (2005) used a curve fit to Deiderichs' data, which gives

$$\varepsilon_{tot,c} - \varepsilon_{th,c} = \varepsilon_{\sigma,c} + \varepsilon_{tr,c} + \varepsilon_{cr,c} = \frac{\sigma_c}{E_{c,20}} \left[1 - \frac{E_{c,20}}{\sigma_{0,20}} f(\theta_c) \right] \tag{5.106}$$

where $f(\theta_c)$ is given by

$$f(\theta_c) = 0,0412(\theta_c - 20) - 0,172 \times 10^{-4} (\theta_c - 20)^2 + 0,33 \times 10^{-6} (\theta_c - 20)^3 \tag{5.107}$$

5.3.2.3 Khoury and Terro

Khoury et al. (1985b) defined the elastic strain as $\sigma_c/E_{0,20}$; thus the transient strain they designated load-induced thermal strain (LITS) that includes any classical creep strain is given by

$$LITS(\theta_c, \sigma) = \varepsilon_{tot,c} - \varepsilon_{th,c} - \frac{\sigma_c}{E_{0,20}} \tag{5.108}$$

Initially Terro (1998) fitted the master curve at a stress level of $0,3\sigma_{0,20}$ to give

$$LITS(\theta_c,0,3\sigma_{0,20}) = 43,87 \times 10^{-6} - 2,73 \times 10^{-6}\theta_c - 6,35 \times 10^{-8}\theta_c^2$$
$$+ 2,19 \times 10^{-10}\theta_c^3 - 2,77 \times 10^{-13}\theta_c^4 \qquad (5.109)$$

For other stress levels, the transient strain is given by

$$LITS(\theta_c,\sigma_c) = LITS(\theta_c,0,3\sigma_{0,20})\left(0,032 + 3,226\frac{\sigma_c}{\sigma_{0,20}}\right) \qquad (5.110)$$

For siliceous aggregates, Equation (5.109) needs modifying to

$$LITS(\theta_c,0,3\sigma_{0,20}) = 1,48 \times 10^{-6}\left(1098,5 - 39,21\theta_c + 0,43\theta_c^2\right)$$
$$- 1,48 \times 10^{-9}\left(2,44\theta_c^3 - 6,27 \times 10^{-3}\theta_c^4 + 5,95 \times 10^{-6}\theta_c^5\right) \qquad (5.111)$$

Additionally the values of LITS may be corrected for volume fractions of aggregate V_a other than the original 65% by using the following equation

$$LITS(\theta_c,\sigma_c)\big|_{V_a} = LITS(\theta_c,\sigma_c)\big|_{65\%}\frac{V_a}{0,65} \qquad (5.112)$$

It should be noted that the model holds up only to temperatures of 590°C.

5.3.2.4 Khennane and Baker

Khennane and Baker (1993) take as a starting point the model of Anderberg and Thelandersson (1976). However, they determine the instantaneous-stress related strain using a stress–strain curve that is initially linear to a yield value that may be taken as 0,45 times the peak concrete strength. The remainder of the characteristic is taken as part of a quarter ellipse (Figure 5.38) with the following equation

$$\frac{(\sigma_{c,\theta} - \sigma_{1,c\theta})^2}{(\sigma_{0,c,\theta} - \sigma_{1,c,\theta})^2} + \frac{(\Delta p - \varepsilon_{p,c,\theta})^2}{(\Delta p)^2} = 1 \qquad (5.113)$$

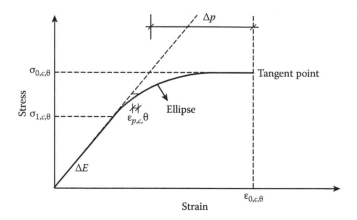

Figure 5.38 Linear elastic-elliptical plastic idealization of the stress–strain curve for concrete. (*Source:* Khennane, A. and Baker, G. (1993) *Journal of Engineering Mechanics*, 119, 1507–1525. With permission.)

The final stress–strain law is in the form of an incremental rule

$$\Delta\varepsilon_{tot,c} = A\Delta\sigma_c - B\sigma_c + \Delta\varepsilon_{tr,c} + \Delta\varepsilon_{th,c} \tag{5.114}$$

where $\Delta\varepsilon_{tot,c}$ is the increment in total strain, σ_c is the stress, $\Delta\sigma_c$ is the increment in stress over the time step, $\Delta\varepsilon_{th,c}$ is the increment of thermal strain, $\Delta\varepsilon_{tr,c}$ is the increment in transient strain defined in the same manner as Anderberg and Thelandersson, and A and B are parameters defined by the following equations

$$A = \frac{E_t}{\left(E_t + \beta\Delta E_t\right)^2} + \frac{H_t}{\left(H_t + \beta\Delta H_t\right)^2} + \beta\frac{k_2}{\sigma_{0,c,20}}\frac{\partial\varepsilon_{th,c}}{\partial\theta}\Delta\theta \tag{5.115}$$

and

$$B = \frac{\Delta E_t}{\left(E_t + \beta\Delta E_t\right)^2} + \frac{\Delta H_t}{\left(H_t + \beta\Delta H_t\right)^2} \tag{5.116}$$

where θ is the temperature, E_t is the slope of the linear portion of the stress strain curve, i.e., the initial tangent modulus, ΔE_t is the change in tangent modulus at time t to $t + \Delta t$, and H_t and ΔH_t are the values of the strain hardening parameter and the change in the strain hardening parameter, β is an interpolation parameter taking a value between zero and unity, and $\Delta\theta$ is the temperature rise. Khennane and Baker (1993) found that the best value for β was 0,5.

The equivalent plastic strain $\varepsilon_{p,c,\theta}$ determined from Equation (5.113) and the strain hardening parameter H are both dependent upon the current stress state. The latter parameter is given by

$$H = \frac{\left(\sigma_{0,c,\theta} - \sigma_{1,c,\theta}\right)^2}{\left(\Delta p\right)^2} \frac{\Delta p - \varepsilon_{p,c,\theta}}{\sigma_{c,\theta} - \sigma_{1,c,\theta}} \tag{5.117}$$

This model does not appear to allow for the instantaneous strain in the concrete to exceed the peak value, and the formulation for transient strain valid for temperatures above 550°C also appears not to be considered. This latter point may be critical since Khennane and Baker appear to produce excellent correlation of experimental results and prediction below temperatures of 550° but far poorer correlation at temperatures above this value. Further, the value of ΔH_t needs to be estimated since it depends on the stress at the end of the incremental time step. Khennane and Baker also indicated that an elaborate algorithm is needed for situations when both the temperature and stress vary during an incremental step. It should be noted that this is a likely case in a full structural analysis of fire-affected concrete members.

5.3.2.5 Schneider

The model used by Schneider (1986b, 1988) is based on a unit stress compliance function, i.e., creep is considered linear with respect to stress. The general background to this approach is detailed in Bažant (1988), and the specific formulation for high temperatures is given in Bažant (1983). An important simplification of the general compliance function approach to creep in a fire is that the duration of between a half and four hours is short compared with the age of the concrete, and thus any time dependence in the model can be ignored. The full background to Schneider's model is given in his other works (1986b; 1988), and only the results will be presented here. The unit stress compliance function $J(\theta,\sigma)$ can be written as

$$J\left(\theta_c, \sigma_c\right) = \frac{1 + \kappa}{E_{c,\theta}} + \frac{\Phi}{E_{c,\theta}} \tag{5.118}$$

where $E_{c,\theta}$ is the temperature-dependent modulus of elasticity and κ is a parameter allowing for nonlinear stress–strain behaviour for stresses exceeding about half the concrete strength, given by Equation (5.119) which is derived from Popovics [1973; see Equation (5.37)],

$$\kappa = \frac{1}{n-1}\left(\frac{\varepsilon_{\sigma,c,\theta}}{\varepsilon_{0,c,\theta}}\right)^n \tag{5.119}$$

with n taking a value of 2,5 for lightweight concrete and 3,0 for normal weight concrete. An alternative formulation for κ is given by

$$\kappa = \frac{1}{n-1}\left(\frac{\sigma(\theta_c)}{\sigma_{0,\theta}}\right)^5 \tag{5.120}$$

The value of $E_{c,\theta}$ is given by

$$E_{c,\theta} = gE_{c,0} \tag{5.121}$$

where the parameter g is given by the following equation

$$g = \begin{cases} 1+\dfrac{f_{c,\theta}}{\sigma_{0,c,20}}\dfrac{\theta_c - 20}{100} & for \quad \dfrac{f_{c,\theta}}{\sigma_{0,c,20}} \leq 0,3 \\[4mm] 1+0,3\dfrac{\theta_c - 20}{100} & for \quad \dfrac{f_{c,\theta}}{\sigma_{0,c,20}} > 0,3 \end{cases} \tag{5.122}$$

where θ_c is the concrete temperature (°C), $f_{c,\theta}/\sigma_{0,c,20}$ is the ratio of the initial stress under which the concrete is heated to the ambient strength, and the creep function Φ is given by

$$\Phi = \begin{cases} g\phi+\dfrac{f_{c,\theta}}{\sigma_{0,c,20}}\dfrac{\theta_c - 20}{100} & for \quad \dfrac{f_{c,\theta}}{\sigma_{0,c,20}} \leq 0,3 \\[4mm] g\phi+0,3\dfrac{\theta_c - 20}{100} & for \quad \dfrac{f_{c,\theta}}{\sigma_{0,c,20}} > 0,3 \end{cases} \tag{5.123}$$

with ϕ given by

$$\phi = C_1 \tanh \gamma_w (\theta_c - 20) + C_2 \tanh \gamma_0 (\theta_c - \theta_g) + C_3 \tag{5.124}$$

and γ_w defined by

$$\gamma_w = 0,001(0,3w + 2,2) \tag{5.125}$$

where w is the moisture content in percent by weight. The stress used in the definitions of g and Φ is that initially applied at the start of the heating period.

The original and modified values of C_1, C_2, C_3, and θ_g and γ_0 proposed by Schneider (1988) and Schneider et al. (2008) are given in Table 5.7,

Table 5.7 Concrete stress–strain model parameters

Concrete type	Parameter				
	C_1	C_2	C_3	$\gamma_0\,(°C)$	$\theta_g\,(°C)$
Quartzite (1988)	2,60	1,40	1,40	0,0075	700
(2008)	2,50	0,70	0,70	0,0075	800
Limestone (1988)	2,60	2,40	2,40	0,0075	650
(2008)	2,50	1,40	1,40	0,0075	700
Lightweight (1988)	2,60	3,00	3,00	0,0075	600
(2008)	2,50	3,00	2,90	0,0075	600

Source: Schneider, U. (1988) *Fire Safety Journal*, 13, 55–68 and Schneider, U. et al. (2008) In *Proceedings of Fifth International Conference on Structures in Fire*, pp. 463–476. With permission.

respectively. However, some evidence indicates that these parameters are likely to be functions of the concrete mix proportions since Purkiss and Bali (1988) report values of 2,1 and 0,7 for C_2 and C_3. A further unpublished analysis by Purkiss of data from Bali (1984) gives slightly different values of 1,5 and 0,95, respectively. An analysis using Anderberg and Thelandersson's data gives values for C_2 and C_3 of 3,27 and 1,78, respectively for the tests with a heating rate of 1°C/min.

By taking into account the effect of the load history before heating on the deformation response to a change in stress and temperature increase, Schneider (1988) aims at incorporating the effect of concrete memory in a model. In recent research, Schneider extended his theory to include the effect of load history during heating on the deformation response of concrete (Schneider et al., 2008; Schneider and Schneider, 2009).

Schneider's model is more complex and comprises a number of important parameters compared with the other models of transient creep strain. In Schneider's model, the relationships of transient creep strain, temperature, and applied stress are nonlinear. In contrast, the formulation by Anderberg and Thelandersson, whilst not justifiable on theoretical grounds, has been successfully used as a model in computer simulations as has the Schneider model.

5.3.2.6 Li and Purkiss

Li and Purkiss (2005) demonstrated that it is possible to determine a numerical apparent stress–strain curve that explicitly includes transient strain. These curves are given in Figure 5.39 with a comparison of Anderberg and Thelandersson and EN 1992-1-2 where one can see there is a close correlation with Anderberg and Thelandersson and EN 1992-1-2 at low temperatures but an increasing divergence from EN 1992-1-2. Li and Purkiss note

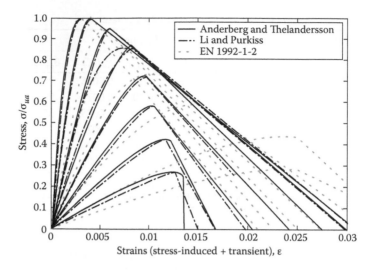

Figure 5.39 Compressions of full stress–strain curves for concrete at temperatures of 40, 100, 200, 300, 400, 500, 600, and 700 °C.

that Khoury and Deiderichs predict lower (but similar) transient strains than Schneider or Anderberg and Thelandersson.

5.3.3 Design code provisions for stress–strain behaviour

Full stress–strain-temperature relationships are needed only when a full elastoplastic analysis is required to determine the fire performance of a concrete structure. Where an end point calculation is sufficient, a stress–strain curve that allows for some creep may be used. EN 1992-1-2 and EN 1994-1-2 give such a set of curves providing data for a stress–strain relationship that may be used for the analysis of concrete sections.

The analytical form of the curve is the same as that proposed by Popovics [Equation (5.37)] with $n = 3$ for normal weight concrete. The parametric variation of data with respect to temperature is given in Table 5.8. However, the strain values at the peak are far higher than those for simple stress–strain curves. It is thought that these higher values allow for transient strain.

Li and Purkiss (2005) note that after 400°C, these peak strains reach around twice the values allowing for transient strains. Anderberg (2005) notes that the values of peak strains used in a structural analysis where axial loads exist may give erroneous results. Although EN 1992-1-2 provides the strength data in a tabular form, Hertz (2005) provides equations for both

Table 5.8 Variations of concrete strength parameters with temperature

Temperature θ^c (°C)	Strength reduction factor ($k_c(\theta_c)$)			$\varepsilon_{cl,\theta}$	$\varepsilon_{cul,\theta}$
	Siliceous	Calcareous	Lightweight		
20	1,000	1,000	1,000	0,0025	0,0200
100	1,000	1,000	1,000	0,0040	0,0225
200	0,950	0,970	1,000	0,0055	0,0250
300	0,850	0,910	1,000	0,0070	0,0275
400	0,750	0,850	0,880	0,0100	0,0300
500	0,600	0,740	0,760	0,0150	0,0325
600	0,450	0,600	0,640	0,0250	0,0350
700	0,300	0,430	0,520	0,0250	0,0375
800	0,150	0,270	0,400	0,0250	0,0400
900	0,080	0,150	0,280	0,0250	0,0425
1000	0,040	0,060	0,160	0,0250	0,0450
1100	0,010	0,020	0,04	0,0250	0,0475
1200	0,000	0,000	0,000	–	–

Note: *The values of $\varepsilon_{cl,\theta}$ and $\varepsilon_{cul,\theta}$ only apply to normal weight concrete.*

Source: Table 3.1 of EN 1992-1-2 and Table 3.3 of EN 1994-1-2.

the hot condition when the concrete is at its weakest and the cold condition when the reinforcement is at its weakest. Equation (5.95) is used with suitable concrete parameters (and with $k = 0$).

This chapter is intended to provide only an overview of the thermal and mechanical behaviours of the main structural materials, before the succeeding chapters and calculations of the performances of structural elements and structures can be undertaken.

Chapter 6

Calculation approach

The analysis of structural fire resistance is a complicated process because it involves many variables such as fire growth and duration, temperature distribution in structural members, interactions of structural members, changes in material properties, and the influence of loads on the structural system. The process generally includes three distinct components: (1) fire hazards analysis to identify fire scenarios and determine the impact of each scenario on adjacent structural members; (2) thermal analysis to calculate temperature history in each member; and (3) structural analysis to determine forces and stresses in each member and whether local or progressive structural collapse would occur during any of the fire hazard scenarios.

The primary objective for conducting such analyses is to determine the length of time that a structure will be able to resist collapse during exposure to a fire, the strength at a pre-determined time, the time lapse in achieving a certain strength reduction, or the time to achieve a given temperature when the structure or structural member is exposed to gas temperatures generated from either a natural fire or the standard furnace test. The integrity of an element, i.e., its ability to resist the passage of flame through gaps in a structure, is not normally calculated; this is best determined using the standard furnace test since this form of failure is mostly applicable to elements such as fire doors and other closure systems. The limit state of integrity will not therefore be considered further as far as calculations are concerned.

For the fire hazards analysis, it is generally not important whether the fire exposure is determined from a fire induced in the standard furnace test or from the effects of a natural or compartment fire. It is sufficient that the compartment temperature–time response is known either as a continuous function or as a series of discrete temperature–time values.

As noted in Chapter 2, the calculation of the temperature response in a structure member can be decoupled generally from the determination of structural response provided that the structure geometry does not undergo significant changes during the period of interest. For steelwork, the boundaries of members are not subject to possible changes during the fire exposure except a total or partial loss of insulation. Therefore it is acceptable to decouple the calculation of the temperature response from that of the structure response.

For concrete, however, the decoupling approach is often not acceptable because spalling is likely to occur. Spalling will change the boundaries of concrete members due to loss of the concrete cover, which is then likely to expose reinforcement to the full effects of the fire, producing a rapid rise of the temperature in the reinforcement, and thus a greater loss in strength (Purkiss and Mustapha, 1996). The problems associated with spalling are covered in more detail in Chapter 7. Since, the mechanisms of spalling are not yet completely understood and its occurrence cannot be quantified, it is not possible to allow for the effects of spalling in the calculation approach. Thus an approach that decouples the calculations of the temperature response and the structural response will of necessity have to be used.

6.1 THERMAL ANALYSIS

The procedure described for the thermal analysis in this section is mainly for concrete, steel or steel–concrete composite members or structures. A much simplified approach is available for timber which will be discussed in Section 6.2. Since the temperature–time curves in fire have been addressed in Chapter 4, this section will focus only on the transfer of heat from a fire to a structural member.

6.1.1 Governing equation and boundary conditions

The analysis of temperature response in a structural member can be subdivided into two parts. One is the heat transfer across the boundary from the furnace or fire into the surface of the structural member via a combination of convection and radiation, usually treated as boundary conditions; the other is heat transfer within the structural member through conduction, treated as a governing equation expressed by the Fourier equation of heat transfer.

Heat conduction is the transfer of thermal energy from one place to another through a solid or fluid due to the temperature difference between the two places. The transfer of thermal energy occurs at the molecular and atomic levels without net mass motion of the material. The rate equation describing this heat transfer mode is Fourier's law, expressed by

$$q = -\lambda \nabla \theta \tag{6.1}$$

where q is the vector of heat flux per unit area, λ is the thermal conductivity tensor, and θ is the temperature. For an isotropic solid such as steel, concrete or masonry, $\lambda = \lambda \mathbf{I}$ where λ is the thermal conductivity that may be a function of the temperature and $\mathbf{I}$ is the identity matrix. The conservation of energy with Fourier's law requires

$$\rho c \frac{\partial \theta}{\partial t} = -\nabla \cdot \mathbf{q} + Q \qquad (6.2)$$

where ρ is the density, c is the specific heat, t is the time, and Q is the internal heat generation rate per unit volume. The specific heat may be temperature dependent. Substituting Equation (6.1) into (6.2) yields

$$\rho c \frac{\partial \theta}{\partial t} = \nabla \cdot (\lambda \nabla \theta) + Q \qquad (6.3)$$

Equation (6.3) is the heat conduction equation and is solved subject to an initial condition and appropriate boundary conditions. The initial condition consists of specifying the temperature throughout the solid at an initial time. The boundary conditions may take the following several forms.

Fire-exposed surface —The surface of the structural member is exposed to a fire, on which the heat transfer involves both convection and radiation, although it is generally accepted that the radiation component is the more dominant after the very early stages of a fire. The net heat flux to the surface of the structural member thus is expressed as

$$\dot{h}_{net} = \dot{h}_{net,c} + \dot{h}_{net,r} \qquad (6.4a)$$

in which, $\dot{h}_{net,c} = \alpha_c (\theta_g - \theta_m) =$ net convective heat flux per unit surface and $\dot{h}_{net,r} = \Phi \varepsilon_m \varepsilon_f \sigma [(\theta_g + 273)^4 - (\theta_m + 273)^4] =$ net radiative heat flux per unit surface where α_c is the coefficient of heat transfer by convection, θ_g is the gas temperature in the vicinity of the fire exposed surface, θ_m is the surface temperature of the structural member, Φ is the configuration factor, ε_m is the surface emissivity of the structural member, ε_f is the emissivity of the fire, and $\sigma = 5{,}67 \times 10^{-8}$ W/m^{2}K^4 is the Stephan Boltzmann constant.

No heat-flow surface — The surface is a thermal symmetric plane or has a large degree of insulation and thus can be assumed as thermally insulated and having no heat flow. Therefore, the net heat flux to the surface of the structural member can be expressed simply as

$$\dot{h}_{net} = 0 \qquad (6.4b)$$

Ambient exposed surface — The surface is exposed to ambient conditions and thus can be treated similarly to that exposed to a fire but replacing the fire temperature with the ambient temperature, θ_a, that is,

$$\dot{h}_{net} = \alpha_c(\theta_a - \theta_m) + \Phi\varepsilon_m\varepsilon_f\sigma[(\theta_a + 273)^4 - (\theta_m + 273)^4] \tag{6.4c}$$

Fixed temperature surface — The surface temperature of the structural member is specified to be constant or a function of a boundary coordinate and/or time.

The boundary condition for the first three surface types described above is called *Neumann* boundary condition and specifies the normal derivative of the temperature, that is,

$$\lambda\frac{\partial\theta_m}{\partial n} = \dot{h}_{net} \tag{6.5a}$$

where n is the normal of the surface. The boundary condition for the fourth surface type is called *Dirichlet* boundary condition and specifies the function of the temperature, that is,

$$\theta_m = \bar{\theta}(t) \tag{6.5b}$$

where $\bar{\theta}(t)$ is the prescribed temperature at the boundary.

Convection is the transfer of thermal energy through a fluid due to motion of the fluid. The energy transfer from one fluid particle to another occurs by conduction, but thermal energy is transported by the motion of the fluid. However, the convection heat transfer coefficient is not a property of the fluid. It is an experimentally determined parameter whose value depends on all the variables influencing convection such as the surface geometry, the nature of fluid motion, the properties of the fluid, and the bulk fluid velocity. EN 1991-1-2 suggests that $\alpha_c = 25$ W/m²K for a fire-exposed surface when the standard temperature–time curve is used and $\alpha_c = 9$ W/m²K for the ambient exposed surface, assuming it contains the effects of heat transfer by radiation.

Unlike the convection that requires a medium to transfer heat, radiation is the transfer of thermal energy between two locations by an electromagnetic wave that requires no medium. The radiation term used here in Equation (6.4) is the traditional one used in textbooks and also implemented in computer packages (Becker et al., 1974; Iding et al., 1977a).

In the literature, several radiation expressions have been suggested. Mooney (1992) proposed a radiation expression based on the concept of the surface radiant energy balance in a fire environment that uses representative temperature and representative emissivity instead of the traditional fire temperature and fire emissivity. Whichever expression is used, experimental data are always required to validate the expression and determine the parameters of the expression. To allow for varying radiative heat flux levels while keeping the surface and fire emissivities as constants, a configuration factor is introduced in the radiative heat flux expression.

A conservative choice for the configuration factor is $\Phi = 1$. A lower Φ value may be obtained from the calculation based on the fraction of the total radiative heat leaving a given radiating surface that arrives at a given receiving surface, as cited in EN 1991-1-2, to take account of so-called position and shadow effects. The theory behind the calculation of the configuration factor is given by Drysdale (1998). Figure 6.1 shows a typical

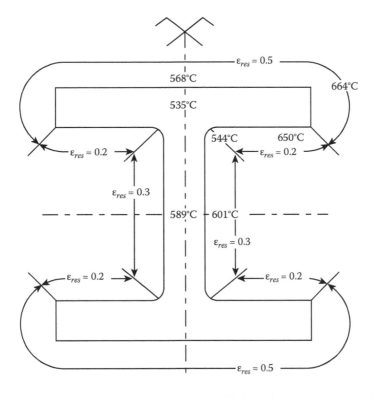

Figure 6.1 Variation of resultant emissivity in prediction of temperatures within steel column. Temperatures given correspond to temperature field calculated at 46 min (temperatures around profile correspond to centre of discretized border elements of 10 mm thickness). (*Source:* Chitty, R., Cox, G., Fardell, P.J. et al. (1992) *Mathematical Fire Modelling and Its Application to Fire Safety Design,* Report 223. Garston: BRE.)

example of the variation of the resultant emissivity in the prediction of temperatures within a steel column when the configuration factor is taken as $\Phi = 1$, i.e., $\varepsilon_{res} = \Phi\varepsilon_m\varepsilon_f = \varepsilon_m\varepsilon_f$ (Chitty et al., 1992). It should be noted that the variation of ε_{res} shown in the figure can be interpreted as the variation of configuration factor due to the difference in positions while taking ε_m and ε_f as constants.

Note that both the governing equation (6.3) and the boundary condition (6.5) are nonlinear. The former is due to the thermal conductivity and specific heat that are temperature dependent, as shown in Chapter 5. The latter is due to the radiative boundary condition that involves a nonlinear term of the temperature. Thus the closed form solution to governing, Equation (6.3) with boundary conditions Equation (6.5), is not possible for even the simplest geometry. Numerical methods such as finite element methods are usually required to solve this kind of heat transfer problem.

6.1.2 Finite element solution of heat transfer problem

The finite element method is a numerical analysis technique for obtaining approximate solutions to engineering problems. It offers a way to solve a complex continuum problem by allowing it to be subdivided into a series of simpler interrelated problems and gives a consistent technique for modelling the whole as an assemblage of discrete parts. The "whole" may be a body of matter or a region of space in which some phenomenon of interest is occurring.

In the heat transfer problem, temperature field is the field variable that is a function of each generic point in the body or solution region. Consequently, the problem is one with an infinite number of unknowns. Finite element analysis reduces the problem to one of a finite number of unknowns by dividing the solution region into elements and by expressing the temperature field in terms of assumed interpolation functions within each element. The interpolation functions are defined in terms of the values of the temperature field at specified points called nodes.

The nodal values of the temperature field and the interpolation functions for the elements completely define the behaviour of the temperature field within the elements. For the finite element representation of the heat transfer problem the nodal values of the temperature field become the unknowns. The matrix equations expressing the properties of the individual elements are determined from the governing equation by using the weighted residual approach. The individual element matrix equations are then combined to form the global matrix equations for the complete system. Once the boundary conditions have been imposed, the global matrix equations can be solved numerically. Once the nodal values of the temperature field are found, the interpolation functions define the temperature field throughout the assemblage of elements.

Assume that the solution domain Ω is divided into M elements and each element has n nodes. Thus, the temperature within each element can be expressed as follows

$$\theta(x,y,z,t) = \sum_{i=1}^{n} N_i(x,y,z)\theta_i(t) = \mathbf{N}(x,y,z)\,\boldsymbol{\theta}_e(t) \tag{6.6}$$

where $N_i(x,y,z)$ is the interpolation function defined at node i, θ_i is the value of the temperature at node i, $\mathbf{N}(x,y,z)$ is the interpolation matrix, and $\boldsymbol{\theta}_e(t)$ is the vector of element nodal temperatures. The element matrix equation is obtained from the governing Equation (6.3) by the method of weighted residuals in which the weighting function is assumed to be the same as the interpolation function, that is,

$$\int_{\Omega_e} \mathbf{N}^T \left\{ \rho c \mathbf{N} \frac{\partial \boldsymbol{\theta}_e}{\partial t} - [\nabla \cdot (\lambda \nabla \mathbf{N} \boldsymbol{\theta}_e)] - Q \right\} d\Omega = 0 \tag{6.7}$$

where Ω_e is the domain of element e. Using Gauss's theorem, for element domain Ω_e of boundary Γ_e the following expression may be derived

$$\int_{\Omega_e} \mathbf{N}^T [\nabla \cdot (\lambda \nabla \mathbf{N})]\, d\Omega = \int_{\Gamma_e} \mathbf{N}^T [(\lambda \nabla \mathbf{N}) \cdot \hat{\mathbf{n}}]\, d\Gamma - \int_{\Omega_e} (\nabla \mathbf{N})^T \cdot (\lambda \nabla \mathbf{N})\, d\Omega \tag{6.8}$$

where $\hat{\mathbf{n}}$ is the normal of the element boundary Γ_e. Using Equation (6.8), Equation (6.7) can be simplified as follows:

$$\int_{\Omega_e} \left(\mathbf{N}^T \rho c \mathbf{N} \right) \frac{\partial \boldsymbol{\theta}_e}{\partial t}\, d\Omega + \int_{\Omega_e} \left(\nabla \mathbf{N}^T \lambda \nabla \mathbf{N} \right) \boldsymbol{\theta}_e\, d\Omega = \int_{\Gamma_e} \mathbf{N}^T \left(\lambda \nabla \mathbf{N} \boldsymbol{\theta}_e \cdot \hat{\mathbf{n}} \right) d\Gamma + \int_{\Omega_e} \mathbf{N}^T Q\, d\Omega \tag{6.9}$$

After some manipulation, the resulting element matrix equation becomes

$$\mathbf{C}_e \frac{\partial \boldsymbol{\theta}_e}{\partial t} + \mathbf{K}_{ce}\, \boldsymbol{\theta}_e = \mathbf{R}_{qe} + \mathbf{R}_{Qe} \tag{6.10}$$

in which,

$$C_e = \int_{\Omega_e} N^T \rho c N d\Omega = \text{element capacitance matrix}$$

$$K_{ce} = \int_{\Omega_e} \nabla N^T \lambda \nabla N d\Omega = \text{element conductance matrix}$$

$$R_{qe} = \int_{\Gamma_e} N^T \left[(\lambda \nabla N \, \theta_e) \cdot \hat{n} \right] d\Gamma = - \int_{\Gamma_e} N^T (q \cdot \hat{n}) d\Gamma$$

$$= \text{element nodal vector of heat flow}$$

$$R_{Qe} = \int_{\Omega_e} (N^T Q) \, d\Omega = \text{element nodal vector of internal heat source}$$

Equation (6.10) is the general formulation of the element matrix equation for transient heat conduction in an isotropic medium. Note that the element nodal temperatures cannot be solved from the element matrix Equation (6.10) because the nodal vector of heat flow on the right side of Equation (6.10) is also an unknown. However, this unknown will be eliminated during the assembly of the element matrix equations or can be identified when applying boundary conditions.

Therefore, the integration in calculating R_{qe} can apply only to the boundaries with the prescribed heat flux. The global finite element matrix equation is obtained by the assembly of element matrix equations, which can be expressed as

$$C \frac{\partial \theta(t)}{\partial t} + K_c \, \theta(t) = R_q + R_Q \tag{6.11}$$

in which,

$$C = \sum_{e=1}^{M} C_e = \text{global capacitance matrix}$$

$$K_c = \sum_{e=1}^{M} K_{ce} = \text{global conductance matrix}$$

$$R_q = \sum_{e=1}^{M} R_{qe} = \text{global nodal vector of heat flow}$$

$$R_Q = \sum_{e=1}^{M} R_{Qe} = \text{global nodal vector of internal heat source}$$

where the summation implies correct addition of the matrix elements in the global coordinates and degrees of freedom. Note that, R_q is other than zero only when it is in the position corresponding to the node that is on a boundary.

For a boundary that has prescribed temperatures, R_q is unknown but θ is known; whereas for a boundary that has prescribed heat fluxes, θ is unknown but R_q is known. Thus, the total number of unknowns in the global finite element equation is always equal to the total number of nodes. It should be noted that, for the prescribed heat flux boundary condition, the expression of R_{qe} may involve the unknown surface temperatures that must be decomposed out from R_{qe}. According to the definition of R_{qe} in Equation (6.10) and noticing that, $-(q \cdot \hat{n}) = \dot{h}_{net} = \dot{h}_{net,c} + \dot{h}_{net,r}$, R_{qe} may be rearranged into

$$R_{qe} = -\int_{\Gamma_e} N^T (q \cdot \hat{n}) d\Gamma = \int_{\Gamma_e} N^T \dot{h}_{net} d\Gamma = \int_{\Gamma_e} N^T \alpha_{eff} (\theta_g - \theta_m) d\Gamma \qquad (6.12)$$

in which,

$$\alpha_{eff} = \frac{\dot{h}_{net}}{\theta_g - \theta_m}$$

$$= \alpha_c + \Phi \varepsilon_m \varepsilon_f \sigma \left[(\theta_g + 273)^2 + (\theta_m + 273)^2 \right] \left[(\theta_g + 273) + (\theta_m + 273) \right]$$

where α_{eff} is the combined convection and radiation coefficient which is temperature dependent. Let

$$R_{qe} = R_{q\theta e} - K_{q\theta e} \theta_e \qquad (6.13)$$

in which

$$R_{q\theta e} = \int_{\Gamma_e} N^T \alpha_{eff} \theta_g d\Gamma$$

$$K_{q\theta e} \theta_e = \int_{\Gamma_e} N^T \alpha_{eff} \theta_m d\Gamma = \int_{\Gamma_e} N^T \alpha_{eff} N \theta_e d\Gamma$$

Similarly, the global nodal vector of heat flow can be rewritten into

$$R_q = R_{q\theta} - K_{q\theta} \theta \qquad (6.14)$$

in which

$$R_{q\theta} = \sum_{e=1}^{M} R_{q\theta e}$$

$$K_{q\theta} = \sum_{e=1}^{M} K_{q\theta e}$$

Thus, Equation (6.11) becomes

$$C\frac{\partial \theta(t)}{\partial t} + (K_c + K_{q\theta})\,\theta\,(t) = R_{q\theta} + R_Q \tag{6.15}$$

Equation (6.15) is the finite element formulation of nonlinear transient heat transfer problems, in which C, K_c, $K_{q\theta}$ and $R_{q\theta}$ are all temperature dependent. The temperature dependence of C is due to the specific heat that is the function of temperature; the temperature dependence of K_c is due to the conductivity that is the function of temperature; while temperature dependence of $K_{q\theta}$ and $R_{q\theta}$ is due to the boundary conditions involving radiation. To solve Equation (6.15), time integration techniques must be employed.

Integration techniques for transient nonlinear solutions typically combine the methods for linear transient solutions and steady-state nonlinear solutions (Huebner et al., 1995; Zienkiewicz and Taylor, 2000). The transient solution of the nonlinear ordinary differential equations is computed by a numerical integration method with iterations at each time step to correct for nonlinearities.

Explicit or implicit one-parameter β schemes are often used as the time integration methods, and Newton-Raphson or modified Newton-Raphson methods are used for the iteration. Let t_k denote a typical time in the response so that $t_{k+1} = t_k + \Delta t$, where $k = 0, 1, 2, ..., N$. A general family of algorithms results by introducing a parameter β such that $t_\beta = t_k + \beta \Delta t$ where $0 \le \beta \le 1$. Equation (6.15) at time t_β can be written as

$$C(\theta_\beta)\frac{\partial \theta_\beta}{\partial t} + K(\theta_\beta, t_\beta)\theta_\beta = R(\theta_\beta, t_\beta) \tag{6.16}$$

where $K = K_c + K_{q\theta}$, $R = R_{q\theta} + R_Q$ are defined at temperature θ_β and time t_β, and the subscript β indicates the temperature vector θ_β at time t_β. By using the following approximations

$$\theta_\beta = (1-\beta)\theta_k + \beta\theta_{k+1}$$

$$\frac{\partial \theta_\beta}{\partial t} = \frac{\theta_{k+1} - \theta_k}{\Delta t} \tag{6.17}$$

$$R(\theta_\beta, t_\beta) = (1-\beta)R(\theta_k, t_k) + \beta R(\theta_{k+1}, t_{k+1})$$

Equation (6.16) can be rewritten into

$$\left(\beta K(\theta_\beta, t_\beta) + \frac{1}{\Delta t} C(\theta_\beta) \right) \theta_{k+1}$$

$$= \left((\beta - 1)K(\theta_\beta, t_\beta) + \frac{1}{\Delta t} C(\theta_\beta) \right) \theta_k + (1 - \beta)R(\theta_k, t_k) + \beta R(\theta_{k+1}, t_{k+1}) \quad (6.18)$$

where θ_{k+1} and θ_β are unknowns and θ_k is known from the previous time step.

Equation (6.18) represents a general family of recurrence relations; a particular algorithm depends on the value of β selected. If $\beta = 0$, the algorithm is the forward difference method, in which, if the capacitance matrices are further lumped, the method becomes explicit and reduces to a set of uncoupled algebraic equations. If $\beta = 1/2$, the algorithm is the Crank-Nicolson method; if $\beta = 2/3$, the algorithm is the Galerkin method; and if $\beta = 1$, the algorithm is the backward difference method.

For a given β, Equation (6.18) is a recurrence relation for calculating the vector of nodal temperatures θ_{k+1} at the end of time step from known values of θ_k at the beginning of the time step. For $\beta > 0$, the algorithm is implicit and requires solution of a set of coupled algebraic equations using iterations because the coefficient matrices K, C and nodal heat vector R are functions of θ. The Newton-Raphson iteration method is often used to solve the nonlinear equations at each time step.

Hughes (1977) shows the algorithm to be unconditionally stable for $\beta \geq 1/2$ as in the corresponding linear algorithm. For $\beta < 1/2$, the algorithm is only conditionally stable, and the time step must be chosen smaller than a critical time step given by

$$\Delta t_{cr} = \frac{2}{1 - 2\beta} \frac{1}{\lambda_m} \quad (6.19)$$

where λ_m is the largest eigenvalue of the current eigenvalue problem. The explicit and implicit algorithms have the same trade-offs as occur for linear transient solutions. The explicit algorithm requires less computational effort, but it is conditionally stable; the implicit algorithm is computationally expensive but unconditionally stable. The nonlinear implicit algorithm requires even greater computational effort than in linear implicit solutions because of the need for iterations at each time step. Thus the selection of a transient solution algorithm for a nonlinear thermal problem is even more difficult than in linear solutions.

A number of finite element computer codes may be used to solve the nonlinear heat transfer equation with the fire boundary condition. Three of the most commonly used codes are FIRES-T3 from the National Institute of

Standards and Technology (NIST) in the U.S., (Iding et al., 1977a), SAFIR from the University of Liège, Belgium (Franssen, 2003), and TASEF from Lund Institute of Technology, Sweden (Sterner and Wickstrom, 1990). In addition to these special codes for structures exposed to fire, general finite element programs such as ABAQUS, ANSYS, DIANA, and Comsol Multiphysics can also be used to conduct heat transfer analysis.

6.2 CALCULATION OF TEMPERATURES IN TIMBER ELEMENTS

The situation for timber is much simpler in that the temperature within a timber element is dependent only on the rate of charring (that defines the depth of charring), the thermal conductivity, and the temperature at the wood–char boundary (Schaffer, 1965). The temperature θ within a semi-infinite element is given by

$$\frac{\theta - \theta_o}{\theta_{cw} - \theta_o} = \exp\left(\frac{\beta_o x}{a_w}\right)$$

(6.20)

where the depth x is measured from the wood–char interface, θ_o is the initial wood temperature which is generally taken as 20°C, $\theta_{cw} = 288\,°C$ is the temperature at the wood–char interface, a_w is the thermal diffusivity of wood which represents how fast heat diffuses through wood material and is defined as $a_w = \lambda/(\rho c)$, and β_o is the rate of charring taken as constant.

6.3 STRUCTURAL ANALYSIS

The analysis of the response of a structure to a fire can be accomplished using the established principles of engineering mechanics. The analysis, however, needs to consider the continuing changes in material properties due to rising temperatures. Those properties that are most significant to structural performance include yield strength, modulus of elasticity, and coefficient of thermal expansion.

The development of numerical techniques and an enhanced knowledge of the thermal and mechanical properties of materials at elevated temperatures have made it possible to determine the fire resistance of various structural members by calculations. It should be emphasised that a structural analysis should examine the fire safety of a whole structure. The response of each member is calculated, and local failures are identified. It is also important to continue the calculations to determine whether these local failures could lead to progressive collapse of the whole structure.

Note that during the course of a fire, plastic deformations may develop in the materials of structural members. Therefore, the structural analysis should be time dependent although the inertia and damping forces may not necessarily be involved in the analysis. The analysis with time dependent based on time steps but without considering the inertia and damping forces is called the quasi-static analysis.

6.3.1 Calculation of structural responses using simple approaches

For some simple structural members such as beams and columns, if fire scenarios are identical or very similar along their longitudinal directions of the member, the temperature field can be assumed to be independent of the longitudinal coordinate and determined based on a two-dimensional heat transfer problem within the cross section of the member. Once the temperature field has been determined, the response of the member to the applied loading can be calculated based on the simple bending theory of Bernoulli beams.

The main assumption of the Bernoulli beam is the linear distribution of the axial strain in the cross section. Under this assumption, the axial strain at any coordinate point of the cross section can be expressed as the sum of a membrane strain and two bending strains as follows

$$\varepsilon(y,z) = \varepsilon_o + y\kappa_{xy} + z\kappa_{xz} \tag{6.21}$$

where ε_o is the membrane strain, and κ_{xy} and κ_{xz} are the curvatures of the beam in the xy- and xz-planes, respectively. On the other hand, the total strain can be decomposed in terms of the components generated by individual actions (Li and Purkiss, 2005)

$$\varepsilon(y,z) = \varepsilon_\sigma(\sigma,\theta) + \varepsilon_{cr}(\sigma,\theta,t) + \varepsilon_{tr}(\sigma,\theta) + \varepsilon_{th}(\theta) \tag{6.22}$$

where σ is the stress, θ is the temperature, t is the time, ε_σ is the stress-induced strain that is the function of stress and temperature, ε_{cr} is the classical creep strain that is the function of stress, temperature, and time, ε_{tr} is the transient strain that is the function of stress and temperature and exists only for concrete material, and ε_{th} is the thermal strain that is the function of temperature. Expressions for ε_σ, ε_{cr}, ε_{tr} and ε_{th} for steel and concrete materials can be found in Chapter 5. When a strain involves a plastic strain, the stress–strain relation is usually expressed in increment form. According to Equation (6.22), the increment of the axial strain can be expressed as

$$\Delta\varepsilon = f_\sigma(\sigma,\theta,t)\Delta\sigma + f_\theta(\sigma,\theta,t)\Delta\theta + f_t(\sigma,\theta,t)\Delta t \tag{6.23}$$

in which

$$f_\sigma(\sigma,\theta,t) = \frac{\partial\varepsilon}{\partial\sigma} = \frac{\partial\varepsilon_\sigma}{\partial\sigma} + \frac{\partial\varepsilon_{cr}}{\partial\sigma} + \frac{\partial\varepsilon_{tr}}{\partial\sigma}$$

$$f_\theta(\sigma,\theta,t) = \frac{\partial\varepsilon}{\partial\theta} = \frac{\partial\varepsilon_\sigma}{\partial\theta} + \frac{\partial\varepsilon_{cr}}{\partial\theta} + \frac{\partial\varepsilon_{tr}}{\partial\theta} + \frac{\partial\varepsilon_{th}}{\partial\theta}$$

$$f_t(\sigma,\theta,t) = \frac{\partial\varepsilon}{\partial t} = \frac{\partial\varepsilon_{cr}}{\partial t}$$

where $\Delta\varepsilon$, $\Delta\sigma$, $\Delta\theta$, and Δt are the increments of strain, stress, temperature, and time, respectively. Similarly, the increment form of Equation (6.21) can be expressed as

$$\Delta\varepsilon = \Delta\varepsilon_o + y\Delta\kappa_{xy} + z\Delta\kappa_{xz} \qquad (6.24)$$

where $\Delta\varepsilon_o$ is the increment of membrane strain and $\Delta\kappa_{xy}$ and $\Delta\kappa_{xz}$ are the increments of curvatures. Substituting Equation (6.23) into (6.24) yields

$$\Delta\sigma = \frac{1}{f_\sigma}(\Delta\varepsilon_o + y\Delta\kappa_{xy} + z\Delta\kappa_{xz} - f_\theta\Delta\theta - f_t\Delta t) \qquad (6.25)$$

Let N_x be the axial membrane force and M_y and M_z be the bending moments about y- and z-axes. Their increments thus can be expressed as

$$\Delta N_x = \int_A \Delta\sigma\, dA$$

$$\Delta M_z = \int_A y\Delta\sigma\, dA \qquad (6.26)$$

$$\Delta M_y = \int_A z\Delta\sigma\, dA$$

Substituting Equation (6.25) into (6.26) yields

$$\Delta N_x = K_{11}\Delta\varepsilon_o + K_{12}\Delta\kappa_{xy} + K_{13}\Delta\kappa_{xz} - \Delta F_1$$

$$\Delta M_z = K_{12}\Delta\varepsilon_o + K_{22}\Delta\kappa_{xy} + K_{23}\Delta\kappa_{xz} - \Delta F_2 \qquad (6.27)$$

$$\Delta M_y = K_{13}\Delta\varepsilon_o + K_{23}\Delta\kappa_{xy} + K_{33}\Delta\kappa_{xz} - \Delta F_3$$

in which

$$K_{11} = \int_A \frac{1}{f_\sigma} dA, \quad K_{12} = \int_A \frac{y}{f_\sigma} dA, \quad K_{13} = \int_A \frac{z}{f_\sigma} dA, \quad \Delta F_1 = \int_A \frac{f_\theta \Delta\theta + f_t \Delta t}{f_\sigma} dA$$

$$K_{22} = \int_A \frac{y^2}{f_\sigma} dA, \quad K_{23} = \int_A \frac{yz}{f_\sigma} dA, \quad \Delta F_2 = \int_A \frac{y(f_\theta \Delta\theta + f_t \Delta t)}{f_\sigma} dA$$

$$K_{33} = \int_A \frac{z^2}{f_\sigma} dA, \quad \Delta F_3 = \int_A \frac{z(f_\theta \Delta\theta + f_t \Delta t)}{f_\sigma} dA$$

Equation (6.27) can be rewritten into the matrix relationship between the increments of generalized strains and generalized forces as

$$\begin{bmatrix} K_{11} & K_{12} & K_{13} \\ K_{12} & K_{22} & K_{23} \\ K_{13} & K_{23} & K_{33} \end{bmatrix} \begin{Bmatrix} \Delta\varepsilon_o \\ \Delta\kappa_{xy} \\ \Delta\kappa_{xz} \end{Bmatrix} = \begin{Bmatrix} \Delta F_1 \\ \Delta F_2 \\ \Delta F_3 \end{Bmatrix} + \begin{Bmatrix} \Delta N_x \\ \Delta M_z \\ \Delta M_y \end{Bmatrix} \qquad (6.28)$$

Equation (6.28) is the generalized form of the bending equation of Bernoulli beams, in which the increments of generalized forces ΔN_x, ΔM_z, and ΔM_y can be expressed in terms of the increments of externally applied mechanical loads and reaction forces at boundaries (for a statically indeterminate structure) through the use of equilibrium equations. The increments ΔF_1, ΔF_2 and ΔF_3 are generated from the temperature and time increments. In the case of ambient temperature, ΔF_1, ΔF_2, and ΔF_3 remain zero, and thus Equation (6.28) reduces to the conventional incremental form of the bending equation of beams. Further, if the material constitutive equation is linear, the stiffness coefficients K_{ij} will be independent of stresses and strains, and thus the relationship between generalized strains and generalized forces will be the same as that between their increments.

   In the case where the temperature and internal forces are independent of the longitudinal coordinate, Equation (6.28) can be solved directly based on increment steps. Examples of this include the column subjected to pure compression and the beam subjected to pure bending. Otherwise, the membrane strain and curvatures must be solved by considering the compatibility along the longitudinal direction of the member with imposed or calculated end conditions (Purkiss and Weeks, 1987; Purkiss, 1990a). Note that

$$\Delta\varepsilon_o = \frac{d(\Delta u)}{dx}, \quad \Delta\kappa_{xy} = -\frac{d^2(\Delta v)}{dx^2}, \quad \Delta\kappa_{xz} = -\frac{d^2(\Delta w)}{dx^2} \qquad (6.29)$$

$$\Delta q_x = \frac{d(\Delta N_x)}{dx}, \quad \Delta q_y = \frac{d^2(\Delta M_z)}{dx^2}, \quad \Delta q_z = \frac{d^2(\Delta M_y)}{dx^2} \qquad (6.30)$$

where Δu is the increment of axial displacement, Δv and Δw are the increments of deflections in y- and z-directions, Δq_x is the increment of axial distributed load (q_x is positive if it is in x-axis direction), and Δq_y and Δq_z are the increments of transverse distributed loads in y- and z-directions (q_y and q_z are positive if they are in y- and z-axis directions). For statically determinate beams, ΔN_x, ΔM_z, and ΔM_y can be determined from Equation (6.30) or from static equilibrium equations directly. For statically indeterminate beams, ΔN_x, ΔM_z, and ΔM_y cannot be determined directly from static equilibrium equations and will involve unknown reaction forces that must be determined from displacement boundary conditions. Substituting Equation (6.29) into (6.28) yields

$$
\begin{bmatrix} K_{11} & K_{12} & K_{13} \\ K_{12} & K_{22} & K_{23} \\ K_{13} & K_{23} & K_{33} \end{bmatrix}
\begin{Bmatrix} \dfrac{d(\Delta u)}{dx} \\[2mm] -\dfrac{d^2(\Delta v)}{dx^2} \\[2mm] -\dfrac{d^2(\Delta w)}{dx^2} \end{Bmatrix}
= \begin{Bmatrix} \Delta F_1 \\ \Delta F_2 \\ \Delta F_3 \end{Bmatrix}
+ \begin{Bmatrix} \Delta N_x \\ \Delta M_z \\ \Delta M_y \end{Bmatrix} \quad (6.31)
$$

Equation (6.31) can be solved using various discrete methods along the longitudinal direction to convert the differentiation equations into algebraic equations. Because of the nonlinearity involved in the material constitutive equation, the stiffness coefficients K_{ij} in Equation (6.31) are temperature dependent and also stress and strain dependent. Thus, iterations are required in solving the equations at each time step to correct for nonlinearities. This finally leads to a complete deformation time history for the given externally applied loads.

Such calculations involved in the analysis are amenable only to computer analysis. Examples of programs include FIRES-RC (Becker and Bresler, 1972), CEFFICOS (Schleich, 1986; 1987), and CONFIRE (Forsén, 1982). Figure 6.2 provides a flowchart for the calculations.

The method can be applied to steel, concrete, and composite steel–concrete members. For timber that chars substantially when subjected to heat, leaving a relatively unaffected core, calculations can be undertaken using normal ambient methods with the use of temperature-reduced strengths if considered appropriate on the core after the parent section has been reduced by the appropriate depth of charring. Hosser et al. (1994) provided a very useful overview and assessment of available simplified, i.e. noncomputer code, design methods.

6.3.2 Calculation of structural responses using finite element analysis packages

The simple approach described in Section 6.3.1 can be applied only to very simple structural members with uniaxial stress states. For frame structures

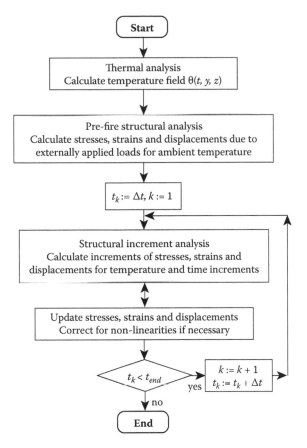

Figure 6.2 Overall calculation procedure for structural behaviour of fire-affected members.

or structural members whose stresses are not uniaxial, structural analysis packages based on finite element approaches should be used. Several such packages are commercially available. However, since a fire involves high temperatures, the software packages to be used must consider the special characteristics of materials at high temperatures and nonlinearities in both geometries and materials.

Several computer packages were specially designed for modelling high-temperature phenomena, including FIRES-RC II (Iding et al., 1977b), FASBUS II (Iding and Bresler, 1987; 1990), SAFIR (Nwosu et al., 1999), and VULCAN (Huang et al. 2003a and b). VULCAN is capable of modelling the global three-dimensional behaviours of composite steel-framed buildings under fire conditions. The analysis considers the whole frame action and includes geometrical and material nonlinearities within its beam, column, and slab elements. It also includes the ability to represent semi-rigid connections that degrade with temperature and partial inter-actions of steel sections and slabs. In addition to these specific software packages, other nonlinear finite element structural analysis programs such as ABAQUS, ANSYS, and DIANA can also be utilised for conducting fire analyses of structures (Sanad et al., 1999).

Most finite element programs require entry of data on the stress–strain temperature behaviour of steel whether reinforcing, pre-stressing, or struc-tural and/or concrete as appropriate. Since the thermal strain is treated separately in most structural analysis programs, the strain used to define the stress–strain temperature behaviour required as an input in the pro-gram is the sum of all other strains.

For steel, this includes the classical creep strain and the strain induced by the mechanical stress. In cases where the classical creep strain is negli-gible, the value reduces to just the strain induced by the mechanical stress, and therefore the stress–strain temperature behaviour can be simply rep-resented by the temperature-dependent stress-strain equations as described in Chapter 5.

For concrete, however, the transient strain is not negligible, and the strain used to define the stress-strain temperature behaviour thus must include both the transient strain and the strain induced by the mechanical stress. Therefore, the temperature-dependent stress–strain equations must be modified to include the transient strain before they can be input to the program (Li and Purkiss, 2005).

Although many computer simulation results appear in the literature, few comparative or benchmark tests have been commissioned to assess ther-mal response and, more importantly, structural response. The latter is very much affected by materials models and the exact formulation of the analy-sis techniques used. Sullivan et al. (1993/4) gave some results following a survey of available computer software packages both for thermal and structural analyses. For the thermal analysis programs, particularly for steel structures, most packages gave comparable answers, and the answers were also in reasonable agreement with experimental data. However, for concrete structures, particularly at temperatures of around 100 to 200°C, agreement was less acceptable because the packages examined did not fully consider moisture transport. It was also noted that the fit between experi-mental and predicted temperatures was often improved by adjusting the values of the parameters defining the thermal diffusivity and/or the thermal boundary conditions such as emissivity.

For the structural analysis packages investigated, a greater spread of acceptability was found. This spread was due partially to the fact that some of the packages were developed for research (and thus lacked adequate doc-umentation or were user unfriendly). Other factors were the assumptions made in the analysis algorithms (e.g., no allowance for large displacements) and inadequate materials models (especially for concrete where transient strain or the effect of stress history was ignored). Many of the programs predicted correct trends but the absolute results did not agree with experimental data. Also, the effect of classical creep on the behaviour of steel was neglected.

6.3.3 Examples

The first example presented is a circular steel tube filled with concrete subjected to pure compression. Its outside surface is exposed to a fire. The problem was solved using the simple approach described in Section 6.3.1 by Yin et al. (2006). Because of the axial symmetry of the problem, the temperature and axial compressive stress are axially symmetric. Figure 6.3 shows the temperature distributions along the radial direction at various times when the composite column was exposed to the fire, the temperature of which is defined by the standard fire curve.

As is seen in Figure 6.3, the variation of the temperature is much smaller in the steel tube than in the concrete core because steel has a much greater thermal diffusivity than concrete. The high temperature in the steel tube

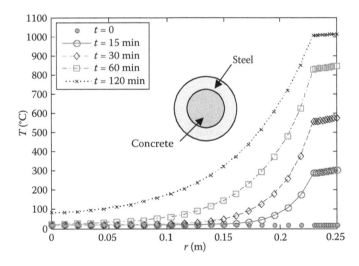

Figure 6.3 Temperature distribution profiles of circular steel tube filled with concrete. Column diameter $D = 500$ mm, steel tube thickness $h_c = 20$ mm, conductivity and specific heat are temperature dependent for both steel and concrete. Standard fire curve is used for fire temperature. [*From:* Yin, J., Zha, X.X., and Li, L.Y. (2006) *Journal of Construction Steel Research*, 62, 723–729. With permission.]

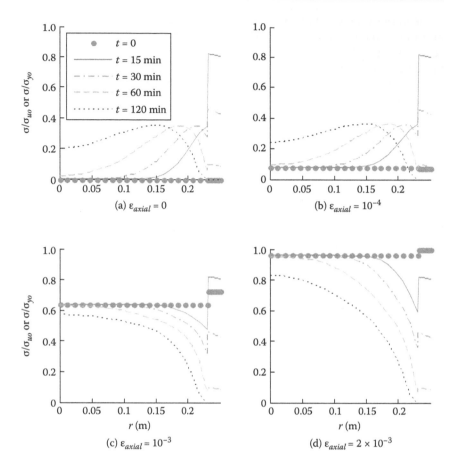

Figure 6.4 Stress distribution profiles along radial direction at various times for different compressive strains (σ_{uo} is the concrete peak compressive stress at ambient temperature and σ_{yo} is the steel yield stress at ambient temperature. (*Source:* Yin, J., Zha, X.X., and Li, L.Y. (2006) *Journal of Construction Steel Research,* 62, 723–729. With permission.)

together with the nonuniform temperature distribution in the concrete core led to a complicated distribution of the axial compressive stress, as demonstrated in Figure 6.4.

Note that the reduction in stress in the steel tube when fire exposure time increases is due to the strength reduction caused by high temperature. Figure 6.5 shows the load displacement curves of the composite column at various fire exposure times. The fire resistance of the column can be obtained by plotting the maximum loads of the load displacement curves against the fire exposure times.

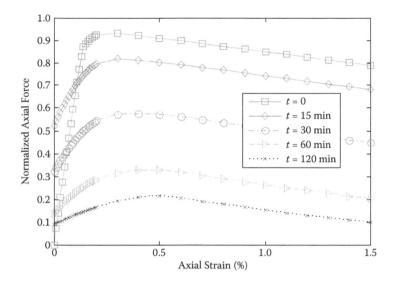

Figure 6.5 Load displacement curves with temperature effects. (*Source:* Yin, J., Zha, X.X., and Li, L.Y. (2006) *Journal of Construction Steel Research*, 62, 723–729. With permission.)

The second example is a two-bay I-section steel frame with columns fixed at the base and beams uniformly loaded on the top. The analysis was performed using the nonlinear finite element analysis package ABAQUS by Ali et al. (2004). Two main steps were followed in the analysis procedure. In the first step, the frame was analyzed under the applied load at room temperature to establish the pre-fire stress and deformation in the frame. In the second step, the history of fire temperature was calculated and imposed on the deformed and loaded structure causing the steel to expand and the mechanical properties to degrade. Both geometric and material nonlinearities were included in the simulations to account for the expected large displacements, plastic deformations, and creep. Five fire scenarios were investigated. The highly nonlinear problem was solved using iterative procedures with automatic time stepping.

Figure 6.6 shows the simulation results. The frame expands slowly toward the wall until it reaches its maximum lateral displacement followed by rapid change in displacement direction and collapse away from the fire wall. The time to collapse is about 45 min. When the fire covers at least one bay of the frame, the time to collapse is largely unaffected by the fire scenario because the plastic hinges and excessive deformations in the span near the wall very much control the failure. The required space between the wall and steel depends on the extent of fire.

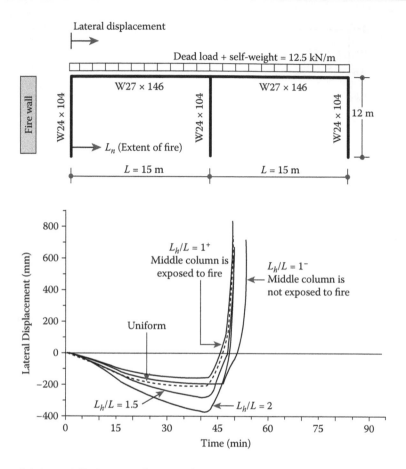

Figure 6.6 Lateral displacement histories for a two-bay steel frame. $L_h/L = 1^-$ = fire local-ized to bay nearest firewall excluding middle column. $L_h/L = 1^+$ = bay closest to wall is exposed to fire including middle column. $L_h/L = 1.5$ = fire extended beyond first bay to cover half of second bay. $L_h/L = 2$ = two bays are heated except for far column. Uniform = fire heats all columns and girders of frame. L_k ($0 \leq L_h \leq 2L$) = length of fuel burning measured from firewall to anywhere within the two bays. [*From:* Ali, F.A., O'Connor, D., and Abu-Tair, A. (2001) *Magazine of Concrete Research*, 53, 197–204. With permission.]

The minimum clearance required between the frame and the wall increases with the length of the fuel burning. This behaviour is consistent with a simple case of uniformly heated steel members that are restrained only at one end in which longitudinal thermal expansion is proportional to the heated length of the member. A significant difference in lateral

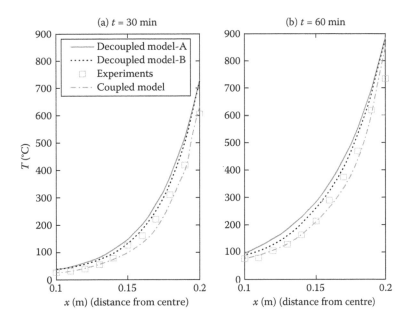

Figure 6.7 Temperature distribution profiles of concrete wall (400 mm) with moisture content 2% of concrete weight at fire exposure times of 30 and 60 min, obtained from different models. Decoupled model A uses specific heat of dry concrete. Decoupled model B uses specific heat recommended in EN 1992-1-2 to account for effects of moisture. Experimental data are taken from Ahmed and Hurst (1997). Coupled model considers the transfer of both heat and moisture (Tenchev, Li and Purkiss, 2001b).

expansion is noticed between the two-bay fire scenario $(L_b/L = 2)$ and the uniform fire case. The uniform fire scenario is mathematically equivalent to a fixed girder at the middle column and results in lateral displacements similar to the one-bay fire case $(L_b/L = 1)$.

Figure 6.7 shows how moisture influences heat transfer and thus the temperature distribution in a concrete structural member. The problem shown here is a simple concrete wall 400 mm thick. The wall has an initial porosity of 0.08, and the corresponding initial moisture content is 2% of concrete weight. The wall is subject to double-side fire exposure. The temperature results for the case where the moisture transfer is considered are taken from Tenchev et al. (2001b). The experimental data are taken from Ahmed and Hurst (1997). The two temperature curves for the case where the moisture transfer is not considered correspond to different specific heat expressions, both of which are given in EN 1992-1-2

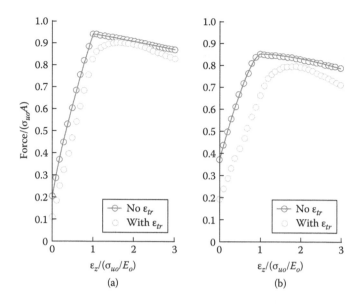

Figure 6.8 Compressive load displacement curves of concrete wall (400 mm): (a) at 20 min and (b) at 60 min. A is the cross section area and σ_{uo} is concrete peak compressive stress. (*Source:* Li. L.Y. and Purkiss, J.A. (2005) *Fire Safety Journal*, 40, 669–686. With permission.)

(one is recommended for use with moisture transfer and the other for use without considering moisture transfer). The comparisons of temperature distribution profiles between models demonstrate that a coupled moisture and heat transfer model provides more accurate results, while the model using an equivalent specific heat to take into account the moisture influence on the heat transfer can slightly improve the prediction but not completely.

Figure 6.8 shows how transient strain influences the fire performance of concrete structures. The problem is about a simple concrete wall of 400 mm thick. The wall is subjected to a uniformly distributed compressive load and heated on both sides as described by the ISO fire curve (Li and Purkiss, 2005). The two sets of results shown in Figure 6.8 correspond to two models; one ignores the transient strain and the other includes the transient strain in the stress–strain temperature behaviour model. Thus, the difference between the two sets of results reflects the influence of the transient strain.

Note from the figure that for short time exposure, the influence of the transient strain on the load displacement curve is not very significant and thus it has little influence on the fire performance of the wall. However,

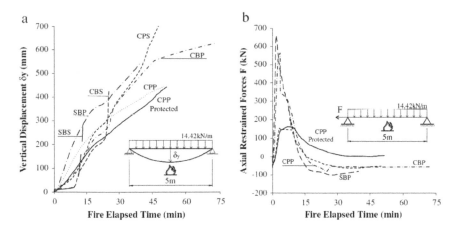

Figure 6.9 Nonlinear analysis results of steel and composite steel–concrete beams and panels under different boundary conditions. (a) Vertical displacements and (b) axial forces. S/C = steel/composite. B/P = beam/panel. S/P = simple (a)/pinned (b) boundary conditions. (*Source:* Silva, J.C. and Landessmann, A. (2013) *Journal of Constructional Steel Research*, 83, 117–126. With permission.)

when the exposure time is not short and the temperature in the concrete is high, the influence of the transient strain on the load displacement curve becomes very significant and thus exerts considerable influence on the fire performance of the wall.

Figure 6.9 demonstrates the importance of considering geometric nonlinearity or second order effects. The results were obtained from the analyses of steel and composite steel–concrete beams and panels under simple (one support can move freely in the axial direction) and pinned boundary conditions (Silva and Landesmann, 2013) during exposure to a fire. The figure indicates that the deflections of the beam or panel obtained from two different boundary conditions are significantly different. This illustrates the effects of axial force on deflection, i.e., the second order effect. Another example for the second-order effects was provided by Forsén (1982) who showed that the deflections of simple and pinned support beams could be qualitatively different.

Both steel and concrete structures can undergo large displacements during a fire due simply to the high temperature that weakens the stiffness of structural members. Therefore, the geometric nonlinearity or the second order effect, i.e., additional moments caused by large deflections or displacements, can substantially affect the results of calculations and the treatment of support conditions, as demonstrated in Figure 6.9.

6.4 COUPLED HEAT AND MASS TRANSFER IN CONCRETE

Unlike steel, concrete is a porous material, whose pores are partially filled with water and air. Under ambient temperature conditions, part of the water is chemically bonded to the cement while the remainder is contained in the concrete pores as free water. When concrete is subject to sufficient heat such as fire, the free water and bound water that release into concrete pores from dehydration will evaporate. If the evaporation rate exceeds the vapour migration rate, pore pressure will build up. This pressure, combined with the effects of temperature and mass concentration gradients, causes the transport of water and a gaseous mixture of dry air and water vapour through the connective pores. The heat absorbed during the continuing evaporation–dehydration processes largely contributes to the inherent fire endurance properties of concrete.

The work related to the study of coupled heat and mass transfer in a porous medium dates back to the early 1960s when Luikov (1966) proposed a system of coupled partial differential equations governing heat and moisture transfer in porous media. Based on Luikov's work, Bažant and Thonguthai (1979), Bažant et al. (1981), Dhatt et al. (1986), and Gong et al. (1991) developed mathematical models to describe the drying of refractory concrete that were solved numerically using finite element methods.

Abdel-Rahman and Ahmed (1996) and Ahmed and Hurst (1997) proposed a model to simulate coupled heat and mass transfer and related processes in one-dimensional porous media exposed to elevated temperatures. The model was based on conservation of mass, momentum and energy with the consideration of the dehydration process of a heated material, and the corresponding effect on material pore size and mass transport mechanisms. However, the model did not take into account the mobility of water.

A more sophisticated model developed by Selih et al. (1994) and Selih and Sousa (1995) considered the migration of free and bound water, diffusion and migration of dry air and water vapour, and the evaporation of free and bound water. However, the individual mass conservations derived in the model were based on a different framework. In contrast to the complex model of Selih et al., Bažant (1997) proposed a simplified model in which the water vapour and free water in liquid phase were treated as a single variable. Thus the model yields only two coupled governing equations; one is the mass transport for the water and the other is the heat transfer for the temperature. The consideration of the evaporation of liquid water was not required, but the dehydration of bound water was taken into account in the model.

The coupled heat and mass transfer model presented in this section is from the work of Tenchev, Li, and Purkiss (2001b) and Tenchev et al.

(2001). One of the distinguishing features of this model, when compared with others, is that dry air, water vapour, and liquid water are treated separately. This enables the modelling of phase changes between solid, liquid, and gaseous mixture more directly and conveniently. The original model in Tenchev, Li, and Purkiss (2001b) was presented using a three-dimensional finite element formulation. However, for the convenience of presentation, here it is presented in the form of a one-dimensional system.

6.4.1 Volumetric fractions of solid, liquid, and gaseous phases

Concrete can be considered as a multiphase system consisting of a solid phase (porous skeleton together with bound water), a liquid phase (free water that partially fills voids of the skeleton), and a gaseous phase (binary mixture of dry air and water vapour filling the residual parts of voids). If ε_i is used to represent the volumetric fraction of phase i :

$$\varepsilon_s + \varepsilon_l + \varepsilon_g = 1 \tag{6.32}$$

where ε_s, ε_l and ε_g are the volumetric fractions of solid, liquid, and gaseous phases, respectively. Furthermore, ε_s can be expressed by:

$$\varepsilon_s = \varepsilon_{ss} + \varepsilon_b \tag{6.33}$$

where ε_{ss} and ε_b are the volumetric fractions of porous skeleton and bound water, respectively. The former can be treated as a constant, while the latter may release into the liquid phase as free water by dehydration when temperature rises to a certain level. Note that the free water can also become water vapour due to evaporation. Therefore ε_b, and thus ε_s, ε_l and ε_g vary with time.

Dehydration of bound water depends on both temperature and type of concrete, which can be experimentally obtained by weight loss measurements on heated specimens. For HSC, the amount of dehydrated water can be approximately expressed as follows (Bažant and Kaplan, 1996):

$$\frac{\varepsilon_d \rho_l}{\rho_{cem}} = \begin{cases} 0 & if & \theta \leq 200°C \\ 7 \times 10^{-4}(\theta - 200) & if & 200°C < \theta \leq 300°C \\ 4 \times 10^{-5}(\theta - 300) + 0{,}07 & if & 300°C < \theta \leq 800°C \\ 0{,}09 & if & 800°C < \theta \end{cases} \tag{6.34}$$

where ε_d is the volumetric fraction of dehydrated bound water, $\rho_l = 1000$ kg/m³ is the specific weight of water, and ρ_{cem} is the mass of anhydrous cement per unit volume concrete, which is a constant for a given type of concrete. For HSC, $\rho_{cem} = 300$ kg/m³. For a given temperature, ε_d can be determined from Equation (6.34), and thus the volumetric fraction of the solid phase can be determined by:

$$\varepsilon_s = \varepsilon_{ss} + \varepsilon_b = \varepsilon_{ss} + \varepsilon_{bo} - \varepsilon_d \tag{6.35}$$

where ε_{bo} is the initial volumetric fraction of bound water, which is equal to the maximum value of ε_d.

The free water content in the liquid phase within concrete can be determined by the so-called equation of state of pore water (Bažant and Thonguthai, 1978). For temperatures above the critical point of water (374,15°C), all free water is assumed to have been vaporised, and thus there is no liquid phase. For any temperature below the critical point of water, the free water content can be determined by temperature and the ratio of water vapour pressure to saturation vapour pressure as follows:

$$\frac{\varepsilon_l \rho_l}{\rho_{cem}} = \begin{cases} \left(\dfrac{\varepsilon_{l,25} \rho_l}{\rho_{cem}} \dfrac{P_v}{P_{sat}} \right)^{1/m(\theta)} & \text{if} \quad \dfrac{P_v}{P_{sat}} \leq 0,96 \\[3ex] \dfrac{\rho_l}{\rho_{cem}} \left[\varepsilon_{l,0,96} + \left(\dfrac{P_v}{P_{sat}} - 0,96 \right) \dfrac{\varepsilon_{l,1,04} - \varepsilon_{l,0,96}}{1,04 - 0,96} \right] & \text{if} \quad 0,96 < \dfrac{P_v}{P_{sat}} < 1,04 \\[3ex] \dfrac{\varepsilon_{lo} \rho_l}{\rho_{cem}} \left[1 + 0,12 \left(\dfrac{P_v}{P_{sat}} - 1,04 \right) \right] & \text{if} \quad 1,04 \leq \dfrac{P_v}{P_{sat}} \end{cases} \tag{6.36}$$

where $\varepsilon_{l,25}$ is the volumetric fraction of saturation water at 25°C, which can be determined if the concrete mix is specified (Neville, 1973), P_v is the water vapour pressure, P_{sat} is the saturation vapour pressure, which can be obtained from the standard water table, ε_{lo} is the initial volumetric fraction of the free water, $\varepsilon_{l,0,96}$ and $\varepsilon_{l,1,04}$ are the volumetric fractions of the free water at $P_v = 0,96 P_{sat}$ and $P_v = 1,04 P_{sat}$, respectively, and $m(\theta)$ is an experimentally determined empirical expression defined as follows:

$$m(\theta) = 1,04 - \frac{(\theta + 10)^2}{22,3(25 + 10)^2 + (\theta + 10)^2} \tag{6.37}$$

The first and third terms in Equation (6.36) originate from Bažant and Thonguthai (1978) and cover free water and water vapour. However, as it will be seen later from the results, if there is free water in an element, the mass of free water will be much larger than that of water vapour. Therefore, the isotherm provides a good estimation of free water content. Otherwise, a different approach need to be developed. The second term of Equation (6.36) is an interpolation between the first and third.

If the temperature and water vapour pressure are known, the volumetric fractions for solid, liquid, and gaseous phases can be determined using Equations (6.34) to (6.37).

6.4.2 Mass transfer of free water and gaseous mixture

In addition to the dehydration of bound water and the evaporation of free water, another feature observed in concrete subjected to fire is the transport of free water, water vapour, and dry air. If it is assumed that the transport and evaporation of bound water occur only after it has released into the liquid phase as free water, the following mass conservation equations for free water, dry air, and water vapour can be obtained:

$$\frac{\partial}{\partial t}(\rho_l \varepsilon_l) = -\nabla[(\rho_l \varepsilon_l)V_l] - \dot{E} + \frac{\partial}{\partial t}(\rho_l \varepsilon_d) \tag{6.38}$$

$$\frac{\partial}{\partial t}(\rho_a \varepsilon_g) = -\nabla\left[(\rho_a \varepsilon_g)V_g - D_g(\rho_a \varepsilon_g)\nabla\left(\frac{\rho_a}{\rho_g}\right)\right] \tag{6.39}$$

$$\frac{\partial}{\partial t}(\rho_v \varepsilon_g) = -\nabla\left[(\rho_v \varepsilon_g)V_g - D_g(\rho_g \varepsilon_g)\nabla\left(\frac{\rho_v}{\rho_g}\right)\right] + \dot{E} \tag{6.40}$$

where ρ_a and ρ_v are the mass concentrations of dry air and water vapour in gaseous phase, $\rho_g = \rho_a + \rho_v$ is the specific weight of gaseous mixture, D_g is the diffusion coefficient of a component in a binary mixture, $\dot{E}$ is the evaporation rate of free water, and V_l and V_g are the migration velocities of free water and gaseous mixture defined by Darcy's law as follows:

$$V_l = -\frac{KK_l}{\mu_l}\nabla P \tag{6.41}$$

$$V_g = -\frac{KK_g}{\mu_g}\nabla P \tag{6.42}$$

where K is the intrinsic permeability of the skeleton, K_l and K_g are the relative permeabilities of the liquid water and gaseous mixture, μ_l and μ_g are the dynamic viscosities of liquid and gaseous mixture, and P is the pore pressure which is the sum of the partial pressures P_a and P_v related to the dry air and water vapour in the gaseous mixture, that is,

$$P = P_a + P_v \tag{6.43}$$

By assuming the gaseous mixture behaves as an ideal mixture of an incondensable ideal gas (dry air) and a condensable ideal water vapour, the partial pressures P_a and P_v can be expressed in terms of the mass concentrations of the dry air and water vapour in the gaseous mixture as follows,

$$P_a = \frac{R(\theta + 273)\rho_a}{M_a} \tag{6.44}$$

$$P_v = \frac{R(\theta + 273)\rho_v}{M_v} \tag{6.45}$$

where $R = 8{,}314$ J/(K mole) is the gas constant, $M_a = 28{,}9$ g/mole and $M_v = 18$ g/mole are the molar masses of the dry air and water vapour. For a given temperature θ, Equations (6.38) to (6.45) can be used to determine $\dot{E}$, ρ_a, ρ_v, V_l, V_g, P_a, P_v and P.

6.4.3 Heat transfer in multiphase medium

The temperature at any point of a multiphase medium can be determined by considering the energy conservation for all the phases present in the medium as follows,

$$\overline{(\rho C)}\frac{\partial \theta}{\partial t} = \nabla\left(k_{eff}\nabla\theta\right) - \overline{(\rho CV)}\nabla\theta - \lambda_e\dot{E} - \lambda_d\frac{\partial}{\partial t}(\rho_l \varepsilon_d) \tag{6.46}$$

in which

$$\overline{(\rho C)} = (\rho_{cem}\varepsilon_{ss})C_s + (\rho_l \varepsilon_l + \rho_l \varepsilon_b)C_l + (\rho_a \varepsilon_g)C_a + (\rho_v \varepsilon_g)C_v$$

$$\overline{(\rho CV)} = (\rho_l \varepsilon_l)C_l V_l + (\rho_a \varepsilon_g)C_a V_g + (\rho_v \varepsilon_g)C_v V_g$$

where C_s, C_l, C_a, and C_v are the specific heats of a porous skeleton, water, dry air, and water vapour, respectively; k_{eff} is the thermal conductivity coefficient; λ_e is the latent heat of evaporation of free water; and λ_d is the specific heat of dehydration of bound water.

The heat transfer Equation (6.46) and mass transfer Equations (6.38) to (6.45) are coupled to each other. These nine equations can be used to solve for the nine unknown variables θ, $\dot{E}$, ρ_a, ρ_v, V_l, V_g, P_a, P_v, and P. By eliminating $\dot{E}$, ρ_a, V_l, V_g, P_a, and P_v algebraically using Equations (6.38) and (6.41) to (6.45), one can finally achieve three independent nonlinear partial parabolic differentiation equations about three independent unknown variables θ, ρ_v, and P. The boundary conditions for θ, ρ_v, and P can be expressed as follows,

$$k_{eff}\frac{\partial\theta}{\partial n}+\alpha J_1 \cdot n + h_{qr}(\theta - \theta_\infty) = 0 \tag{6.47}$$

$$J_v \cdot n - \beta(\rho_v - \rho_{v\infty}) = 0 \tag{6.48}$$

$$P = P_\infty \tag{6.49}$$

where n is the unit outward normal of the surface, α is the isobaric heat of evaporation of water from liquid state, J_l is the flux of free water at the boundary, h_{qr} is the combined convective and radiation heat transfer coefficient, J_v is the flux of water vapour at the boundary, β is the transfer coefficient, and the parameters with subscripts ∞ imply that they are defined in the surrounding environment.

The nonlinear partial parabolic differentiation equations with boundary conditions defined in Equations (6.47) to (6.49) can be solved numerically using finite element methods. Detailed finite element formulations can be found in Tenchev et al. (2001b). Temperature-dependent material constants can be found in Tenchev et al. (2001) for normal concrete and in Tenchev et al. (2001b) for HSC.

6.4.4 Numerical results

Figure 6.10 shows the temperature distributions of a HSC column of 400 mm × 400 mm when two opposite sides of the column are exposed to a standard fire curve temperature, while the other two sides are fully insulated. Good agreement between the numerical results and test data is demonstrated at different times and also at different positions.

The numerical results for pore pressure P, mass concentration of water vapour ρ_v, mass of water vapour in unit volume of concrete $\varepsilon_g\rho_v$, and mass of liquid water in unit volume of concrete $\varepsilon_l\rho_l$ are plotted in Figure 6.11. The figure shows that high pore pressure develops during the first hour of a

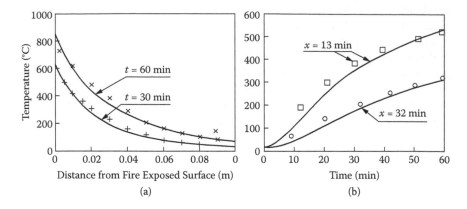

Figure 6.10 Distribution of temperatures at (a) given times and (b) given locations. Cross, plus and square symbols are experimental data. (From Tenchev, R.T., Li, L.Y. and Purkiss, J.A. (2001b) *Numerical Heat Transfer,* 39, 685–710. With permission.)

fire. The zone of increased pore pressure (Figure 6.11a) corresponds to the increased mass concentration of water vapour (Figure 6.11b).

A comparison of Figure 6.11b and Figure 6.11c clearly indicates that the sudden drop in the mass of water vapour in unit volume of concrete is due to a drop in the volumetric fraction of gaseous phase that corresponds to the transition between wet and dry zones (Figure 6.11d).

In the literature, reported pore pressures in concrete under intense heating vary from less than 1 MPa for normal concrete (Dayan and Gluekler, 1982; Ahmed and Hurst, 1997) to 16 MPa for HSC (Consolazio et al., 1997; Ahmed and Hurst, 1997; Ju and Zhang, 1998). This indicates that pore pressure plays an important role in the spalling of concrete.

The distribution of liquid water (Figure 6.11d) shows several well defined zones, including a dry zone between $x = 0$ and line a, a zone of evaporation between lines a and b, a zone of increased water content between lines b and c, and a zone of initial state between line c and $z = 0{,}1$ m. The increased water content results from the condensation of water vapour transferred into the cooler interior of the concrete. It corresponds to the so-called moisture clog (Ulm et al., 1999). This finding is similar to what was reported in literature (Majumdar et al., 1995; Majumdar and Marchertas, 1997; Ahmed and Hurst, 1997).

A rapid increase of pore pressure as a function of time at a given location (see, for example, $x = 30$ mm in Figure 6.11e) starts with the formation of a moisture clog (dotted line d going to Figure 6.11h). The moisture

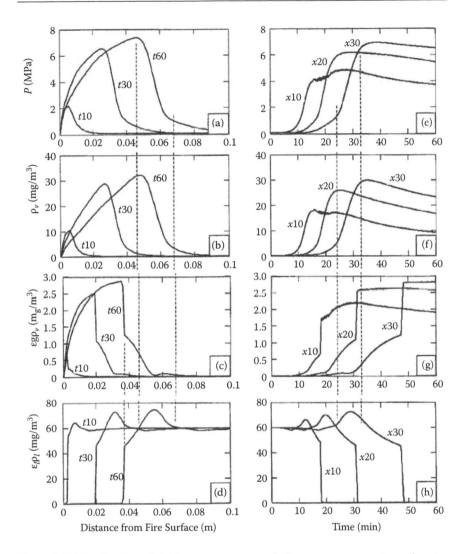

Figure 6.11 Distribution of (a/e) pore pressure, (b/f) mass concentration of water vapour, (c/g) mass of water vapour in unit volume of concrete, and (d/h) mass of liquid water in unit volume of concrete at (a–d) three times (*t*10 = 10 min, *t*20 = 20 min and *t*30 = 30 min) and (e–h) three locations (*x*10 = 10 mm, *x*20 = 20 mm, and *x*30 = 30 mm). (*Source:* Tenchev, R.T., Li, L.Y., and Purkiss, J.A. (2001b) *Numerical Heat Transfer*, 39, 685–710. With permission.)

clog blocks the inward transfer of water vapour. During the time interval between lines d and e, the mass concentration of water vapour increases and causes an increase of pore pressure.

As the pores are almost saturated with liquid water (Figure 6.11h), the volumetric fraction of gaseous phase and thus the mass of water vapour in unit volume of concrete are very small (Figure 6.11g). The time when the maximum saturation at $x = 30$ mm occurs is 30 min, which corresponds to a temperature around 200°C. After that, reduction of liquid water follows because of evaporation. The increase in the mass of water vapour is mainly due to increased volumetric fraction of the gaseous phase (Figure 6.11f). After its maximum value, the pore pressure decreases only slightly. After the evaporation of all liquid water, the relatively high value of pore pressure is maintained by the thermal expansion of the gaseous mixture and water vapour transferred from the interior.

Figure 6.12 plots the maximum pore pressure and its corresponding time, location, and temperature. The figure indicates a rapid increase of the maximum pore pressure around 30 min after the start of the fire. At that time, the maximum pore pressure is located about 0,02 m away from the fire-exposed surface, and the corresponding temperature is about 220°C. If spalling is caused by high pore pressure, this is the most likely time and place where it will occur. This is in good agreement with experimental observations (Copier, 1979).

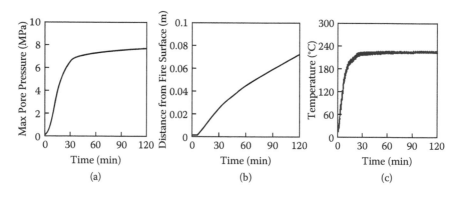

Figure 6.12 Maximum pore pressures at (a) different times, (b) corresponding locations, and (c) temperatures. (*Source:* Tenchev, R.T., Li, L.Y., and Purkiss, J.A. (2001b) *Numerical Heat Transfer*, 39, 685–710. With permission.)

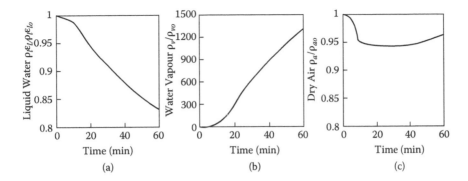

Figure 6.13 Evolution of (a) relative liquid water, (b) relative water vapour, and (c) relative dry air during first hour of fire. (*Source:* Tenchev, R.T., Li, L.Y., and Purkiss, J.A. (2001b) *Numerical Heat Transfer*, 39, 685–710. With permission.)

After that time, both the maximum pore pressure and temperature values remain almost constant although the corresponding location moves steadily toward the interior where evaporable water is still available.

Figure 6.13 plots the evolution of the liquid water, water vapour, and dry air during the first hour of a fire. As expected, the liquid water decreases when evaporation takes place, and the total moisture (liquid water plus water vapour) decreases as water vapour escapes from the surface. Dry air also decreases because it is displaced by water vapour and expands thermally. The total change of the moisture content may be measured experimentally and thus may be used to check the accuracy of the numerical model indirectly.

Chapter 7

Design of concrete elements

The design of concrete elements affected by exposure to fire may be undertaken in one of two ways: a prescriptive approach or a calculation approach. The prescriptive approach, in which tables of minimum dimensions, minimum axis distance, and other parameters corresponding to a given standard fire endurance are consulted, was covered in Chapter 3. It will not be discussed further except to note that even in the calculation approach minimum dimensions are conveniently determined using such tables when the exposure is taken as the standard furnace temperature-time curve. It should be noted that the minimum axis distance required for adequate fire resistance for low periods of fire resistance is likely to be less than that required for durability, and thus a calculation approach may not be viable.

Because of the complexity of the stress–strain relationships of concrete in compression where the elastic strains, unrestrained thermal expansion, and transient strains must be considered, it is not generally possible in simple design methods to consider the deformation history of a structural element. It is normally sufficient to consider only the resistance to the applied forces and thus the analytical approach reduces to an end-point calculation, where the inequality between the resistance effect and the load effect is determined at a given time. However, by using numerical stress–strain curves such as those proposed by Li and Purkiss (2005), it may be possible to determine deformation history using spreadsheets. The time to failure for a structural element may be determined by calculating the moment capacity or axial capacity at a series of discrete time steps.

The previous chapter dealt with matrix techniques for both temperature distribution and stresses within structural elements. This chapter will concentrate on hand calculation methods, although the use of spreadsheets may be found advantageous. The methods presented will assume that the thermal analysis and structural analysis may be decoupled.

7.1 CALCULATION OF TEMPERATURES

The basic theory for calculating the temperature distribution in any element was given in Chapter 6, but it should be noted that empirical or graphical solutions are available as design tools. It is generally accurate enough to use such data for the end-point designs of concrete members. Where exposure is to a parametric or real fire curve, the temperatures within a concrete element, including those of any steel reinforcement, continue to rise for a period after the maximum fire or gas temperature has been reached. Thus the mechanical response or resistance of any concrete element so exposed will need calculating during the early stages of the cooling period as the critical design strength may be attained after the maximum fire temperature has occurred.

7.1.1 Graphical data

The three main sources of graphical data are the *ISE and Concrete Society Design Guide* (1978), the *FIP/CEB Report* (1978), and EN 1992-1-2.

7.1.1.1 ISE and Concrete Society design guide (1978)

This publication gives temperature profiles for both flat soffit slabs and beams exposed to the standard furnace temperature–time response. Note that although the flat soffit slab data in the design guide are given for both normal and lightweight concrete, the beam data are for siliceous aggregate normal weight concrete only. The guide suggests that for lightweight concrete beams, the temperatures may be taken as 80% of those for normal-weight concrete.

7.1.1.2 FIP/CEB report (1978)

This report provides temperature data on more varied types of concrete, including limestone aggregates, but only for exposure to the standard furnace curve.

7.1.1.3 EN 1992-1-2

Temperature profiles for standard fire resistance periods for slabs and beam–column sections cast from normal weight concrete are given in Annex A. Annex A implies that these profiles are results of calculation. However, whether these profiles were calibrated against actual test data and what values of thermal properties were used are not known.

7.1.2 Empirical methods

These in general are based on curve fitting techniques from data derived from furnace tests or the superposition of simple solutions to the Fourier heat transfer equation. Two such methods are available: (1) A method based on the analysis of results from TASEF-2 (Wickström, 1985a and 1986); and (2) the method of Hertz (1981a and b). Both methods can be applied to exposure to an actual compartment temperature–time curve (for Wickström's method, the parametric curve in Annex A of EN 1991-1-2 is used) or the standard furnace test curve. Both methods are applicable to different concretes because thermal diffusivity values are entered as data. Both methods yield a temperature rise $\Delta\theta$ above ambient. The explanations of both methods below are limited to exposure to the standard furnace curve.

7.1.2.1 Wickström method

The temperature rise in a normal weight concrete element $\Delta\theta$ is given by:
For uniaxial heat flow:

$$\Delta\theta = n_x n_w \Delta\theta_f \tag{7.1}$$

For biaxial heat flow:

$$\Delta\theta_{xy} = (n_w(n_x + n_y - 2n_x n_y) + n_x n_y)\Delta\theta_f \tag{7.2}$$

where n_w is given by

$$n_w = 1 - 0,0616t^{-0,88} \tag{7.3}$$

and assuming constant thermal properties n_x (or n_y with y substituted for x) by

$$n_x = 0,18\ln u_x - 0,81 \tag{7.4}$$

where

$$u_x = \frac{a}{a_c}\frac{t}{x^2} \tag{7.5}$$

and a is the thermal diffusivity of the concrete under consideration and a_c is a reference value of $0,417 \times 10^{-6}$ m²/sec. For $a = a_c$, Equation (7.4) reduces to

$$n_x = 0,18\ln\frac{t}{x^2} - 0,81 \tag{7.6}$$

with x (or y) subject to the limit

$$x \geq 2h - 3,6\sqrt{0,0015t} \tag{7.7}$$

where t is the time (hr), x and y are the depths into the member (m), and h is the overall depth of the section (m). Equation (7.6) only applies to a concrete whose thermal conductivity reduces approximately linearly from around 1,25 W/m°C at 100°C to 0,5 W/m°C at 1200°C. This roughly corresponds to the lower bound curve in EN 1992-1-2. However, Clause 10.3 of PD 7974-3 allows the use of this method without apparent restrictions. $\Delta\theta_g$ is given by the rise in furnace temperature above ambient at time t. Note that the temperature rise above ambient on the surface of the element is given by $n_w\Delta\theta_g$.

7.1.2.2 Hertz method

The unidimensional time-dependent temperature $\theta(x,t)$ is given by

$$\Delta\theta(x,t) = f_1(x,t) + f_2(x,t) + f_3((x,t)) \qquad (7.8)$$

where the functions $f_1(x,t)$, $f_2(x,t)$ and $f_3(x,t)$ are solutions to the heat transfer equation for specific boundary conditions. These functions are given by

$$f_1(x,t) = E\left(1 - \frac{x}{3,363\sqrt{at}}\right)^2 \qquad (7.9)$$

$$f_2(x,t) = De^{-x\sqrt{\frac{\pi}{2Ca}}} \sin\left(\frac{\pi t}{C} - x\sqrt{\frac{\pi}{2Ca}}\right) \qquad (7.10)$$

$$f_3(x,t) = \frac{D+E}{2(e^{LC}-1)}\left(1 - e^{\left(L(t-C)-x\sqrt{\frac{L}{a}}\right)}\right) \qquad (7.11)$$

where a is the thermal diffusivity. Note that f_1 is set equal to zero if

$$1 - \frac{x}{3,363\sqrt{at}} \leq 0 \qquad (7.12)$$

f_2 is set equal to zero if

$$\frac{\pi t}{C} - x\sqrt{\frac{\pi}{2Ca}} \leq 0 \qquad (7.13)$$

and f_3 is set equal to zero if

$$L(t-C) - x\sqrt{\frac{L}{a}} \leq 0 \qquad (7.14)$$

The parameters E, D and C are dependent upon the heating régime and L is dependent upon the temperature curve during cooling and is given by

$$L = \frac{2}{C}\ln\left(\frac{3D}{E-2D}\right)$$ (7.15)

If $E - 2D$ is negative, then $E - 2D$ is set to 0,02. The temperature rise on the surface during heating is given by $D + E$. Note that for exposure to the standard furnace temperature–time curve, L does not need calculating, since f_3 is always zero.

For exposure to the standard furnace curve, values of C, D, and E are given in Table 7.1. Note that C is equal to twice the required time period. For the values of these parameters when exposure is to a parametric compartment, temperature–time response reference should be made to Hertz. For two-dimensional heat flow, the above method needs modification in that the temperature $\theta(x,y,t)$ is given by:

$$\Delta\theta(x,y,t) = \Delta\theta_0\left(\xi_{\theta,x} + \xi_{\theta,y} - \xi_{\theta,x}\xi_{\theta,y}\right)$$ (7.16)

where $\Delta\theta_0$ is the surface temperature rise at time t and $\xi_{\theta,x}\Delta\theta_0$ is the temperature rise at the point considered, assuming unidimensional heat flow on the x direction and $\xi_{\theta,y}\Delta\theta_0$ is the temperature rise for heat flow on the y direction. It is instructive to compare the surface temperatures predicted by Wickström, Hertz and Figure A2 of EN 1992-1-2. This comparison is carried out in Table 7.2, where all three sets of results are within about 40°C of each other. Thus it would appear acceptable to use any of the methods to predict temperature rise.

Table 7.1 Temperature analysis parameters for concrete members under standard conditions

Time (hr)	C (hr)	D (°C)	E (°C)
0,5	1,0	150	600
1,0	2,0	220	600
1,5	3,0	310	600
2,0	4,0	360	600
3,0	6,0	410	600
4,0	8,0	460	600

Note: Where the element at the top of a fire compartment has a web narrower than the bottom flange, the values of D and E should be multiplied by 0,9.

Source: Hertz, K.D. (1981b) *Simple Temperature Calculations of Fire Exposed Concrete Constructions.* Report 159, Lyngby: Technical University of Denmark. With permission.

Table 7.2 Comparison of surface temperature rises of Wickström, Hertz, and EN 1992-1-2

Standard fire duration (hr)	Furnace temperature	Surface temperatures (°C)		
		Wickström	Hertz	EN 1992-1-2[a]
0,5	842	749	770	730
1,0	945	888	840	880
1,5	1006	963	930	950
2,0	1049	1015	980	1010
3,0	1110	1084	1030	1080
4,0	1153	1132	1080	1120

[a] Figure A.2.

7.1.3 Values of thermal diffusivity

Both calculations of internal temperatures require the value of thermal diffusivity. EN 1992-1-2 gives no guidance on this. EN 1993-1-2 gives values of the thermal conductivity λ_c and specific heat c_c that may be used in simple calculations. These are $\lambda_c = 1,6$ W/mK and $c_c = 1000$ J/kg°C. If a value of ρ_c of 2400 kg/m3 is assumed, then from Equation (5.2):

$$a_c = \frac{\lambda_c}{\rho_c c_c} = \frac{1,6}{1000 \times 2400} = 0,67 \times 10^{-6} \ \text{m}^2/\text{sec}$$

Hertz (1981b) quotes values of $0,35 \times 10^{-6}$ m²/sec for Danish sea gravel or granite concrete and $0,52 \times 10^{-6}$ m²/sec for a quartzite concrete. Wickström (1986) gives a value of $0,417 \times 10^{-6}$ m²/sec for normal weight concrete. It would therefore appear that for siliciceous aggregate concretes a_c should lie in the range 0,417 to $0,67 \times 10^{-6}$ m²/sec. PD 7974-3 implies the use of a value of $a_c = 0,417 \times 10^{-6}$ m²/sec

7.1.4 Position of 500°C isotherm

This is required for the method first proposed by Anderberg (1978b). From Wickström (1985a, 1986) for uniaxial heat flow, the position x for a temperature rise $\Delta\theta_x$ at time t and furnace temperature rise $\Delta\theta_f$ is given by

$$x = \left[\frac{\dfrac{a}{0,417 \times 10^{-6}} t}{\exp\left(4,5 + \dfrac{\Delta\theta_x}{0,18 n_w \Delta\theta_f} \right)} \right]^{0,5} \tag{7.17}$$

Table 7.3 Comparison of depths of 500°C isotherm (mm) using Wickström, Hertz, and EN 1992-1-2 methods

Time (hr)	Thermal diffusivity (× 10⁻⁶ m²/sec)						
	Wickström			Hertz			EN 1992-1-2
	0,417	0,52	0,67	0,417	0,52	0,67	
0,5	12	13	14	17	18	21	10
1,0	23	25	29	27	30	34	21
1,5	31	35	40	36	41	46	30
2,0	39	44	49	44	49	56	37
3,0	52	58	66	57	63	72	48
4,0	64	71	81	68	76	86	60

For the 500°C isotherm, $\Delta\theta_x = 480°C$ and $x = x_{500}$. Values of x_{500} were determined for values of a_c of 0,415, 0,52, and 0,67 × 10⁻⁶ m²/sec. A closed form solution cannot be obtained from Hertz. Thus a spreadsheet was used to obtain values of x_{500} for the same values of a_c. Values of x_{500} obtained from Figure A2 of EN 1992-1-2 are tabulated in Table 7.3.

It appears from Table 7.3 that Wickström overestimates the value of x_{500} for all values of a_c with the least differences for $a_c = 0,417 × 10⁻⁶$ m²/sec. Hertz consistently overpredicts the value of x_{500} and is thus conservative. In all calculations for the 500°C isotherm, Wickström's approximation will be adopted with a value of $a_c = 0,417 × 10⁻⁶$ m²/sec.

7.2 SIMPLE CALCULATION METHODS

7.2.1 Calculation of load effects

This section applies to whichever method is used to determine section capacity. Load effect may be calculated using (1) load combinations and load factors given in EN 1990 to determine the effects of the actions on the structure, or (2) the actions to be considered during a fire may be taken as η_{fi} times those at the ultimate limit state.

7.2.1.1 Direct calculation

The combination of loading is determined from the unfavourable (and favourable) permanent actions with a partial safety factor of 1,0 and the variable actions multiplied by ψ_2 from EN 1990 (Clause 4.3.1(2), EN 1991-1-2). From Table A1.1 of EN 1990, ψ_2 takes a value of 0,3 for domestic, residential, office, and high level traffic areas, 0,6 for shopping, congregation, and low level traffic areas, and 0,8 for storage areas.

7.2.1.2 Indirect calculation

The design effect in fire $E_{d,fi}$ should be taken as $\eta_{fi} E_d$, where E_d is the design effect at ambient and η_{fi} is given by

$$\eta_{fi} = \frac{G_k + \psi_{fi} Q_{k,1}}{\gamma_G G_k + \gamma_{Q,1} Q_{k,1}} \tag{7.18}$$

$$\eta_{fi} = \frac{G_k + \psi_{fi} Q_{k,1}}{\gamma_G G_k + \gamma_{Q,1} \psi_{0,1} Q_{k,1}} \tag{7.19}$$

$$\eta_{fi} = \frac{G_k + \psi_{fi} Q_{k,1}}{\xi \gamma_G G_k + \gamma_{Q,1} Q_{k,1}} \tag{7.20}$$

depending upon which set of combination rules is used to determine E_d. As a conservative simplification, a value of $\eta_{fi} = 0{,}7$ may be adopted.

7.2.2 Partial safety factors

The recommended value for material strength partial safety factors $\gamma_{M,fi}$ is 1,0 for both steel and concrete.

7.2.3 Methods for determining section capacity

EN 1992-1-2 allows two methods for determining section resistance: the 500°C isotherm method (Anderberg, 1978b) and the method of slices of Hertz (1981a and b, 1985).

7.2.3.1 Reduced section method (500°C Isotherm)

This method was proposed by Anderberg (1978b), following the analysis of a number of fire tests carried out on flexural reinforced concrete elements. Some limitations are placed on the use of the method and involve minimum thicknesses for either standard exposure times or fire load densities (Table B1, EN 1992-1-2). If used with parametric curves, the opening factor must exceed 0,14 m$^{\frac{1}{2}}$. The calculations are carried out by assuming (1) the concrete within the 500°C isotherm remains unaffected by heat and (2) the reduction factors for the reinforcement (assuming Class N) are for compression reinforcement and tension reinforcement as shown below.

Compression reinforcement and tension reinforcement with strain in reinforcement $\varepsilon_{s,fi} < 2$ % (Clause 4.2.4.3) —

$20° \leq \theta \leq 100°C$:

$$k_s(\theta) = 1{,}0 \tag{7.21}$$

$100°C \leq \theta \leq 400°C$:

$$k_s(\theta) = 0,7 - 0,3\frac{\theta - 400}{300} \tag{7.22}$$

$400°C \leq \theta \leq 500°C$:

$$k_s(\theta) = 0,57 - 0,13\frac{\theta - 500}{100} \tag{7.23}$$

$500°C \leq \theta \leq 700° \ C$:

$$k_s(\theta) = 0,1 - 0,47\frac{\theta - 700}{200} \tag{7.24}$$

$700°C \leq \theta \leq 1200°C$:

$$k_s(\theta) = 0,1\frac{1200 - \theta}{500} \tag{7.25}$$

Tension reinforcement with $\varepsilon_{s,fi} \geq 2\%$ (Table 3.2a, hot rolled) —
$20° \leq \theta \leq 400°C$:

$$k_s(\theta) = 1,0 \tag{7.26}$$

$400°C \leq \theta \leq 500°C$:

$$k_s(\theta) = 0,78 - 0,22\frac{\theta - 500}{200} \tag{7.27}$$

$500°C \leq \theta \leq 600°C$:

$$k_s(\theta) = 0,47 - 0,31\frac{\theta - 600}{100} \tag{7.28}$$

$600°C \leq \theta \leq 700° \ C$:

$$k_s(\theta) = 0,23 - 0,24\frac{\theta - 700}{100} \tag{7.29}$$

$700°C \leq \theta \leq 800° \ C$:

$$k_s(\theta) = 0,11 - 0,12\frac{\theta - 800}{100} \tag{7.30}$$

$800°C \le \theta \le 1200°C$:

$$k_s(\theta) = 0,11\frac{1200 - \theta}{400} \tag{7.31}$$

In accordance with EN 1992-1-1, the depth of the stress block is taken as λx, where x is the depth to the neutral axis where λ is given by:

$$\lambda = 0,8 - \frac{f_{ck} - 50}{200} \le 0,8 \tag{7.32}$$

and the concrete strength is taken as ηf_{cd} where η is given by:

$$\eta = 1,0 - \frac{f_{ck} - 50}{200} \le 1,0 \tag{7.33}$$

All concrete in tension is ignored. In both methods, the value of the load duration factor α_{cc} is taken as 1,0, and therefore will not be included in the calculations. The temperatures in the reinforcing are determined by considering the heat flow in both the x and y directions as appropriate. The moment capacity of the section M_u is given by:

$$M_u = M_{u1} + M_{u2} \tag{7.34}$$

where M_{u1} is due to the tension reinforcement and M_{u2} is due to the compression reinforcement and its balancing tension reinforcement. M_{u1} is given by:

$$M_{u1} = A_{s1}f_{sd,fi}(\theta_m)z \tag{7.35}$$

and the mechanical reinforcement ratio ω_k is given by:

$$\omega_k = \frac{A_{s1}f_{sd,fi}(\theta_m)}{b_{fi}d_{fi}f_{cd,fi}(20)} \tag{7.36}$$

M_{u2} is given by:

$$M_{u2} = A_{s2}f_{scd,fi}(\theta_m)z' \tag{7.37}$$

The total tension steel area A_s is given by:

$$A_s = A_{s1} + A_{s2} \tag{7.38}$$

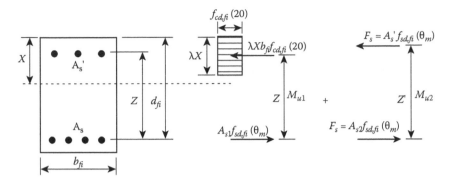

Figure 7.1 Stress distribution at ultimate limit state for a rectangular concrete cross section with compression reinforcement (Source: EN 1992-1-2. © British Standards Institution.).

Symbols are defined in Figure 7.1. The values of $f_{scd,fi}(\theta_m)$ and $f_{sd,fi}(\theta_m)$ are the temperature-reduced strengths of the reinforcement at a mean temperature θ_m in a given layer. Where the reinforcement is in layers, the mean temperature reduced strength $k(\varphi)f_{sd,bi}$ is given by:

$$k(\varphi)f_{sd,fi} = \frac{\Sigma[k_s(\theta_i)f_{sd,i}A_1]}{\Sigma A_i} \qquad (7.39)$$

and the effective axis distance a by:

$$a = \frac{\Sigma[a_i k_s(\theta_i)f_{sd,i}A_i]}{k_s(\theta_i)f_{sd,i}A_i} \qquad (7.40)$$

For the more usual case where all the reinforcement has the same strength, Equations (7.39) and (7.40) reduce to:

$$k(\phi) = \frac{\Sigma[k_s(\theta_i)A_1]}{\Sigma A_i} \qquad (7.41)$$

and the effective axis distance a to:

$$a = \frac{\Sigma[a_i k_s(\theta_i)A_i]}{\Sigma k_s(\theta_i)A_i} \qquad (7.42)$$

7.2.3.2 Method of slices (Zone Method)

The heat-affected concrete is divided into a series of slices and the temperature θ determined at the mid-depth of each slice. If the section has dimensions $h \times b$, where $h > b$, the temperatures are calculated based on the heat flow in smaller dimension and, if necessary, considering the heat flow from both opposite faces. Where $h = b$, the heat flow in either direction need only be considered. For reinforcing, the temperatures are determined by considering the heat flow in both the x and y directions as appropriate. The concrete strength reduction factor for siliceous aggregate concrete $k_c(\theta)$ is given by (Table 3.1, EN 1992-1-2):

For $20°C \leq \theta \leq 100°C$:

$$k_c(\theta) = 1,0 \tag{7.43}$$

For $100°C \leq \theta \leq 200°C$:

$$k_c(\theta) = 0,95 - 0,05\frac{\theta - 200}{100} \tag{7.44}$$

For $200°C \leq \theta \leq 400°C$:

$$k_c(\theta) = 0,75 - 0,2\frac{\theta - 400}{200} \tag{7.45}$$

For $400°C \leq \theta \leq 800°C$:

$$k_c(\theta) = 0,15 - 0,6\frac{\theta - 800}{400} \tag{7.46}$$

For $800°C \leq \theta \leq 900°C$:

$$k_c(\theta) = 0,08 - 0,07\frac{\theta - 900}{100} \tag{7.47}$$

For $900°C \leq \theta \leq 1000°C$:

$$k_c(\theta) = 0,04 - 0,04\frac{\theta - 1000}{100} \tag{7.48}$$

For $1000°C \leq \theta \leq 1100°C$:

$$k_c(\theta) = 0,01 - 0,03\frac{\theta - 1100}{100} \tag{7.49}$$

For $1100°C \leq \theta \leq 1200°C$:

$$k_c(\theta) = 0,1\frac{1200-\theta}{100} \tag{7.50}$$

The mean concrete strength reduction factor $k_{c,m}$ is given by:

$$k_{c,m} = \frac{1-\dfrac{0,2}{n}}{n}\sum_{i=1}^{n}k_c(\theta_i) \tag{7.51}$$

The factor $1 - 0,2/n$ in Equation (7.51) compensates for the fact that $k_c(\theta_i)$ is determined at the centre of a strip. The effective width of a uniform stress block is determined by calculating the width of the damage zone a_z given by:

For columns:

$$a_z = w\left[1-\left(\frac{k_{c,m}}{k_c(\theta_M)}\right)^{1,3}\right] \tag{7.52}$$

For beams and slabs:

$$a_z = w\left[1-\left(\frac{k_{c,m}}{k_c(\theta_M)}\right)\right] \tag{7.53}$$

The strength reduction factor $k_c(\theta_M)$ is determined at the centre of the member, and w is the half width for exposure on opposite faces. For columns, $2w$ is the lesser cross-sectional dimension. In the original work by Hertz, the equations for beams and slabs were identical, each with an index of 1,3. EN 1992-1-2 provides graphical data for reduction of concrete strength and section width in Figure B.5 of Annex B. The basis of the curves in EN 1992-1-2, however, is not known.

The reinforcement strength reduction factors are the same as those used in the 500°C isotherm method, and the method of analysis for beams is also similar.

EN 1992-1-2 imposes a restriction on the zone method: it may be used only for exposure to the standard furnace curve. The reason for this is unclear as Hertz implied that there was no restriction as he tabulated mean strength and damage zone width parameters for exposure to parametric fire curves. The zone method is not applicable to slabs. In all subsequent examples, $a_c = 0,417 \times 10^{-6}$ m²/sec will be used where appropriate.

7.2.3.3 Calibration of 500°C isotherm and zone methods

It should be noted that both methods were calibrated against the results from standard furnace tests that involved little or no axial restraint to the member. This is not of great importance for beams and slabs as restraint forces in a structure are likely to be low. For columns, this will not be the general case as restraint forces will be induced by the stiffness of the surrounding structure to the thermal expansion of the columns (Dougill, 1966). Thus in a stiff structure, column loads will change during the course of a fire with the internal stresses in the column redistributing due to the effects of transient strain. It is therefore recommended that simple calculation methods for columns should be used only for isolated columns or columns in a flexible structure.

Example 7.1: Concrete slab design

A simply supported slab in a multispan structure has been designed and detailed in Figure 7.2. Continuity steel in the top face has been provided purely to resist the effect of cracking over the support. The cover is to satisfy durability only. The slab is to be checked for a 2-hr fire resistance period when exposed to the standard furnace test. Assume the usage of the structure is as an office. The structural loading in the fire limit state is given by:

$$0,125 \times 25 + 1,0 + 0,3 \times 2,5 = 4,875 \text{ kPa}$$

The ψ factor has been set as $0,3$ and the partition loading has been treated as totally permanent.

$$M_{Ed,fi} = 4,875 \times 2,7^2/8 = 4,44 \text{ kNm/m}.$$

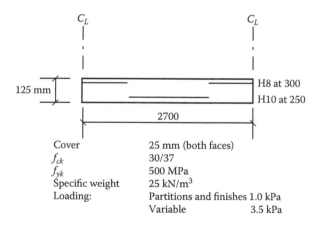

Cover	25 mm (both faces)
f_{ck}	30/37
f_{yk}	500 MPa
Specific weight	25 kN/m^3
Loading:	Partitions and finishes 1.0 kPa
	Variable 3.5 kPa

Figure 7.2 Design data for a reinforced concrete slab (Example 7.1).

Only the 500°C isotherm method may be used where the 500°C isotherm is approximately 40 mm from the bottom of the slab, thus for sagging moments the concrete strength is unaffected (Table 7.3, Wickström). Use Wickström, Equations (7.1) to (7.6) to determine temperatures. For temperature in the steel:

Centroid of the reinforcement is at $20 + 10/5 = 25$ mm
$n_w = 1 - 0,0616\ t^{-0,88} = 1 - 0,0616 \times 2^{-0,88} = 0,967$
$n_x = 0,18\ln(t/x^2) - 0,81 = 0,18\ln(2/0,025^2) - 0,81 = 0,643$
$\Delta\theta_g = 345\log(480t + 1) = 345\log(480 \times 2 + 1) = 1029°C$
$\Delta\theta = n_x n_w \Delta\theta_g = 0,643 \times 0,967 \times 1029 = 640°C$
or
$\theta_s = 640 + 20 = 660°C$ if ambient temperature of 20°C is assumed

As the strength reduction will be high (and the neutral axis depth small), use the values of strength reduction factors for $\varepsilon_{s,fi} > 2\%$. From Equation (7.29):

$k_s(\theta) = 0,23 - 0,24(\theta-700)/100 = 0,23 - 0,24(660 - 700)/100 = 0,326$

For tension force, F_s:

$F_s = 0,326 \times 500 \times 314 = 51182$ N
For Grade 30/37 concrete, $\lambda = 0,8$ and $\eta = 1,0$
so
$F_c = \eta f_{cd,fi}(20)\lambda x b = 0,8(30/1,0) \times 0,8x \times 1000 = 19200x$.
Equating F_s and F_c gives $x = 2,67$ mm.
If $\varepsilon_{cu} = 3500$ µstrain, $\varepsilon_s > 2\%$
$M_{u1} = F_s(d - 0,5\lambda x) = 51182(125 - 25 - 0,5 \times 0,8 \times 2,67)$
$\quad\quad = 5,06$ kNm/m

The moment capacity of 5,06 kNm/m exceeds the applied moment of $M_{Ed,fi}$ of 4,44 kNm/m. Thus the slab is satisfactory. If a factor of η_{fi} of 0,7 were used, M_{Ed} would be $0,7(0,125 \times 25 + 1,0 + 2,5) \times 2,7^2/8 = 4,23$ kNm/m, which would also be satisfactory. The reason the use of $\eta_{fi} = 0,7$ produces a slightly lower value of $M_{Ed,fi}$ is due to the relatively low value of the variable load of 2,5 kPa.

Example 7.2: Concrete beam design

Determine the load carrying capacity history over a complete range of standard furnace exposures, and check the duration the beam can last. The data for the example are given in Figure 7.3.

Permanent load = $0,45 \times 0,85 \times 25 = 9,56$ kNm/m
Permanent load moment = $9,56 \times 11^2/8 = 145$ kNm
Assuming office loading, $\psi_2 W = 0,3 \times 260 = 78$ kN
Moment due to variable loading = $78 \times 4 = 312$ kNm.
Total applied moment $(M_{Ed,fi}) = 145 + 312 = 457$ kNm

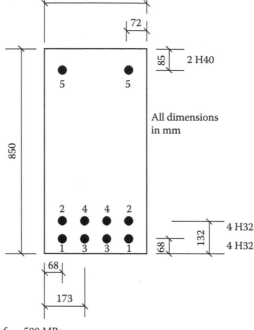

f_{yk} = 500 MPa
Concrete grade c 30/37
Span 11 m
Specific weight 25 kN/m³
Variable loading 2 × 260 kN at 4m from support

Figure 7.3 Design data for a concrete beam (Example 7.2).

500°C isotherm method — The position of the 500°C isotherm is only needed parallel to the side faces. These values are obtained from Table 7.3, and the reduced width is then given by $450 - 2x_{500}$. The determination of reinforcement temperatures is based on data in Table 7.4 and Table 7.5. As all the reinforcement is the same size, Equation (7.41) reduces to

$$k(\phi) = \frac{1}{n}\sum k_s(\theta_i)$$ (7.54)

Equation (7.54) expands to:

$$k(\phi) = \frac{k_s(\theta_1) + k_s(\theta_2) + k_s(\theta_3) + k_s(\theta_4)}{4}$$

Table 7.4 Temperatures and strength reduction factors for Bar 5

t (hr)	n_w	72 (mm)	68 (mm)	132 (mm)	173 (mm)	θ_s (°C)	$k_s(\theta)$
				n_x		Bar 5	
0,5	0,887	0,012	0,033	0	0	29	1,0
1,0	0,938	0,137	0,158	0	0	139	0,961
1,5	0,957	0,210	0,231	0	0	218	0,882
2,0	0,967	0,262	0,283	0,044	0	281	0,819
3,0	0,977	0,335	0,356	0,117	0,019	376	0,724
4,0	0,982	0,380	0,407	0,169	0,071	450	0,547

and Equation (7.42) reduces to:

$$a = \frac{\sum[a_i k_s(\theta_i)]}{\sum k_s(\theta_i)} \tag{7.55}$$

Equation (7.55) expands to:

$$a = \frac{68\,[k_s(\theta_1) + k_s(\theta_3)] + 132[k_s(\theta_2) + k_s(\theta_4)]}{4k(\varphi)}$$

$b_{fi} = b - 2x_{500} = 450 - 2x_{500}$
$F_{sc} = A_s' f_{scd,fi}(\theta_m) = A_{sc}k(\theta_s)f_{yk} = 2513 \times 500k(\theta_s) = 1256500k(\theta_s)$
$A_{si}f_{s,fi}(\theta) = A_s f_{Sd,fi}(\theta_m) - F_{sc} = 6434 \times 500k(\varphi) = 3{,}217k(\varphi)$ MN
$x = A_{si}f_{s,fi}(\theta)/(0,8 \times f_{cd}(20) \times b_{fi}) = A_{si}f_{s,fi}(\theta)/(0,8 \times 30 \times b_{fi})$
  $= A_{si}f_{s,fi}(\theta)/(24 \times b_{fi})$
$M_{u1} = A_{si}f_{s,fi}(\theta)[h - a - 0,4x] = A_{si}f_{s,fi}(\theta)[850 - a - 0,4x]$
$M_{u2} = F_{sc}[h - a - 85] = F_{sc}[765 - a]$

The values of M_{u1}, M_{u2} and M are determined in Table 7.6.

Table 7.5 Temperatures and strength reduction factors for bars 1 to 4

	Bar 1 (68, 68)		Bar 2 (68, 132)		Bar 3 (68, 173)		Bar 4 (132, 173)	
t (hr)	θ_s (°C)	$k_s(\theta)$	θ_s (°C)	$k_s(\theta)$	θ_s (°C)	$k_s(\theta)$	θ_s (°C)	$k_s(\theta)$
0,5	67	1,0	44	1,0	44	1,0	20	1,0
1,0	273	1,0	157	1,0	157	1,0	20	1,0
1,5	408	0,881	238	1,0	238	1,0	20	1,0
2,0	506	0,761	333	1,0	302	1,0	64	1,0
3,0	646	0,360	480	0,802	412	0,878	163	1,0
4,0	744	0,177	586	0,513	520	0,718	274	1,0

Table 7.6 Determination of M_{u1}, M_{u2} and M

t (hr)	k(φ)	a (mm)	b_{fi} (mm)	F_s (MN)	$F_{s,tot}$ (MN)	F_{sl} (MN)	x (mm)	M_{u1} (kNm)	M_{u2} (kNm)	M (kNm)
0,5	1,0	100	426	1,257	3,217	1,960	192	1319	836	2155
1,0	1,0	100	404	1,143	3,217	2,074	214	1378	760	2138
1,5	0,97	101	388	1,108	3,120	2,012	216	1333	736	2069
2	0,94	102	372	1,029	3,204	2,175	244	1415	682	2097
3	0,76	106	346	0,910	2,445	1,535	185	1028	600	1628
4	0,60	109	322	0,687	1,930	1,243	161	975	451	1426

Method of slices — Nine vertical strips of 50-mm width are taken parallel to the vertical faces of the beam cross section and the temperatures determined at the mid-depth of each strip using Equations (7.8) to (7.10). The results are given in Table 7.7. The sum of the concrete strength reduction factor $\Sigma k_c(\theta)$ is given by

$$\Sigma k_c(\theta) = 2k_c(\theta_1) + 2k_c(\theta_2) + 2k_c(\theta_3) + 2k_c(\theta_4) + k_c(\theta_5)$$

and $k_{c,m}$ by

$$k_{c,m} = \frac{1 - \dfrac{0,2}{n}}{n}\sum k_c(\theta_1) = \frac{1 - \dfrac{0,2}{9}}{9}\sum k_c(\theta_1) = 0,109\sum k_c(\theta_i)$$

Table 7.7 Determination of concrete temperatures, strength reduction factors, and section width reduction

				Strip no (x)							
	1 25		2 75		3 125		4 175		5 225		
t (hr)	θ_1(°C)	$k_c(\theta_1)$	θ_2(°C)	$k_c(\theta_2)$	θ_3(°C)	$k_c(\theta_3)$	θ_4(°C)	$k_c(\theta_4)$	θ_5(°C)	$k_c(\theta_5)$	Σk_c (θ) a_z
0,5	385	0,765	41	1,00	20	1,00	20	1,00	20	1,00	8,530 16
1,0	516	0,576	128	0,986	21	1,00	20	1,00	20	1,00	8,124 26
1,5	620	0,420	201	0,949	48	1,00	20	1,00	20	1,00	7,738 35
2	689	0,317	268	0,882	82	1,00	22	1,00	20	1,00	7,398 44
3	773	0,191	371	0,779	139	0,981	50	1,00	20	1,00	6,902 56
4	842	0,121	452	0,672	198	0,951	85	1,00	31	1,00	6,488 66

Table 7.8 Determination of temperature and strength reduction factors for bar 5

			$\xi_{\theta,x}$			Bar 5	
t (hr)	D + E (°C)	72 (mm)	68 (mm)	132 (mm)	173 (mm)	θ_s (°C)	$k_s(\theta)$
0,5	750	0,038	0,055	0	0	49	1,0
1,0	820	0,146	0,168	0	0	140	0,960
1,5	910	0,219	0,246	0,020	0	219	0,881
2,0	960	0,278	0,305	0,050	0,002	287	0,813
3,0	1010	0,367	0,393	0,102	0,032	390	0,710
4,0	1060	0,426	0,451	0,144	0,064	471	0,539

The value of $k_c(\theta_M)$ is given by $k_c(\theta_S)$ which equals 1,0 for all time steps. The value of a_z is given by:

$$a_z = w\left[1 - \frac{k_{c,m}}{k_c(\theta_M)}\right] = \frac{450}{2}\left[1 - \frac{0,109\sum k_c(\theta_i)}{1,0}\right]$$

The next stage is to determine the $\xi_{\theta,x}$ (or $\xi_{\theta,y}$) factors to calculate steel temperatures. The temperatures corresponding to depths of 68, 132, and 173 mm are given in Table 7.8 along with the $\xi_{\theta,x}$ values derived from the calculated temperatures. The determinations of reinforcement temperatures and strength reduction factors are tabulated in Table 7.8 and Table 7.9. The values of $k(\phi)$ and a are calculated as before and given in Table 7.10.

In conclusion, it should be noted that there is little difference in the results from Anderberg (500°C isotherm method) and Hertz (zone method) up to

Table 7.9 Temperatures and strength reduction factors for bars 1 to 4

	Bar 1 (68, 68)		Bar 2 (68, 132)		Bar 3 (68, 173)		Bar 4 (132, 173)	
t (hr)	θ_s (°C)	$k_s(\theta)$	θ_s (°C)	$k_s(\theta)$	θ_s (°C)	$k_s(\theta)$	θ_s (°C)	$k_s(\theta)$
0,5	100	1,0	41	1,0	41	1,0	20	1,0
1,0	272	1,0	158	1,0	158	1,0	20	1,0
1,5	413	0,876	258	1,0	244	1,0	20	1,0
2,0	516	0,730	346	1,0	314	1,0	70	1,0
3,0	658	0,331	479	0,803	437	0,849	152	1,0
4,0	761	0,157	582	0,526	535	0,671	231	1,0

Table 7.10 Determination of M_{u1}, M_{u2}, and M

t (hr)	k(φ) (mm)	a (mm)	b_{fi} (MN)	F_s (MN)	$F_{s,tot}$ (MN)	F_{sl} (mm)	x (mm)	M_{u1} (kNm)	M_{u2} (kNm)	M (kNm)
0,5	1,0	100	418	1,257	3,217	1,960	195	1317	836	2155
1,0	1,0	100	398	1,206	3,217	2,011	211	1339	802	2141
1,5	0,969	101	380	1,107	3,117	2,010	220	1329	735	2066
2	0,933	102	362	1,022	3,001	1,979	228	1300	678	1978
3	0,748	106	338	0,892	2,406	1,514	187	1013	588	1601
4	0,589	109	318	0,677	1,895	1,218	160	825	444	1269

2 hr, but Hertz indicates a slightly more rapid decrease after 2 hr. In both cases, the beam will last more than 4 hr.

7.3 COLUMNS

This section covers only cases where columns are not subjected to bending moments or where buckling need not be considered. Annex B3 of EN 1992-1-2 gives a method of handling this situation, but it is iterative as the column curvature(s) must be taken into account. Also if buckling is not critical at the normal ambient limit state, it need not be considered at the fire limit state. This is in contradiction to Hertz (1985) who used the Rankine equation to determine the load capacity of any column under a fire limit state.

Example: 7.3: Concrete column

Determine the fire resistance of a short reinforced concrete column 400 × 800 mm with 8 H25 bars having a cover of 35 mm (Figure 7.4). The concrete is Grade 50/60.

$A_c = 320000$ mm², $A_s = 3927$ mm²

At ambient, the load carrying capacity N_{Rd} is given as

$$N_{Rd} = A_c \frac{\alpha_{cc}f_{ck}}{\gamma_m} + A_s \frac{f_{yk}}{\gamma_m} = 320000\frac{0,85 \times 50}{1,5} + 3927\frac{500}{1,15} = 10,77 \text{ MN}$$

As the use of the structure and the relative magnitudes of permanent and variable loads are not known, take the fire loading as $\eta_{fi}N_{Ed}$, so

$$N_{Ed,fi} = \eta_{fi}N_{Ed} = 0,7 \times 10,77 = 7,54 \text{ MN}$$

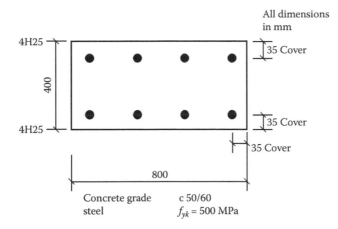

All dimensions
in mm

4H25

35 Cover

400

35 Cover
4H25

35 Cover

800

Concrete grade   c 50/60
steel $f_{yk} = 500$ MPa

Figure 7.4 Design data for a concrete column (Example 7.3).

Although EN 1992-1-2 suggests that the method of slices is more accurate, use both methods as an illustration.

500°C isotherm method — Figure B.1 of EN 1992-1-2 suggests that
rounding of the isotherm at corners should be considered or that a rectangular area equal to that contained within rounding at the corners
should be taken. As an approximation, draw a 45-degree line through
the point producing 500°C through bi-directional heating (Figure 7.5).
With $a_c = 0,417 \times 10^{-6}$ m²/sec, the distance from the face $x_{500,2}$ is given
from Equations (7.1) through (7.6) as:

$$x_{500,2} = \sqrt{\frac{t}{\exp\left[\dfrac{n_w - \sqrt{n_w^2 - (2n_w - 1)\dfrac{\Delta\theta}{\Delta\theta_g}}}{0,18(2n_w - 1)} + 4,5\right]}} \qquad (7.56)$$

The loss of area at each corner $A_{c,loss}$ is given by:

$$A_{c,loss} = 2(x_{500,2} - x_{500})^2 \qquad (7.57)$$

and the total loss $A_{c,loss,\,total}$ by:

$$A_{c,loss,total} = 8(x_{500,2} - x_{500})^2 \qquad (7.58)$$

This represents a lower bound as the actual loss of area is less than
this. The values of the loss in concrete area due to isotherm rounding are given in Table 7.11. Carry out the calculations twice, first not

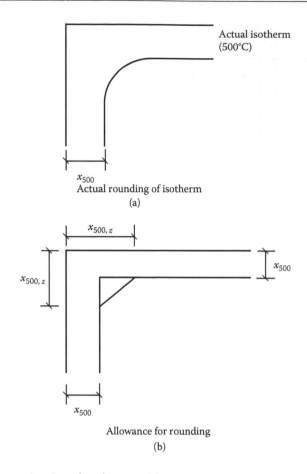

Figure 7.5 Approximation of isotherm position at corner.

Table 7.11 Determination of loss in area from isotherm rounding

t (hr)	x_{500} (mm)	$x_{500,2}$ (mm)	$A_{c,loss,\,total}$ (mm^2)
0,5	12	25	1352
1,0	23	43	3200
1,5	31	57	5408
2,0	39	69	7200
3,0	52	89	10952
4,0	64	107	14792

Table 7.12 Determination of reinforcement temperatures
and strength reduction factors

				Bar 1		Bar 2	
t (hr)	n_w	$\Delta\theta_g$ (°C)	n_x	θ_s (°C)	$k_s(\theta_s)$	θ_s (°C)	$k_s(\theta_s)$
0,5	0,887	822	0,162	138	0,962	239	0,861
1,0	0,938	925	0,287	269	0,831	451	0,548
1,5	0,957	1006	0,360	360	0,740	594	0,349
2,0	0,967	1049	0,412	429	0,557	690	0,124
3,0	0,977	1090	0,485	536	0,485	808	0,078
4,0	0,982	1133	0,536	617	0,295	899	0,060

considering loss of area due to isotherm rounding and second with
consideration of isotherm rounding.

Ignoring rounding: the values of x_{500} are taken from Table 7.3. To
determine the temperatures in the reinforcement x (=y) = 35 + 25/2 =
47,5 mm, the results of the calculation are given in Table 7.12. The
results of the determination of load carrying capacity are to be found
in Table 7.13. As both α_{cc} and γ_{mc} are equal to 1,0 in the fire limit state,
the concrete capacity in compression effectively becomes:

$$N_{c,fi} = b_{fi}d_{fi}f_{ck} \tag{7.59}$$

The compression capacity of the reinforcement with $\gamma_{m,s} = 1,0$ becomes:

$$N_{s,fi} = 1964 \times 500(k_s(\theta_1) + k_s(\theta_2)) = 982000\ (k_s(\theta_1) + k_s(\theta_2)) \tag{7.60}$$

The structural fire load $N_{fi,Ed}$ is 7,54 MN; thus the column would last
more than 4 hr. The results considering rounding of the isotherms are
given in Table 7.14. As both α_{cc} and γ_{mc} are equal to 1,0 in the fire limit
state, the concrete capacity in compression effectively becomes

$$N_{c,fi} = (b_{fi}d_{fi} - A_{c,loss,total})f_{ck} \tag{7.61}$$

Table 7.13 Determination of load carrying capacity

t (hr)	x_{500} (mm)	b_{fi} (mm)	d_{fi} (mm)	$N_{c,fi}$ (kN)	$N_{s,fi}$ (kN)	$N_{Rd,fi}$ (kN)
0,5	12	376	776	14,59	1,79	16,38
1,0	23	354	754	13,35	1,16	14,51
1,5	31	338	738	12,47	1,10	13,57
2,0	39	322	722	11,62	0,67	12,29
3,0	52	296	696	10,30	0,55	10,85
4,0	64	272	672	9,14	0,35	9,49

Table 7.14 Load carrying capacity allowing for isotherm rounding

t (hr)	x_{500} (mm)	b_{fi} (mm)	d_{fi} (mm)	$A_{c,loss,total}$ (mm²)	$N_{c,fi}$ (kN)	$N_{s,fi}$ (kN)	$N_{Rd,fi}$ (kN)
0,5	12	376	776	1352	14,52	1,79	16,31
1,0	23	354	754	3200	13,19	1,16	14,35
1,5	31	338	738	5408	12,20	1,10	13,30
2,0	39	322	722	7200	11,26	0,67	11,83
3,0	52	296	696	10952	9,75	0,55	10,30
4,0	64	272	672	14792	8,40	0,35	8,75

The compression capacity of the reinforcement is determined using Equation (7.60). The structural axial fire load $N_{fi,Ed}$ is 7,54 MN, so the column would still last more than 4 hr. Note that the effect of corner rounding is not significant until around 2 hr, and even at 4 hr the axial capacity is reduced only by around 8%.

Zone method — Use 50 mm wide slices (i.e., four per half width). As the number of strips is even, the mean concrete temperature at the centre will need separate calculation. The heat transfer is considered through the narrower dimension of 400 mm, and the results are given in Table 7.15. Strictly for the point $x = 200$, the heat flow is bi-directional, but considering uniaxial flow is satisfactory because the temperature rises are small. The mean concrete strength reduction factor $k_{c,m}$ is given by

$$k_{c,m} = \frac{1 - \frac{0,2}{n}}{n} \sum_1^4 k_c(\theta_i) = \frac{1 - \frac{0,2}{4}}{4} \sum_1^4 k_c(\theta_i) = 0,2375 \sum_1^4 k_c(\theta_i) \qquad (7.62)$$

The temperatures and strength reduction factors are given in Table 7.16 (noting the centroid of the reinforcement is at 47,5 mm from the surface). As $k_c(\theta_M) = 1,0$ for all cases, Equation (7.52) reduces to:

$$a_z = w\left[1 - k_{c,m}^{1,3}\right] = 200\left[1 - k_{c,m}^{1,3}\right] \qquad (7.63)$$

Table 7.15 Concrete zone temperatures and strength reduction factors

t (hr)	x = 25 (mm) θ (°C)	$k_c(\theta)$	x = 75 (mm) θ (°C)	$k_c(\theta)$	x = 125 (mm) θ (°C)	$k_c(\theta)$	x = 175 (mm) θ (°C)	$k_c(\theta)$	x = 200 (mm) θ (°C)	$k_c(\theta)$
0,5	385	0,765	41	1,00	20	1,00	20	1,00	20	1,00
1,0	516	0,576	128	0,986	21	1,00	20	1,00	20	1,00
1,5	620	0,420	201	0,949	48	1,00	20	1,00	20	1,00
2,0	689	0,317	268	0,883	82	1,00	22	1,00	20	1,00
3,0	773	0,191	371	0,779	139	0,981	50	1,00	27	1,00
4,0	842	0,121	452	0,672	198	0,951	85	1,00	52	1,00

Table 7.16 Determination of reinforcement temperatures and strength losses

			Bar 1		Bar 2	
t (hr)	D+E (°C)	$\xi_{\theta,x}$	$\theta(°C)$	$k_s(\theta)$	$\theta(°C)$	$k_s(\theta)$
0,5	750	0,189	162	0,938	277	0,823
1,0	820	0,338	297	0,803	481	0,535
1,5	910	0,414	397	0,703	618	0,293
2,0	960	0,469	470	0,540	709	0,098
3,0	1010	0,546	571	0,403	822	0,076
4,0	1060	0,594	650	0,218	905	0,059

Table 7.17 Determination of load carrying capacity

	Strips 1 through 4						
t (hr)	$\Sigma k_c(\theta_i)$	$k_{c,m}$	a_z (mm)	b_{fi} (mm)	d_{fi} (mm)	$N_{c,fi}$ (kN)	$N_{s,fi}$ (kN)	$N_{Rd,fi}$ (kN)
0,5	3,765	0,894	27	346	746	12,91	1,73	14,64
1,0	3,562	0,846	39	322	722	11,62	1,31	12,93
1,5	3,369	0,800	50	300	700	10,50	0,98	11,48
2,0	3,200	0,760	60	280	680	9,52	0,63	10,15
3,0	2,951	0,701	74	252	652	8,22	0,47	8,69
4,0	2,744	0,652	85	230	630	7,25	0,27	7,54

The determination of the load carrying capacity using Equations (7.55) and (7.56) is shown in Table 7.17. Using the zone method, the column would last 4 hr as $N_{Ed,fi} = 7,54$ MN.

7.4 COMPARISONS OF METHODS OF CALCULATION

For the two examples where both methods are applicable, the zone method would appear slightly more conservative than the 500°C isotherm method, although this may in part be due to the different methods of calculating temperature rise as the method derived by Hertz produces slightly higher temperatures than those derived by Wickström.

7.5 DESIGN AND DETAILING CONSIDERATIONS

7.5.1 Shear

For simply supported or continuous reinforced concrete construction, shear is rarely a problem (Krampf, undated). However, this will not be the case for prestressed concrete due to the moments induced in sections by the prestress.

Bobrowski and Bardhan-Roy (1969) indicated that the critical section for shear was between 0,15 and 0,2 L from the support where L is the span. Shear is unlikely to be critical in conventional precast prestressed concrete floor units (Lennon, 2003; van Acker, 2003/4; Fellinger, 2004), provided the precast units are constrained to act as a diaphragm by adequate tying in the plane of the floor. The tests reported by Lennon (2003) involved a natural fire of a time equivalent of approximately 1 hr and indicated no spalling.

7.5.2 Bond

This is generally not a problem even though bond strengths are severely reduced in a fire. The problem is more likely to be worse in pre-stressed concrete construction where bond in the anchorage length is needed to transfer the pre-stress force into the concrete. However, few failure of pre-stressed concrete were reported directly attributable to loss in bond. It is not a general practice to check bond strengths in fire design. Fellinger (2004) indicates it can be considered a good practice to insulate floor units over the transfer length of the prestressing.

7.5.3 Spalling

Spalling occurs in one of two forms in a fire. The first is explosive spalling that occurs very early in a fire and is likely to lead to loss of cover of the main reinforcing and hence to more rapid rises in temperature and resultant strength loss leading to reduced fire performance. The second form is known as sloughing, whereby the concrete gradually comes away due to loss of effective bond and strength loss. This mode tends to occur toward the end of a fire or late during a standard furnace test and is rarely critical.

The magnitude of the effects of spalling is demonstrated by both test results and computer simulation. Results are given by Aldea et al. (1997; quoted in Table 3.1) and Purkiss et al. (1996) who recorded the results from two sets of tests carried out by a fire research station from 1964 to 1976. Most of the columns suffered losses around 30% of the cross-sectional area and failed to achieve levels of fire endurance that would have been anticipated from relevant design guides.

Computer simulation (Mustapha, 1994; Purkiss and Mustapha, 1995) indicated that a loss of cross-sectional area can lead to reduction of fire endurance around 40 to 50%. The exact mechanism of explosive spalling is still not understood, but it is affected by the factors (Malhotra, 1984; Connolly, 1995, 1997) to be discussed below.

7.5.3.1 Moisture content

A concrete with high moisture content is more likely to spall since one of the possible mechanisms of spalling is the build-up of high vapour pressures near the surface leading to tensile failures in the concrete caused by moisture clog (Shorter and Harmathy, 1965). However, it is now recognized that the critical isotherm for pore pressure build-up is the 200°C isotherm and not the 100°C (Khalafallah, 2001).

The blanket limit of a moisture content of 3%, below which EN 1992-1-2 indicates spalling will not occur, should be questioned. Lennon et al. (2007) perpetuate this. They also indicate without substantiation that exposure to low humidity inside a building will satisfy this criterion. The original proposal for control of spalling by Meyer-Ottens (1975) also suggested stress limits.

7.5.3.2 Concrete porosity and permeability

A more porous concrete, and therefore one with a high permeability, will allow the dissipation of vapour pressure, and thus relieve any build-up within a section. However, a porous concrete will give a poor performance with respect to durability. It has also become clear that the combination of moisture content and permeability is critical (Tenchev and Purnell, 2005) as indicated in Figure 7.6.

The values of water content W (Figure 7.6) are defined in respect to an initial water content of 80 kg/m³ at $W = 50\%$. The water content of 80 kg/m³ is equivalent to a percentage moisture content by weight of 3,3. It is possible in a homogeneous concrete to determine pore pressures using a coupled heat and mass transfer model (Tenchev, Li, and Purkiss, 2001a and b;

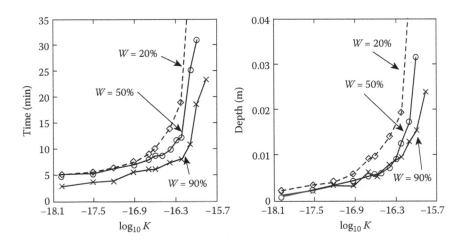

Figure 7.6 Time and depth of spalling as a function of permeability K (m2) for three levels of initial free water content ρ_L (Source: Tenchev, R. and Purnell, P. (2005) *International Journal of Solids and Structures*, 42, 6550–6565. With permission.)

Tenchev et al., 2001) and as a result to predict levels and times to spalling. The following results are from Tenchev, Purkiss and Li (2001). The time to spalling, t_{spall} and the depth of spalling x_{spall} are given by

$$t_{spall} = (382 - 3,34S + 0,00538S^2 - 0,00054S^3)p_{or} \qquad (7.64)$$

$$x_{spall} = (1,09 - 0,0085S)p_{or} \qquad (7.65)$$

where p_{or} is the porosity of the concrete and S is the initial water saturation in percent. In deducing Equations (7.64) and (7.65), it has been assumed that an initial porosity of 0,08 corresponds to an initial permeability of 8×10^{-17} m^2.

7.5.3.3 Stress conditions

From evidence of fire tests and observations in fires, it has been noted that spalling is likely to be more severe in areas where a concrete cross section is in compression, i.e., areas of hogging moments in beams, slabs, or columns. This may also be true for pre-stressed slabs, whether pre- or post-tensioned. The evidence for this is contradictory. Van Herbergen and Van Damme (1983) in tests on post-tensioned unbonded pre-stressed slabs report spalling. Unfortunately, neither concrete strengths nor permeability are reported.

Kelly and Purkiss (2008) report that spalling occurred in a single test on a post-tensioned slab. It should be noted that the moisture content at the time of test was 4,6%, which may be considered high. Bailey and Ellobody (2009) report tests on unbonded and bonded post-tensioned concrete slabs. In all cases, the moisture content was below 2,5% with concrete strengths in the range of 30 to 48 MPa at test. No spalling was reported.

The influence of comprehensive stress is probably due to the issue that in areas of compressive stress cracks cannot open up to relieve internal pressure. This does not mean that spalling cannot occur in areas of sagging moments where tensile cracks exist since it is possible that pressure build-up will still occur because tension cracks are generally discrete and not parts of a continuum.

7.5.3.4 Aggregate type

The evidence available suggests that the aggregate most likely to produce spalling is siliceous aggregate, with limestone producing less spalling and lightweight concrete the least. Spalling is likely to be linked to the basic porosity of the aggregate in that siliceous aggregate is impermeable compared to the others and moisture transport has to occur through the mortar matrix. However, evidence now indicates that limestone and lightweight aggregates may create problems, especially in younger concretes as the pore structure of the aggregate may provide convenient reservoir storage for free water (Connolly, 1995).

7.5.3.5 Section profile and cover

Some evidence suggests that sharp profiles produce more spalling than rounded or chamfered edges. Spalling is also exacerbated in thin sections because the depth of spalling constitutes more of the section dimension and also because there is less of a cool reservoir for moisture to migrate toward (Kalafallah, 2001).

High covers are also likely to produce greater amounts of spalling. Thus, design codes frequently place restrictions when high covers are needed at high fire resistance periods in order to maintain low temperatures in reinforcing. These restrictions often concern the placement of a light mesh with 4-mm wires at a spacing of 100 mm at the surface of the concrete cover when the axis distance exceeds 70 mm to retain the cover (EN 1992-1-2). This supplementary mesh is often difficult to place and on site is often omitted. Tests have shown that even with high covers mesh is not absolutely necessary to give fire resistance periods up to 4 hr (Lawson, 1985).

7.5.3.6 Heating rate

The higher the heat flux, the less chance pore pressures have to dissipate to the relatively cool internal regions of a concrete element. The rate of heating is therefore critical to an assessment of the likelihood of spalling. A hydrocarbon-type curve will therefore be far more critical than a cellulosic-type fire. The influence of heating rate was demonstrated in tests on high strength concrete columns by Ali et al. (2001) who demonstrated that the level of spalling was lower at low heating rates compared to high rates, and that this effect appeared not to depend on load level. The high heating rate in the tests corresponded to BS 476 Part 20 with the low heating rate corresponding to BS 476 up to a temperature of around 300°C and then approximately linear with a rate of 75°C/min.

7.5.3.7 Concrete strength

In spite of the above, normal strength concretes ($f_{ck} \leq 60$ MPa) may not spall. However, if a concrete designed as normal strength has a much higher strength than designed for, problems may ensue. In the Cardington fire test on a concrete frame structure, the floor slab with flint aggregate was designed to be Grade C30/37. The actual strength at the time of test was 61 MPa (cube) or approximately 50 MPa (cylinder). With a moisture content of 3,8% and a permeability of $6,75 \times 10^{-17}$ m^2, the slab suffered severe spalling in the test (Bailey, 2002). The problems are exacerbated for high strength concretes.

7.5.4 High strength and self-compacting concretes

These concretes can produce worse symptoms of spalling as the pore structure porosity is lower (i.e., the permeability is lower) and build-up of pore pressures is much greater. The higher tensile strengths of such concretes do not remove the problem as tiny pores will act as stress raisers and hence reduce the effective tensile strength.

EN 1992-1-2 allows a number of methods to reduce the effect of spalling on high strength concrete. However, the most effective is to add 2% of polypropylene fibres to the concrete (Lennon and Clayton, 1999; Clayton and Lennon, 2000; Persson, 2003; Boström, 2002 and 2004; Jansson and Boström, 2004). It is not recommended to use steel fibres to attempt to control spalling as Hertz (1992) found they had little or no effect.

7.5.5 Detailing

Where beams and slabs are designed to act as continuous members in a fire or where advantage, as in Example 7.1, is taken of anti-crack steel to give continuity in a fire but not at ambient, it is absolutely essential that the hogging reinforcement is detailed so that anchorage forces needed to generate those hogging moments may be sustained. It is thus essential that such reinforcement be fully anchored beyond the point of contraflexure. For continuous members, the ISE/Concrete Society Report (1978) and EN 1992-1-2 both give detailing requirements that should be followed.

Chapter 8

Design of steel elements

The determination of thermal and structural responses may be decoupled in a similar manner to that adopted in the design of concrete elements. The calculations may be simplified even further for steelwork as the temperature gradient across a member may be neglected. Since there is no transient strain component, it is possible to use empirically modified steady state stress–strain data to allow for the effects of classical creep. As the thermal properties of steel are sensibly independent of the steel strength, it is also possible to utilize empirical equations to determine the temperature rise within an element. The simplifications due to thermal gradient and stress behaviour mean that it is possible to consider calculation methods for steelwork over a broader base than concrete.

This chapter covers both the calculation of temperatures within an element and the methods by which the load carrying capacity at elevated temperatures may be determined.

8.1 CALCULATION OF TEMPERATURES

8.1.1 Basic principles

As stated above, it is possible to ignore the effect of any thermal gradient through a member which means that the temperature rise is entirely a function of time. Account is, however, taken of the proportion of the element exposed to the effects of the fire, i.e., a column or stanchion exposed on four sides or a beam element on three sides. Since the temperature régime is independent of the spatial coordinate system, the Fourier heat diffusion is much simplified. The resultant heat flow $\dot{h}_{net}$ is now given by:

$$\dot{h}_{net} = KF(\theta_t - \theta_{a,t}) \tag{8.1}$$

where K is the coefficient of total heat transfer, F is the surface area of the element exposed to the fire, $\theta_{a,t}$ is the temperature of the steel, and θ_t is the

211

gas temperature at time t. The coefficient of total heat transfer has three components due to convection, radiation, and insulation. A full discussion of the evaluation of the convection and radiation boundary conditions is given in Chapter 6. The three components of the heat flow boundary conditions are discussed below:

Convection (α_c) — This mode other than in the very early stages of a fire is not dominant, and α_c may be taken as 25 W/m²C for cellulosic fires and 50 W/m²C for hydrocarbon fires.

Radiation (α_r) — This is calculated using the Stefan-Boltzmann law for radiation:

$$\alpha_{cr} = \phi\left(\frac{5,67\times10^{-8}\varepsilon_{res}}{\left(\theta_g+273\right)-\left(\theta_a+273\right)}\right)\left(\left(\theta_g+273\right)^4-\left(\theta_a+273\right)^4\right) \qquad (8.2)$$

where ϕ is the configuration factor that conservatively takes a value of unity and ε_{res} is the resultant emissivity that may be taken as $\varepsilon_f\varepsilon_m$, where ε_f is the emissivity of the fire compartment and ε_m is the surface emissivity. For all furnace tests, the emissivity of the fire compartment ε_f equals 1,0. For steel (and concrete), the surface emissivity ε_m equals 0,7 and thus the resultant emissivity ε_{res} is also 0,7.

Insulation — This is given by d_p/λ_p, where d_p is the thickness of the insulation (m) and λ_p is the effective thermal conductivity of the insulation (W/m°C). The effective thermal conductivity of the insulation is temperature dependent, but it is usual practice to adopt a constant value that is correct around 550°C. The total coefficient of heat transfer K is given by:

$$K = \frac{1}{\dfrac{1}{\alpha_c+\alpha_r}+\dfrac{d_p}{\lambda_p}} \qquad (8.3)$$

The rate of heat flow into the element may also be written in incremental form as:

$$\dot{h}_{net} = c_a\rho_a V_i \frac{\Delta\theta_{a,t}}{\Delta t} \qquad (8.4)$$

where c_a is the specific heat of steel, ρ_a is the density of steel, and V_i is the volume per unit length. Equating the values of heat flow in Equations (8.1) and (8.4) gives:

$$\Delta\theta_{a,t} = \frac{K}{c_a\rho_a}\frac{F}{V_i}\left(\theta_t-\theta_{a,t}\right)\Delta t \qquad (8.5)$$

Changing notation from F/V_i to A_m/V gives Equation (8.5) as:

$$\Delta\theta_{a,t} = \frac{K}{c_a\rho_a}\frac{A_m}{V}(\theta_t - \theta_{a,t})\Delta t \qquad (8.6)$$

The calculation of the heated perimeter per unit volume of steelwork A_m/V is considered in Section 8.1.6. Modifications to Equation (8.6) may now be made to deal with the specific cases of uninsulated and insulated steelwork.

8.1.2 Heat flow in uninsulated steelwork

With no insulation, Equation (8.6) reduces to:

$$\Delta\theta_{a,t} = \frac{\alpha_c + \alpha_r}{c_a\rho_a}\frac{A_m}{V}(\theta_t - \theta_{a,t})\,\Delta t \qquad (8.7)$$

where $\theta_{a,t}$ is the steel temperature, θ_t is the gas temperature at time t, and A_m/V is the exposed surface area per unit volume. EN 1993-1-2 Clause 4.2.5.1(5) indicates that Δt should not exceed 5 sec. However, this is likely to be conservative and Equation (8.11) should be noted. Example 8.1 indicates the difference between using a Δt of 5 sec and a value calculated from Equation (8.11). The use of Equation (8.7) is demonstrated in Examples 8.1 and 8.2.

8.1.3 Heat flow in insulated steelwork

In early work published by the European Community for Coal and Steel (ECCS, 1983; 1985), the heat flow to insulated steelwork was subdivided into cases where the insulation had either negligible or substantial heat capacity. The reason for this subdivision was that the equation used for calculations where the insulation had substantial heat capacity was unstable when the other situation was operative. It was later shown by Melinek and Thomas (1987) and Melinek (1989) that the ECCS method for heavy insulation could not be justified on theoretical grounds. However, for historical completeness, the ECCS method is included.

8.1.3.1 ECCS method of calculation

As indicated above, both the cases need consideration. The first is where the insulation has negligible heat capacity because it is relatively thin or dry. The external face of the steelwork may then be considered to have the same temperature as the furnace gases. The second case is where the insulation has substantial heat capacity or substantial moisture content. The insulation temperature can then be taken as the mean of the steel temperature and the gas temperature. The insulation is deemed to have substantial heat capacity when the parameter Φ is greater than 0,50, where Φ is defined by:

$$\Phi = \frac{c_p \rho_p}{c_a \rho_a} d_p \frac{A_m}{V} \tag{8.8}$$

where c_p is the specific heat of the insulation and ρ_p its density. In both these cases, the heat transfer term due to the insulation is dominant. The component of the total heat transfer coefficient due to convection and radiation may be neglected and K can be taken equal to λ_p/d_p. The derivation of the ECCS equations [(8.9) and (8.10)] is given in Malhotra (1982a).

Heat flow in insulated members with negligible heat capacities — Equation (8.6) is rewritten as:

$$\Delta\theta_{a,t} = \frac{\dfrac{\lambda_p}{d_p}}{c_a \rho_a} \frac{A_m}{V} (\theta_t - \theta_{a,t}) \tag{8.9}$$

Heat flow in members with substantial heat capacities — Equation (8.6) is modified to:

$$\Delta\theta_{a,t} = \frac{\dfrac{\lambda_p}{d_p}}{c_a \rho_a} \frac{A_m}{V} \frac{(\theta_t - \theta_{a,t})\Delta t}{1 + \dfrac{\Phi}{2}} - \frac{\Delta\theta_t}{1 + \dfrac{2}{\Phi}} \tag{8.10}$$

Note that the original notation of the ECCS equations has been modified to agree with EN 1993-1-2 and EN 1994-1-2. The ECCS rules give the following equation to determine the critical value of Δt to ensure convergence of the finite difference form of the heat transfer equation:

$$\Delta t \leq \frac{25000}{\dfrac{A_m}{V}} \tag{8.11}$$

8.1.3.2 EN 1993-1-2 approach

EN 1993-1-2 (and EN 1994-1-2) use the method derived by Wickström (1985b) with the governing equation written as:

$$\Delta\theta_{a,t} = \frac{\dfrac{\lambda_p}{d_p}}{c_a \rho_a} \frac{A_m}{V} \frac{(\theta_t - \theta_{a,t})\Delta t}{1 + \dfrac{\Phi}{3}} - (e^{\Phi/10} - 1)\Delta\theta_t \tag{8.12}$$

EN 1993-1-2 places a limit on the value of Δt as 30 sec (Clause 4.2.3.2). Wickström, however, suggests that the limit on Δt should be taken as:

$$\Delta t \le \frac{c_a \rho_a}{\lambda_p} \frac{V}{A_m} \left(1 + \frac{\Phi}{3}\right) < 60s \qquad (8.13)$$

Wickström also suggests that a time shift $\bar{t}$ needs to be introduced at the commencement of heating to allow for the thermal capacity of the insulation and thus improve the accuracy of Equation (8.12). This time shift is given by:

$$\bar{t} = c_a \rho_a \frac{V}{A_m} \frac{d_p}{\lambda_p} \left(1 + \frac{\Phi}{3}\right) \frac{\Phi}{8} \qquad (8.14)$$

Melinek and Thomas (1987) derive the time shift term differently and give the following equation:

$$\bar{t} = c_a \rho_a \frac{V}{A_m} \frac{d_p}{\lambda_p} \left(1 + \frac{\Phi}{3}\right) \left(\frac{\Phi}{2\Phi + 6}\right) \qquad (8.15)$$

EN 1993-1-2 makes no direct reference to the time shift to allow for the thermal capacity of the insulation. The heat flow calculations for an insulated section are covered in Example 8.3. It should be noted that the calculations for both uninsulated and insulated sections are best performed on a spreadsheet, as done for the examples in this chapter and the next.

8.1.4 Effect of moisture

As noted above, the effect of moisture in insulation is to slow down the rate of temperature rise and cause a dwell in temperature rise around 100°C as the water is vaporized. Two approaches may be used to deal with this effect. The first is to use a moisture-dependent effective density for the insulation in the temperature calculations. The second is to introduce a dwell or delay time t_v when the steel temperature reaches 100°C. The latter approach is indicated in EN 1993-1-2. While no explicit method is included, reference is made to EN 13381-4.

8.1.4.1 Effective density of insulation

The effective density of the insulation ρ'_p is given by Equation (8.16)

$$\rho'_p = \rho_p (1 + 0,03p) \qquad (8.16)$$

where p is the moisture content in percent by weight.

8.1.4.2 Delay time

The delay time t_v in minutes may be calculated from the following equation (ECCS, 1983; PD 7974-2):

$$t_v = \frac{0,2p\rho_p d_p^2}{\lambda_p}$$

(8.17)

It is easier to use an effective density for the insulation than to employ a delay time as it is iterative; the time delay depends on the insulation thickness.

8.1.5 Empirical approach for temperature calculation

Equations for such an approach have been derived from a study of experimental results.

8.1.5.1 Bare steelwork

Twilt and Witteveen (1986) give the following equation for the temperature rise on bare steelwork:

$$t_{fi,d} = 0,54(\theta_{a,t} - 50)\left(\frac{A_m}{V}\right)^{-0,6}$$

(8.18)

where $\theta_{a,t}$ is the temperature in the steel reached at a time t (min). Note that Equation (8.18) only holds for $10 \le t_{fi,d} \le 80$ min; $400 \le \theta_{a,t} \le 600°C$, and $10 \le A_m/V \le 300$ m^{-1}.

8.1.5.2 Protected steelwork

A similar curve fitting exercise on test results from times to failure for steelwork protected with dry insulation indicated that time to failure depends solely on the term $(d_p/\lambda_p)(V_i/A_p)$ with the following resultant equation:

$$t_{fi,d} = 40(\theta_{a,t} - 140)\left(\frac{d_p}{\lambda_p}\frac{V}{A_m}\right)^{0,77}$$

(8.19)

This equation was fitted to data determined from tests in which the insulation material was light. Thus when insulation has substantial heat capacity, Equation (8.19) takes the following form (Wickström, 1985b; Melinek and Thomas, 1987):

$$
t_{fi,d} = 40(\theta_{a,t} - 140)\left(\frac{d_p}{\lambda_p}\left(\frac{V}{A_m} + \frac{d_p \rho_p}{\rho_a}\right)\right)^{0,77}
\tag{8.20}
$$

Equation (8.20) can be rewritten as:

$$
\frac{\rho_p}{\rho_a}\frac{1}{\lambda_p}d_p^2 + \frac{1}{\lambda_p}\frac{V}{A_m}d_p - \left[\frac{t_{fi,d}}{40(\theta_{a,t} - 140)}\right]^{1,3} = 0
\tag{8.21}
$$

or

$$
d_p = \frac{-\dfrac{1}{\lambda_p}\dfrac{V}{A_m} + \sqrt{\left(\dfrac{1}{\lambda_p}\dfrac{V}{A_m}\right)^2 + 4\dfrac{\rho_p}{\rho_a}\dfrac{1}{\lambda_p}\left[\dfrac{t_{fi,d}}{40(\theta_{a,t}-140)}\right]^{1,3}}}{2\dfrac{\rho_p}{\rho_a}\dfrac{1}{\lambda_p}}
\tag{8.22}
$$

8.1.6 Calculation of A_m/V

The cross-sectional area V is always that of the basic steel section. The heated perimeter A_m depends on the type of insulation, e.g., sprayed insulation or intumescent paint applied to the section profile or board insulation that boxes the section, and on the number of sides of the member exposed to the fire effect. The calculation of A_m/V may be performed using basic principles (Figure 8.1) or determined from data tables applicable to the sections produced in a given country such as those published by Tata Steel Construction (2012) in the UK. The units of A_m/V are m^{-1} and usually expressed to the nearest 5 m^{-1}.

8.1.7 Thermal properties of insulation materials

To apply this method, knowledge of the various thermal properties of the insulation materials is needed. Currently, specific values for any particular material may not be available. The generic data in Table 8.1 may be used (Malhotra, 1982a). Similar data were given by Lawson and Newman

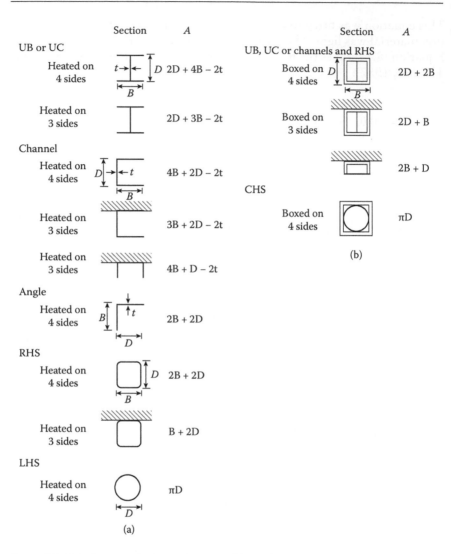

Figure 8.1 Calculations of values of ratios of heated perimeter per unit volume (A/V) for steel elements: (a) values of A for unprotected steelwork or steelwork protected by sprayed insulation or intumescent paint and (b) values of A for boxed sections.

(1996). More recently, a more comprehensive set of data including variation of thermal properties with temperature and the effects of moisture content was compiled by Wang et al. (2013).

Three examples for calculating temperature rise will be carried out, two on uninsulated steelwork and the other on insulated steelwork. The

Table 8.1 Generic thermal data for insulation materials

Material	Density (kg/m³)	Specific heat (J/kg°C)	Thermal Conductivity (W/m°C)	Moisture Content (percent by weight)
Sprayed mineral fibre	250-350	1050	0,1	1,0
Vermiculite slabs	300	1200	0,15	7,0
Vermiculite/gypsum slabs	800	1200	0,15	15,0
Gypsum plaster	800	1700	0,20	20,0
Mineral fibre sheets	500	1500	0,25	2,0
Aerated concrete	600	1200	0,30	2,5
Lightweight concrete	600	1200	0,80	2,5
Normal-weight concrete	2200	1200	1,70	1,5

Source: Malhotra, H.L. (1982a) *Design of Fire-Resisting Structures*. Glasgow: Surrey University Press. With permission.

examples on insulated steelwork use data from Table 8.1 for the values of thermal properties of insulation.

Example 8.1: Temperature rise on unprotected steelwork exposed to standard furnace curve

Calculate the temperature–time response curve for a $203 \times 203 \times 52$ UC heated to the standard furnace temperature–time curve for 30 min. For a bare $203 \times 203 \times 52$ UC, $A_m/V = 180$ m⁻¹. The resultant emissivity ε_{res} is taken as 0,70. Calculate maximum value of Δt from Equation (8.11):

$$\Delta t = 25000/(A_i/V_i) = 25000/180 = 139 \text{ sec}$$

Use $\Delta t = 2$ min (120 sec) and a constant value of $c_a = 600$ J/kg°C. The heat transfer coefficient α is given by:

$$\alpha = 25 + \frac{0,7 \times k_{sh} \times 0,56 \times 10^{-8}}{\theta_t - \theta_{a,t}} [(\theta_t + 273)^4 - (\theta_{a,t} + 273)^4]$$

The k_{sh} is the shielding parameter [equivalent to the configuration factor ϕ in Equation (8.2)]. The value of k_{sh} will initially be set equal to 1. The governing equation for this case is Equation (8.7) which with numerical values substituted for c_a, ρ_a, A_i/V_i and Δt becomes

$$\Delta\theta_{a,t} = \frac{\alpha}{c_a\rho_a} \frac{A_m}{V} (\theta_t - \theta_{a,t})\Delta t = \frac{\alpha}{600 \times 7850} 180 \, (\theta_t - \theta_{a,t}) \, 120 = \frac{\alpha}{218} (\theta_t - \theta_{a,t})$$

At $t = 0$, the base temperature is taken as 20°C. The gas temperature θ_t is given by the standard curve

$$\theta_t = 20 + 345\log(8t + 1)$$

The values of θ_t, α, $\theta_t - \theta_{a,t}$, and $\Delta\theta_a$ are calculated at 1, 3, 5, 7 min, etc., with the resultant temperature in the steel $\theta_{a,t}$ given at 0, 2, 4, min, etc. The calculations are presented in Table 8.2. In addition to the calculations with $\Delta t = 120$ sec, the final calculated results with $\Delta t = 5$ sec as recommended by EN 1993-1-2 Clause 4.2.3.1 are given in Table 8.2. Note the small difference between both sets of results. Also, opportunities were taken to evaluate the effects of using values of c_a that vary with temperature, given by Equations (5.4) to (5.7) and the effect of the shielding factor k_{sh}. For I-sections, EN 1993-1-2 Clause 5.2.5.1(2) gives the shielding factor k_{sh} as:

$$k_{sh} = 0.9\frac{\left[\dfrac{A_m}{V}\right]_b}{\dfrac{A_m}{V}} \tag{8.23}$$

where $[A_m/V]_b$ is the box value of the section factor. For a 203 × 203 × 52 UC, $[A_m/V]_b = 125$ m^{-1},

$$k_{sh} = 0.9\frac{\left[\dfrac{A_m}{V}\right]_b}{\dfrac{A_m}{V}} = 0.9\frac{125}{180} = 0.625$$

The final results from using k_{sh} with constant c_a, ignoring the shielding factor and incorporating variable c_a, are also given in Table 8.2. The results are plotted in Figure 8.2 along with the mean temperatures measured in a furnace test (Data Sheet 41 from Wainman and Kirby, 1988). The test terminated at 23 min with an average temperature in the column of 688°C. The maximum temperature measured at 0,61 m up the column was 723°C in the flange and 713°C in the web. The furnace temperature was 23°C below that from the standard furnace curve.

The calculations with a constant specific heat of 600 J/kg°C gave 788°C. Calculations with variable specific heat of steel at 761°C were made at 23 min. The results of these calculations are also plotted in Figure 8.2. Note, however, that around 20 min, the predicted

Table 8.2 Bare steel results

t (min)	θ_t (°C)	α (W/m²°C)	θ_t − θ (°C)	Δθ_a,t (°C)	θ_a,t (°C)	Δt = 5 (°C)	Variable c_a (°C)	k_sh = 0,625 (C°)
		Original Calculations						
0					20	20	20	20
	349	42,2	329	636				
2					84	77	107	74
	502	57,7	419	111				
4					195	181	238	163
	576	74,1	382	130				
6					325	301	372	265
	626	94,1	301	130				
8					455	421	489	371
	663	118	208	113				
10					568	525	579	472
	693	142	125	82				
12					650	607	640	561
	717	164	68	51				
14					701	666	683	634
	739	180	38	32				
16					733	708	713	688
	757	192	25	22				
18					755	738	732	727
	774	202	20	18				
20					773	760	740	755
	789	211	16	16				
22					789	779	751	776
	802	218	14	14				
24					803	794	771	793
	814	226	12	13				
26					816	808	797	808
	826	233	11	12				
28					828	821	819	821
	837	239	10	11				
30					839	832	835	832
	847	245	9	10				
32					849	843	846	843

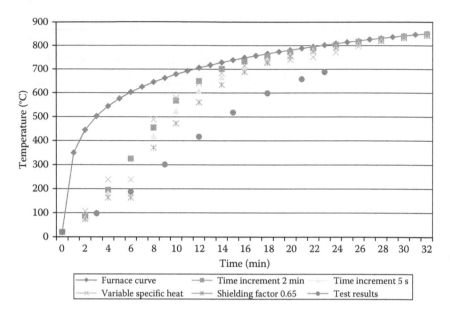

Figure 8.2 Response of unprotected steelwork.

temperatures are sensibly independent of assumptions. However all the calculations predict temperatures higher than those in the test (Figure 8.2) with a difference of around 100°C. If the assumption of constant specific heat is made, a value of $\varepsilon_{res}k_{sh} = 0,15$ would have to be employed. This would give a temperature of 692°C at 23 min. If $k_{sh} = 0,625$, ε_{res} would have to be taken as 0,24. This is not unreasonable if Figure 6.1 (from Chitty et al., 1992) is examined where resultant emissivities as low as 0,2 are required to give reasonable prediction of temperatures in columns.

Example 8.2: Determination of temperature response of bare steelwork to parametric fire curve

Calculate the temperature-time response curve for a 203 × 203 × 52 UC heated to the parametric temperature–time curve determined in Example 8.1 for both heating and cooling. For a bare 203 × 203 × 52 UC, $A_i/V_i = 180$ m⁻¹. The resultant emissivity ε_{res} is taken as 0,70 (shielding factor is ignored). The recommended value of Δt of 5 sec is used. The governing equation is identical to that of Example 8.1.

The gas temperature θ_t is given by the parametric fire curve of Example 4.1. The results are plotted in Figure 8.3a. After about 15 min,

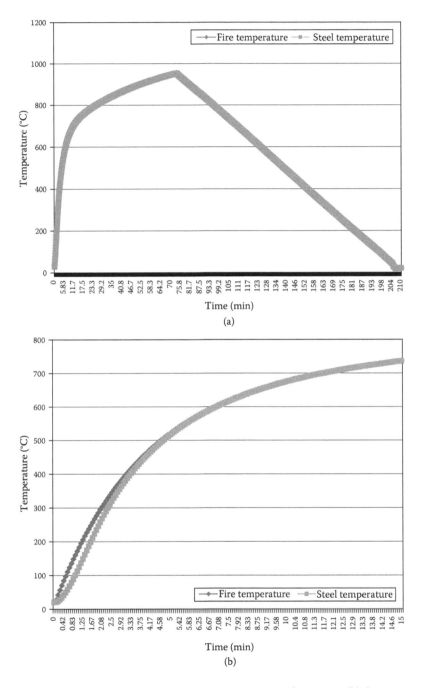

Figure 8.3 (a) Response of bare steelwork to parametric fire curve. (b) Response to parametric fire curve for first 15 min.

the steel temperatures and fire temperatures are virtually indistinguishable. On the cooling section of the curve, the steel temperatures lag the fire temperature by 2 or 3°C. At the maximum fire temperature, the steel is around 0,5°C cooler. The results from the first 15 min of the calculations are plotted in Figure 8.3b.

Example 8.3: Determination of temperature response of insulated specimen subjected to standard furnace curve

Determine the temperature history for a 203 × 203 × 46 UC with fire protection formed by 30-mm thick mineral fibre boarding encasing the column on all four sides (Figure 8.4) using the heat transfer equations of EN 1993-1-2 (Table 8.3). Material data are as follows:

$$\lambda_p = 0.25 \text{ W/m°C}, \rho_p = 500 \text{ kg/m}^3, p = 2\%, c_p = 1500 \text{ J/kg°C},$$
$$A_p/V_i = 140 \text{ m}^{-1}, \rho_a = 7850 \text{ kg/m}^3, c_a = 600 \text{ J/kg°C}.$$

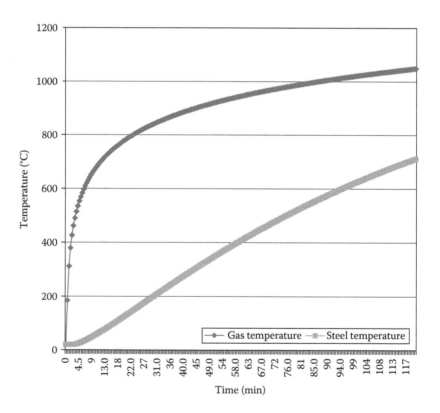

Figure 8.4 Temperature response of protected member.

Table 8.3 Calculation of temperature rises of protected steelwork exposed to standard furnace curve (Example 8.3)

Time (sec)	θ_t (°C)	$\theta_t - \theta_{a,t}$ (°C)	$\Delta\theta_{a,t}$ (°C)	$\theta_{a,t}$ (°C)
0				20
	184,6	164,6	−11,1	
30				20
	311,6	291,6	−7,6	
60				20
	379,3	359,3	−2,8	
90				20
	425,8	405,8	−1,0	
120				20
	461,2	441,2	0	
150				20
	489,8	469,7	0,7	
180				20,8
	513,8	493,1	1,2	
210				22,0
	534,5	512,6	1,6	
240				23,5

The calculation is continued to

7080				703,6
	1046,2	344,6	2,0	
7110				705,7
	1046,8	343,2	2,0	
7140				707,7
	1047,5	340,4	2,0	
7170				709,7
	1048,7	339,1	2,0	
7200				711,7

From Equations (8.8) and (8.12):

$$\Phi = \frac{c_p\rho_p}{c_a\rho_a}d_p\frac{A_m}{V} = \frac{1500\times500(1+0,03\times2)}{600\times7850}0,030\times140 = 0,709$$

$$\Delta\theta_{a,t} = \frac{\frac{\lambda_p}{d_p}\frac{A_m}{V}}{c_a\rho_a}\frac{(\theta_t - \theta_{a,t})\Delta t}{1+\frac{\Phi}{3}} - \left(e^{\left(\frac{\Phi}{10}\right)} - 1\right)\Delta\theta_t$$

$$= \frac{\frac{0,25}{0,030}}{600\times7850}140\frac{30(\theta_t - \theta_{a,t})}{1+\frac{0,709}{3}} - \left(e^{\frac{0,709}{10}} - 1\right)\Delta\theta_t$$

$$= 0,00601(\theta_t - \theta_{a,t}) - 0,0735\Delta\theta_t$$

EN 1993-1-2 gives a recommended value of Δt of 30 sec used in this example. The recommendation by Wickström is 60 sec and Equation (8.11) gives a Δt of 178.6 sec (about 3 min in practice). As in Example 8.1, the reference temperature is taken as 20°C and the gas temperature θ_t is calculated from the standard furnace curve.

Note that in the early stages of heating, the heat flux is adsorbed by the insulation and thus there appears to be negative heat transfer to the section with $\Delta\theta_{a,t}$ taking values less than zero. These negative values were ignored and set to zero when calculating the steel temperatures. Typical calculations for the start and end of the heating period are found in Table 8.3.

If the time shift calculations are carried out according to Equations (8.14) and (8.15), a value of 6,9 min is obtained from the formulation proposed by Wickström and 7,5 min from that by Melinek and Thomas. These values are not significantly different, and thus the temperature rise at 120 min can be taken as the value reached in 113 min, i.e., a delay period of about 7 min. The results are plotted in Figure 8.4.

Having established methods of determining the temperature rise in steelwork, it is now possible to consider the design of such members in a fire. It is convenient because of the different approaches used to consider noncomposite and composite construction separately. Composite construction is where a second material, generally concrete, acts in conjunction with steelwork so that each carries part of the load. Concrete floor units sitting on the top flange of a beam do not act compositely even though the presence of the floor units increases the carrying capacity of the beam by reducing the effect of lateral torsional buckling and by acting as a heat sink to reduce the temperatures attained in the steelwork. Composite steelwork is covered in the next chapter.

8.2 DESIGN OF NONCOMPOSITE STEEL WORK

Historically, the results from furnace tests on members loaded to their full design strengths led to the concept of a critical failure temperature around 550°C for both columns and beams. The beams were tested with concrete slabs resting on the top flanges and were noncomposite. A research programme was set up in the mid-1980s to evaluate the effects of varying load patterns and partial or total shielding of the web and exposed flanges of members with no additional protection.

This research showed that for certain categories of sections, the temperatures were below those required to cause failure. Shielding of the web and exposed flanges that had the effect of inducing thermal gradients in the section and thus allowing a redistribution of carrying capacity from the hotter

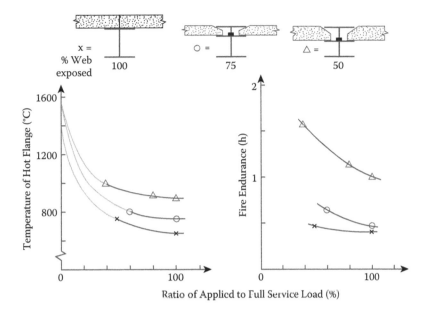

Figure 8.5 Fire test performance of partially protected beams. (*Source:* Robinson, J.T. and Latham, D.J. (1986) In *Design of Structures against Fire.* London: Elsevier, pp. 225–236. With permission.)

to the cooler parts of the section and reducing the applied stresses had the effect of increasing the inherent fire resistance of unprotected steelwork (Kirby, 1986; Robinson and Latham, 1986; Robinson and Walker, 1987). The complete test data are given in Wainman and Kirby (1988 and 1989). Results from unprotected beams and columns with partial or total shielding and varying loadings are given in Figure 8.5 and Figure 8.6.

8.2.1 Determination of structural load in fire limit state

This follows an identical procedure to that for concrete (Section 7.2.1) except that η_{fi} may be taken as 0,65.

8.2.2 EN 1993-1-2 approach for determining structural fire capacity

8.2.2.1 Background of Eurocode method

The method used here was first formulated in the ECCS Recommendations (1983) and the ECCS design guide (1985). The original approach was to calculate the ratio of the required strength at elevated temperature to

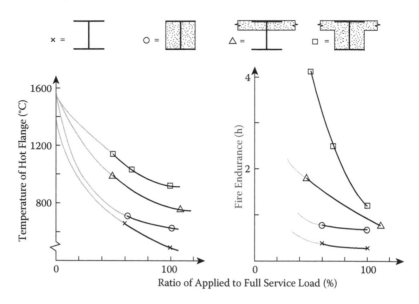

Figure 8.6 Fire test performance of partially protected columns. (*Source*: Robinson, J.T. and Latham, D.J. (1986) In *Design of Structures against Fire*. London: Elsevier, pp. 225–236. With permission.)

that at ambient to ensure the element would not collapse. Thus for beams designed elastically

$$k_{y,\theta} = \frac{f_{a\max,\theta_{cr}}}{f_{ay,20}} = \frac{\kappa}{\theta} \frac{W_{el}}{W_{pl}} \frac{q_{fi,d}}{q_{Sd,el}}$$
(8.24)

where $f_{a\max,\theta cr}/f_{ay,20°C}$ is the stress ratio ($= k_{y,\theta}$), κ is a factor allowing for the nonuniform temperature distribution, geometric imperfections and strength variations, θ is a factor, greater than unity, allowing for redistribution between the elastic ambient moment distribution and the plastic distribution under fire, W_{el}/W_{pl} is the ratio between the elastic and plastic section moduli (known as the shape factor), and $q_{fi,d}/q_{Sd,el}$ is the ratio of the design load (action) in the fire to the elastic design load (action). For beams designed plastically

$$k_{y,\theta} = \frac{f_{a\max,\theta_{cr}}}{f_{ay,20}} = \kappa \frac{q_{fi,d}}{q_{Sd}}$$
(8.25)

where $q_{fi,d}/q_{Sd}$ is the ratio of the fire load (action) to the ultimate design load (action). As a corollary, note that a beam designed elastically has a greater reserve of strength when exposed to a fire than a beam designed plastically. The basis behind the derivation of parameter κ and the values thereof are given in Pettersson and Witteveen (1979/80) who showed how the variations in material strengths, transverse and longitudinal temperature gradients, and structural imperfections affected the calculated strengths of members based on the simplifying assumption that these variations did not exist.

8.2.2.2 Eurocode methods

The Eurocode gives two methods for the design of steelwork members in a fire. The first method is to satisfy the load carrying criterion and the second is a limiting or critical temperature approach.

8.2.2.2.1 Load carrying method

This can simply be stated as the satisfaction of the following criterion:

$$E_{fi,d} \leq R_{fi,d,t} \tag{8.26}$$

where $E_{fi,d}$ is the design value of the internal force to be resisted and $R_{fi,d,t}$ is the design resistance at time t and should be calculated in accordance with the principles of EN 1993-1-1.

- Section classification

This should be carried out in accordance with EN 1993-1-1, except the value of ε is modified to allow for the effects of temperature increase to:

$$\varepsilon = 0,85 \left[\frac{235}{f_y} \right]^{0,5} \tag{8.27}$$

- Tension members (Clause 4.2.21)

With a uniform temperature distribution, the axial tensile capacity $N_{fi,\theta,Rd}$ may be calculated as

$$N_{fi,\theta,Rd} = k_{y,\theta} N_{Rd} \left[\frac{\gamma_{M,1}}{\gamma_{M,fi}} \right] \tag{8.28}$$

Since $\gamma_{M,1} = \gamma_{M,fi} = 1,0$, Equation (8.28) reduces to

$$N_{fi,\theta,Rd} = k_{y,\theta} N_{Rd} \tag{8.29}$$

where $k_{y,\theta}$ is the normalized strength reduction at a temperature of θ_a and N_{Rd} is the ambient design resistance. For tension members with a nonuniform temperature distribution, the axial capacity may be obtained by summing the contributions of incremental areas or conservatively using the maximum steel temperature reached and assuming constant temperature.

- Compression members

The axial force of compression members for Classes 1, 2, and 3 cross sections is calculated as follows (Clause 4.2,3.2):

$$N_{b,fi,t,Rd} = \chi_{fi} A k_{y,\theta} \frac{f_y}{\gamma_{M,fi}}$$

(8.30)

where the buckling strength reduction factor χ_{fi} is determined from

$$\chi_{fi} = \frac{1}{\phi_\theta + \sqrt{\phi_\theta^2 - \bar{\lambda}_\theta^2}}$$

(8.31)

with

$$\phi_\theta = 0,5 \left[1 + 0,65 \sqrt{\frac{235}{f_y}} \bar{\lambda}_\theta + \bar{\lambda}_\theta^2 \right]$$

(8.32)

where the normalized slenderness ratio $\bar{\lambda}_\theta$ is defined as

$$\bar{\lambda}_\theta = \bar{\lambda} \left[\frac{k_{y,\theta}}{k_{E,\theta}} \right]^{0,5}$$

(8.32)

where $k_{E,\theta}$ is the temperature-dependent reduction factor for Young's modulus and $\bar{\lambda}$ is the normalized slenderness ratio, except that buckling length for a continuous column at other than the top storey may be taken as 0,5 times the column length (0,7 for the top storey), provided the frame is braced and that the fire resistance of the compartments above and below have a resistance not less than that required for the column.

- Beams

Class 1 or 2 section classification (Clause 4.2.3.3) — With a uniform temperature distribution, the moment capacity $M_{fi,\theta,Rd}$ with no lateral torsional buckling may be calculated as:

$$M_{fi,\theta,Rd} = k_{y,\theta} M_{Rd} \left[\frac{\gamma_{M,1}}{\gamma_{M,fi}} \right]$$

(8.34)

Since $\gamma_{M,1} = \gamma_{M,fi} = 1,0$, Equation (8.34) reduces to

$$M_{fi,\theta,Rd} = k_{y,\theta}M_{Rd} \qquad (8.35)$$

where $k_{y,\theta}$ is the normalized strength reduction at a temperature of θ_a and M_{Rd} is the ambient design resistance. For flexural members with a nonuniform temperature distribution, the moment capacity may be obtained by summing the contributions of incremental areas or the design moment of resistance $M_{fi,t,d}$ as follows,

$$M_{fi,\theta,Rd} = \frac{M_{fi,t,d}}{\kappa_1\kappa_2} \qquad (8.36)$$

where κ_1 allows for nonuniform temperature distribution within the cross section and should be taken as 1,0 for a beam exposed on all four sides, 0,85 for a protected beam exposed on three sides with a composite or concrete slab on the fourth, and 0,70 for an unprotected beam exposed on three sides with a composite or concrete slab on the fourth. The parameter κ_2 should be taken as 1,0 except at the supports of a statically indeterminate beam when it should be taken as 0,85. Where lateral-torsional buckling can occur, the moment capacity $M_{b,fi,t,d}$ is given by

$$M_{b,fi,t,d} = \chi_{LT,fi}W_{pl,y}k_{y,\theta,com}\frac{f_y}{\gamma_{M,fi}} \qquad (8.37)$$

where $k_{y,\theta,com}$ is the strength reduction factor for the temperature in the compression flange (that may be based conservatively on the uniform temperature θ_a), and $\chi_{LT,fi}$ is determined from

$$\chi_{LT,fi} = \frac{1}{\varphi_{LT,\theta,com} + \sqrt{\left[\varphi_{LT,\theta,com}\right]^2 - \left[\overline{\lambda}_{LT,\theta,com}\right]^2}} \qquad (8.38)$$

with

$$\varphi_{LT,\theta,com} = 0,5\left[1 + 0,65\sqrt{\frac{235}{f_y}}\,\overline{\lambda}_{LT,\theta,com} + \left(\overline{\lambda}_{LT,\theta,com}\right)^2\right] \qquad (8.39)$$

$$\overline{\lambda}_{LT,\theta,com} = \overline{\lambda}_{LT}\left[\frac{k_{y,\theta,com}}{k_{E,\theta,com}}\right]^{0,5} \qquad (8.40)$$

where $k_{E,\theta,com}$ is the temperature-dependent reduction factor for Young's modulus for the compression flange, and $\overline{\lambda}_{LT}$ is the normalized lateral-torsional buckling slenderness ratio.

Class 3 (Clause 4.2.3.4) — These are dealt with exactly as Class 1 or 2 beams except that the moment capacity M_{Rd} is determined using the elastic section modulus W_{el} and not the plastic section modulus W_{pl}.

Shear (Class 1, 2 or 3) — The design shear resistance $V_{fi,t,d}$ is determined from

$$V_{fi,t,d} = k_{y,\theta,web} V_{Rd} \left[\frac{\gamma_{M,1}}{\gamma_{M,fi}} \right] \tag{8.41}$$

where $k_{y,\theta,web}$ is the strength reduction factor based on the temperature within the web.

- Members subject to combined bending and axial compression

The design buckling resistance $R_{fi,t,d}$ should satisfy the following interaction equations:

Class 1 or 2:

$$\frac{N_{fi,Ed}}{\chi_{min,fi} A k_{y,\theta} \dfrac{f_y}{\gamma_{M,fi}}} + \frac{k_y M_{y,fi,Ed}}{W_{pl,y} k_{y,\theta} \dfrac{f_y}{\gamma_{M,fi}}} + \frac{k_z M_{z,fi,Ed}}{W_{pl,z} k_{y,\theta} \dfrac{f_y}{\gamma_{M,fi}}} \leq 1,0 \tag{8.42}$$

$$\frac{N_{fi,Ed}}{\chi_{z,fi} A k_{y,\theta} \dfrac{f_y}{\gamma_{M,fi}}} + \frac{k_{LT} M_{y,fi,Ed}}{\chi_{LT,fi} W_{pl,y} k_{y,\theta} \dfrac{f_y}{\gamma_{M,fi}}} + \frac{k_z M_{z,fi,Ed}}{W_{pl,z} k_{y,\theta} \dfrac{f_y}{\gamma_{M,fi}}} \leq 1,0 \tag{8.43}$$

Class 3:

$$\frac{N_{fi,Ed}}{\chi_{min,fi} A k_{y,\theta} \dfrac{f_y}{\gamma_{M,fi}}} + \frac{k_y M_{y,fi,Ed}}{W_{el,y} k_{y,\theta} \dfrac{f_y}{\gamma_{M,fi}}} + \frac{k_z M_{z,fi,Ed}}{W_{el,z} k_{y,\theta} \dfrac{f_y}{\gamma_{M,fi}}} \leq 1,0 \tag{8.44}$$

$$\frac{N_{fi,Ed}}{\chi_{z,fi} A k_{y,\theta} \dfrac{f_y}{\gamma_{M,fi}}} + \frac{k_{LT} M_{y,fi,Ed}}{\chi_{LT,fi} W_{el,y} k_{y,\theta} \dfrac{f_y}{\gamma_{M,fi}}} + \frac{k_z M_{z,fi,Ed}}{W_{el,z} k_{y,\theta} \dfrac{f_y}{\gamma_{M,fi}}} \leq 1,0 \tag{8.45}$$

where

$$k_{LT} = 1 - \frac{\mu_{LT} N_{fi,Ed}}{\chi_{z,fi} A k_{y,\theta} \dfrac{f_y}{\gamma_{M,fi}}} \leq 1,0 \tag{8.46}$$

$$\mu_{LT} = 0,15\overline{\lambda}_{z,\theta}\beta_{M,LT} - 0,15 \leq 0,9 \tag{8.47}$$

$$k_y = 1 - \frac{\mu_y N_{fi,Ed}}{\chi_{y,fi} A k_{y,\theta} \dfrac{f_y}{\gamma_{M,fi}}} \leq 3,0 \tag{8.48}$$

$$\mu_y = (1,2\beta_{M,y} - 3)\overline{\lambda}_{y,\theta} + 0,44\beta_{M,LT} - 0,29 \leq 0,8 \tag{8.49}$$

$$k_z = 1 - \frac{\mu_z N_{fi,Ed}}{\chi_{z,fi} A k_{y,\theta} \dfrac{f_y}{\gamma_{M,fi}}} \leq 3,0 \tag{8.50}$$

$$\mu_z = (2\beta_{M,z} - 5)\overline{\lambda}_{z,\theta} + 0,44\beta_{M,z} - 0,29 \leq 0,8 \tag{8.51}$$

Equations (8.46) to (8.51) are subject to the limit that $\overline{\lambda}_{z,\theta} \leq 1,1$. Figure 4.2 of EN 1993-1-2 gives values of the equivalent moment factor β_M for various types of moment diagrams due to lateral in-plane loading cases and also for cases of end moments only where β_M is given by

$$\beta_M = 1,8 - 0,7\psi \tag{8.52}$$

where ψ is the ratio between the end moments such that $-1 > \psi > 1$. Note that where lateral-torsional buckling or strut buckling can occur, the procedure to determine the critical temperature in the member is iterative as the buckling coefficients are also temperature dependent. Where columns carry moments, the situation worsens in that some coefficients in the interaction equations are also temperature dependent.

The tabular data for $k_{E,\theta}$ from Table 3.1 of EN 1993-1-2 are given by the following equations:

For $20°C \leq \theta_a \leq 100°C$:

$$k_{E,\theta} = 1,0 \tag{8.53}$$

For $100°C \leq \theta_a \leq 500°C$:

$$k_{E,\theta} = 0,6 - 0,4\frac{\theta_a - 500}{400} \tag{8.54}$$

For $500°C \leq \theta_a \leq 600°C$:

$$k_{E,\theta} = 0,31 - 0,29\frac{\theta_a - 600}{100} \tag{8.55}$$

For $600°C \leq \theta_a \leq 700°C$:

$$k_{E,\theta} = 0,13 - 0,18\frac{\theta_a - 700}{100} \qquad (8.56)$$

For $700°C \leq \theta_a \leq 800°C$:

$$k_{E,\theta} = 0,09 - 0,04\frac{\theta_a - 800}{100} \qquad (8.57)$$

For $800°C \leq \theta_a \leq 1100°C$:

$$k_{E,\theta} = 0,0225 - 0,0675\frac{\theta_a - 1100}{300} \qquad (8.58)$$

For $1100°C \leq \theta_a \leq 1200°C$:

$$k_{E,\theta} = 0,0225\frac{1100 - \theta_a}{100} \qquad (8.59)$$

The tabular data for $k_{y,\theta}$ from Table 3.1 of EN 1993-1-2 are given by the following equations:

For $20°C \leq \theta_a \leq 400°C$:

$$k_{y,\theta} = 1,0 \qquad (8.60)$$

For $400°C \leq \theta_a \leq 500°C$:

$$k_{y,\theta} = 0,78 - 0,22\frac{\theta_a - 500}{100} \qquad (8.61)$$

For $500°C \leq \theta_a \leq 600°C$:

$$k_{y,\theta} = 0,47 - 0,31\frac{\theta_a - 600}{100} \qquad (8.62)$$

For $600°C \leq \theta_a \leq 700°C$:

$$k_{y,\theta} = 0,23 - 0,24\frac{\theta_a - 700}{100} \qquad (8.63)$$

For $700°C \leq \theta_a \leq 800°C$:

$$k_{y,\theta} = 0,11 - 0,12\frac{\theta_a - 800}{100} \qquad (8.64)$$

For $800°C \leq \theta_a \leq 900°C$:

$$k_{y,\theta} = 0,06 - 0,05\frac{\theta_a - 900}{100} \qquad (8.65)$$

For $900°C \leq \theta_a \leq 1100°C$:

$$k_{y,\theta} = 0,02 - 0,04\frac{\theta_a - 1100}{200} \qquad (8.66)$$

For $1100°C \leq \theta_a \leq 1200°C$:

$$k_{y,\theta} = 0,02\frac{1100 - \theta_a}{100} \qquad (8.67)$$

8.2.2.2.2 Limiting temperature criterion

For a member to perform adequately where deflection or instability (buckling) is not critical in a fire EN 1993-1-2 requires that

$$\theta_a \leq \theta_{a,cr} \qquad (8.68)$$

The determination of the actual steel temperature θ_a was covered earlier in this chapter and in Examples 8.1 to 8.3 and will not be discussed further.

The determination of the critical steel temperature $\theta_{a,cr}$ is dependent on the degree of utilization μ_0. The relationship between $\theta_{a,cr}$ and μ_0 has been determined from elementary plasticity theory and the reduction in steel strength with temperature and is given empirically by:

$$\theta_{a,cr} = 39,19\ln\left[\frac{1}{0,9674\mu_0^{3,833}} - 1\right] + 482 \qquad (8.69)$$

subject to the limit $\mu_0 > 0,013$. The degree of utilization is defined as

$$\mu_0 = \frac{E_{fi,d}}{R_{fi,d,0}} \qquad (8.70)$$

where $R_{fi,d,0}$ is the resistance of the member at time $t = 0$ determined in accordance with the principles outlined above, and $E_{fi,d}$ is the design effect of the structural fire actions. Alternatively μ_0 may be defined conservatively as

$$\mu_0 = \eta_{fi}\left[\frac{\gamma_{M,fi}}{\gamma_{M1}}\right] \tag{8.71}$$

The use of $\theta_{a,cr}$ only holds for tension members and Class 1, 2, or 3 beams and compression members.

Example 8.4: Determination of fire protection requirements for lateral-torsionally restrained beam.

A Grade S275 beam is simply supported over a span of 8 m. It carries permanent loading from 125-mm thick precast concrete units (205 kg/m²), 40-mm concrete screed, 20-mm wood blocks, and a suspended ceiling of mass 40 kg/m². The lightweight partitions comprise 1,0 kN/m² and the variable loading is 2,5 kPa. The beams are 457 × 152 × 60 UB at 3,75-m pitch. Design mineral board protection to give 60 min standard fire resistance.

Permanent Load	Component kg/m²	kPa
Beam self weight	16	
Precast units	205	
Screed (2400 × 0,040)	96	
Wood blocks (900 × 0,020)	18	
Suspended ceiling	40	
Total	375	3,75
Partitions		1,00
Total		4,75

Variable load in the fire limit state is $0,3 \times 2,5 = 0,75$ kPa. $M_{fi,Ed} = 3,75(4,75 + 0,75)8^2/8 = 163$ kNm. $M_{Rd} = 275 \times 1287 \times 10^{-3}/1,0 = 354$ kNm.

Check section classification:

$$\varepsilon = 0,85\left[\frac{235}{f_y}\right]^{0,5} = 0,85\left[\frac{235}{275}\right]^{0,5} = 0,786$$

Flange classification:

$$c = 0,5[b - 2r - t_x] = 0,5[152,9 - 2 \times 10,2 - 8,1] = 62,6$$

$$\frac{c}{t_f} = \frac{62,6}{13,3} = 4,71$$

Limiting value for Class 1: $9\varepsilon = 9 \times 0,786 = 7,07$. So flange is Class 1.
Web check where c = depth between fillets = 407,6 mm:

$$\frac{c}{t_w} = \frac{407,6}{8,1} = 50,3$$

Limiting value for Class 1 is $9\varepsilon = 72 \times 0,786 = 56,6$. So web is Class 1.

For section resistance, set $M_{fi,\theta,Rd}$ equal to $M_{fi,Ed}$, then from Equation
(8.34),

$$k_{y,\theta} = \frac{M_{fi,Ed}}{M_{Rd}} = \frac{163}{354} = 0,460$$

As the temperature is nonuniform, this value may be modified by $1/\kappa_1$ ($\kappa_2 = 1,0$ as the beam is simply supported). For this case (beam pro-
tected on three sides with concrete slab on the top flange, $\kappa_1 = 0,85$) from
Equation (8.36), the effective value of $k_{y,\theta}$ is $0,46/0,85 = 0,541$, and from
Equation (8.62) $\theta_a = 577°C$.

Fire protection data are mineral fibre boarding as in Example 8.3
(i.e., $\rho_p = 500$ kg/m³, $p = 2\%$, $c_p = 1500$ J/kg°C, $\lambda_p = 0,25$ W/m°C). For
a four-sided box, $A_m/V - 140$ m^{-1}. Using Equation (8.22):

$$d_p = \frac{-\dfrac{1}{\lambda_p}\dfrac{V}{A_m} + \sqrt{\left(\dfrac{1}{\lambda_p}\dfrac{V}{A_m}\right)^2 + 4\dfrac{\rho_p}{\rho_a}\dfrac{1}{\lambda_p}\left[\dfrac{t_{fi,d}}{40\left(\theta_{a,t} - 140\right)}\right]^{1,3}}}{2\dfrac{\rho_p}{\rho_a}\dfrac{1}{\lambda_p}}$$

$$= \frac{-\dfrac{1}{0,25 \times 140} + \sqrt{\left(\dfrac{1}{0,25 \times 140}\right)^2 + 4\dfrac{530}{7850 \times 0,25}\left[\dfrac{60}{40\left(577 - 140\right)}\right]^{1,3}}}{2\dfrac{530}{7850 \times 0,25}}$$

$$= 0,019 m$$

From Equation (8.12) using the heat transfer equations, $d_p = 0,0195$ m
($\theta_{a,t} = 576°C$).

For the critical temperature approach:

$$\mu_0 = \frac{E_{fi,d}}{R_{fi,d,0}} = \frac{163}{354} = 0,460$$

$$\theta_{a,cr} = 39,19\ln\left[\frac{1}{0,9674\mu_0^{3,833}} - 1\right] + 482$$

$$= 39,19\ln\left[\frac{1}{0,9674 \times 0,46^{3,833}} - 1\right] + 482 = 598°C$$

It should be observed that the two temperatures differ only by around 20°C. From Equation (8.22)

$$
d_p = \frac{-\dfrac{1}{\lambda_p}\dfrac{V}{A_m} + \sqrt{\left(\dfrac{1}{\lambda_p}\dfrac{V}{A_m}\right)^2 + 4\dfrac{\rho_p}{\rho_a}\dfrac{1}{\lambda_p}\left[\dfrac{t_{fi,d}}{40(\theta_{a,t}-140)}\right]^{1,3}}}{2\dfrac{\rho_p}{\rho_a}\dfrac{1}{\lambda_p}}
$$

$$
= \frac{-\dfrac{1}{0,25\times140} + \sqrt{\left(\dfrac{1}{0,25\times140}\right)^2 + 4\dfrac{530}{7850\times0,25}\left[\dfrac{60}{40(598-140)}\right]^{1,3}}}{2\dfrac{530}{7850\times0,25}}
$$

$$
= 0,018 \text{ m}
$$

From Equation (8.12) using heat transfer calculations, $d_p = 0{,}0185$ m ($\theta_{a,t} = 597°C$).

The use of the quadratic equation for determination of protection thicknesses and the heat transfer calculations give almost identical results. The limiting temperature approach is slightly less conservative, although the resultant differences in fire protection thicknesses are negligible.

Example 8.5: Design of fire protection for beam

Determine the thickness of mineral fibre board protection required to give 90-min fire resistance for a 406 × 178 × 74 UB (Grade S355 JR) whose design data are given in Figure 8.7a.

As the beam can suffer lateral-torsional buckling, the critical temperature approach cannot be used. Also it will be assumed conservatively that the compression flange temperature is equal to the uniform temperature. As the solution to Equation (8.37) is iterative, a spreadsheet was used in which the temperature varied until the moment capacity was just greater than the applied moment.

From the ambient design $\overline{\lambda}_{LT} = 0{,}715$.
At a temperature $\theta_a = 565°C$:
From Equation (8.62) $k_{y,\theta} = 0{,}579$
From Equation (8.55) $k_{E,\theta} = 0.411$

From Equation (8.40):

$$
\overline{\lambda}_{LT,\theta,com} = \overline{\lambda}_{LT}\left[\frac{k_{y,\theta,com}}{k_{E,\theta,com}}\right]^{0,5} = 0{,}715\left[\frac{0{,}579}{0{,}411}\right]^{0,5} = 0{,}848
$$

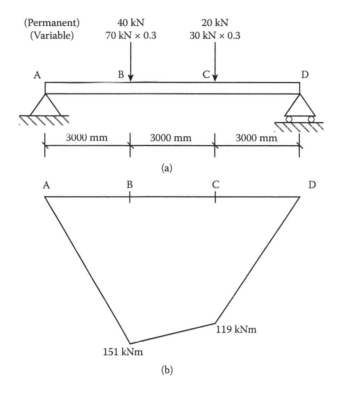

Figure 8.7 (a) Beam loading data. (b) Bending moment diagram for Example 8.4.

From Equation (8.39):

$$\phi_{LT,\theta,com} = 0,5\left[1+0,65\sqrt{\frac{235}{f_y}}\;\bar{\lambda}_{LT,\theta,com} + \left(\bar{\lambda}_{LT,\theta,com}\right)^2\right]$$

$$= 0,5\left[1+0,848\times0,65\sqrt{\frac{235}{355}} + 0.848^2\right] = 1,084$$

From Equation (8.38):

$$\chi_{LT,fi} = \cfrac{1}{\phi_{LT,\theta,com} + \sqrt{\left[\phi_{LT,\theta,com}\right]^2 - \left[\bar{\lambda}_{LT,\theta,com}\right]^2}} = \cfrac{1}{1,084 + \sqrt{1,084^2 - 0,848^2}}$$

$$= 0,568$$

From Equation (8.37):

$$M_{b,fi,t,d} = \chi_{LT,fi} W_{pl,y} k_{y,\theta,com} \frac{f_y}{\gamma_{M,fi}} = 0,568 \times 1301 \times 0,579 \times 355 \times 10^{-3} = 152 \text{ kNm}$$

This is greater than the applied moment $M_{fi,Ed}$ of 151 kNm. For a three-sided box, $A/V = 105 \text{ m}^{-1}$:

From Equation (8.22):

$$d_p = \frac{-\dfrac{1}{\lambda_p}\dfrac{V}{A_m} + \sqrt{\left(\dfrac{1}{\lambda_p}\dfrac{V}{A_m}\right)^2 + 4\dfrac{\rho_p}{\rho_a}\dfrac{1}{\lambda_p}\left[\dfrac{t_{fi,d}}{40(\theta_{a,t}-140)}\right]^{1,3}}}{2\dfrac{\rho_p}{\rho_a}\dfrac{1}{\lambda_p}}$$

$$= \frac{-\dfrac{1}{105\times0,25} + \sqrt{\left(\dfrac{1}{105\times0,25}\right)^2 + 4\dfrac{530}{7850\times0,25}\left[\dfrac{90}{40(565-140)}\right]^{1,3}}}{2\dfrac{530}{7850\times0,25}}$$

$$= 0,025 \text{ m}$$

From Equation (8.12), $d_p = 0,0255$ m $(\theta_{a,t} = 565°C)$.

Example 8.6: Column heated under axial load only

A 254×254×107 UC Grade 275 carries a permanent load and a variable axial load both equal to 1000 kN at ambient limit state. The effective length of the column is 3,5 m. Design box protection to give 90 min fire performance.

The ambient design gives $\bar{\lambda} = 0,612$ with $N_{Rd} = 3112$ kN. At ambient, $N_{Ed} = 2850$ kN.

Check the section classification:

$$\varepsilon = 0,85\left[\frac{235}{275}\right]^{0,5} = 0,85\left[\frac{235}{275}\right]^{0,5} = 0,786$$

Flange classification:

$$c = 0,5[b - 2r - t_w] = 0,5[258,8 - 2\times12,7 - 12,8] = 110,3 \text{ mm}$$

$$\frac{c}{t_f} = \frac{110,3}{20,5} = 5,38$$

Limiting value for Class 1 is $9\varepsilon = 9 \times 0,786 = 7,07$. So flange is Class 1. For web check, c = depth between fillets = 200,3 mm:

$$\frac{c}{t_w} = \frac{200,3}{12,8} = 15,6$$

Limiting value for Class 1 is $9\varepsilon = 72 \times 0,786 = 56,6$. So web is Class 1. As the exact end conditions are not known, take the buckling length in the fire limit state as 3,5 m. Taking $\psi = 0,3$ on the variable load gives

$$N_{Ed,fi} = 1,0 \times 1000 + 0,3 \times 1000 = 1300 \text{ kN}$$

For a steel temperature of 565°C:
From Equation (8.62):

$$k_{y,\theta} = 0,5785$$

From Equation (8.55):

$$k_{E,\theta} = 0,4115$$

From Equation (8.33):

$$\bar{\lambda}_\theta = \bar{\lambda} \left[\frac{k_{y,\theta}}{k_{E,\theta}} \right]^{0,5} = 0,612 \left[\frac{0,5785}{0,4115} \right]^{0,5} = 0,725$$

From Equation (8.32):

$$\phi_\theta = 0,5 \left[1 + 0,65 \sqrt{\frac{235}{f_y}} \, \bar{\lambda}_\theta + \bar{\lambda}_\theta^2 \right] = 0,5 \left[1 + 0,725 \times 0,65 \sqrt{\frac{235}{275}} + 0,725^2 \right]$$

$$= 0,981$$

From Equation (8.31):

$$\chi_{fi} = \frac{1}{\phi_\theta + \sqrt{\phi_\theta^2 - \bar{\lambda}_\theta^2}} = \frac{1}{0,981 + \sqrt{0,981^2 - 0,725^2}} = 0,609$$

From Equation (8.30):

$$N_{b,fi,t,Rd} = \chi_{fi} A k_{y,\theta} \frac{f_y}{\gamma_{M,fi}} = 0,609 \times 13600 \times 0,5785 \times 275 \times 10^{-3} = 1318 \text{ kN}$$

This exceeds the design load $N_{fi,Ed}$ of 1300 kN and is therefore satisfactory. For a four-sided box $A_m/V = 75 \text{ m}^{-1}$.

From Equation (8.22):

$$d_p = \frac{-\dfrac{1}{\lambda_p}\dfrac{V}{A_m} + \sqrt{\left(\dfrac{1}{\lambda_p}\dfrac{V}{A_m}\right)^2 + 4\dfrac{\rho_p}{\rho_a}\dfrac{1}{\lambda_p}\left[\dfrac{t_{fi,d}}{40(\theta_{a,t}-140)}\right]^{1,3}}}{2\dfrac{\rho_p}{\rho_a}\dfrac{1}{\lambda_p}}$$

$$= \frac{-\dfrac{1}{75\times 0,25} + \sqrt{\left(\dfrac{1}{75\times 0,25}\right)^2 + 4\dfrac{530}{7850\times 0,25}\left[\dfrac{90}{40(565-140)}\right]^{1,3}}}{2\dfrac{530}{7850\times 0,25}}$$

$$= 0,019 \text{ m}$$

From Equation (8.12):

$d_p = 0,0197$ m ($\theta_{a,t} = 566°C$).

The ASCE Standard Calculation Methods for Structural Fire Protection (2007) give the following formulae for the use of gypsum wall board for fire protection of columns:

$$t_{fi,requ} = 1,60\left[\frac{d_p \dfrac{W'}{H_p}}{2}\right]^{0,75} \tag{8.72}$$

where $t_{fi,requ}$ is the fire resistance period based on standard classification (hr), H_p is the heated perimeter (mm), d_p is thickness of the board, and W' is mass per unit length of the column and wall boarding and is given by

$$W' = W + 0,0008d_p H_p \tag{8.73}$$

where W is the mass per unit length of the steel section (kg/m). The formulae are based on a critical temperature around 525°C and are valid only for W/H_p less than or equal to 0,215. Equations (8.72) and (8.73) may be solved as a quadratic in d_p to give

$$d_p = \frac{\sqrt{\left(\dfrac{W}{H_p}\right)^2 + 0,0064\left(\dfrac{t_{fi,requ}}{1,6}\right)^{\frac{4}{3}}} - \dfrac{W}{H_p}}{0,0016} \tag{8.74}$$

In this example, $W = 107$ kg/m and $H_p = 2(h + b) = 2(266,7 + 258,8) =$ 1051 mm, which satisfies the criterion $W/H_p = 0,102 < 0,215$. Hence

$$d_p = \frac{\sqrt{\left(\dfrac{W}{H_p}\right)^2 + 0,0064\left(\dfrac{t_{fi,requ}}{1,6}\right)^{\frac{4}{3}}} - \dfrac{W}{H_p}}{0,0016} = \frac{\sqrt{0,102^2 + 0,0064\left(\dfrac{1,5}{1,6}\right)^{\frac{4}{3}}} - 0,102}{0,0016}$$

$$= 16,0 \text{ mm}$$

This is comparable to the value of 19,7 mm from the more rigorous calculation.

Example 8.7: Design of fire protection for column under moment and axial force

Determine the thickness of plaster board protection required to give 90-min fire resistance for a column ($305 \times 305 \times 137$ Grade S 275) with an axial force from permanent actions of 600 kN and a moment about the major axis due to variable actions of 300 kNm. The buckling length of the column in both the ambient and fire limit states is 3,5 m.

From the ambient design, $\bar{\lambda}_y = 0,294$; $\bar{\lambda}_z = 0,514$, and $\bar{\lambda}_{LT} = 0,276$.

For the ambient design $\bar{\lambda}_{LT} < 0,4$ which indicates lateral-torsional buckling will not occur, there is no indication whether this restriction still applies in the fire limit state. Thus checks with and without lateral-torsional buckling should be carried out. Also assume that the column is an edge type toward the mid-height of a multi-storey structure when the moments at the ends of the column will be of opposite sign and approximately equal. Thus the ratio ψ can be taken as –1. Thus from Equation (8.52):

$$\beta_{M,\psi} = 1,8 - 0,7\psi = 1,8 - 0,7(-1) = 2,5$$

This value will also apply in the determination of k_y, k_z and k_{LT}. For a value of $\theta_a = 640°C$,

From Equation (8.63):

$$k_{y,\theta} = 0,374$$

From Equation (8.56):

$$k_{E,\theta} = 0,238$$

From Equation (8.33):

$$\bar{\lambda}_{\theta,y} = \lambda_y \left[\frac{k_{y,\theta}}{k_{E,\theta}}\right]^{0,5} = 0,294 \left[\frac{0,374}{0,238}\right]^{0,5} = 0,369$$

From Equation (8.32):

$$\phi_{\theta,y} = 0,5\left[1+0,65\sqrt{\frac{235}{f_y}}\,\overline{\lambda}_\theta + \overline{\lambda}_\theta^2\right] = 0,5\left[1+0,369\times0,65\sqrt{\frac{235}{275}}+0,369^2\right]$$

$$= 0,679$$

From Equation (8.31):

$$\chi_{fi,y} = \frac{1}{\phi_\theta + \sqrt{\phi_\theta^2 - \overline{\lambda}_\theta^2}} = \frac{1}{0,679+\sqrt{0,679^2-0,369^2}} = 0,801$$

From Equation (8.33):

$$\overline{\lambda}_{\theta,z} = \overline{\lambda}_z\left[\frac{k_{y,\theta}}{k_{E,\theta}}\right]^{0,5} = 0,514\left[\frac{0,374}{0,238}\right]^{0,5} = 0,644$$

From Equation (8.32):

$$\phi_{\theta,z} = 0,5\left[1+0,65\sqrt{\frac{235}{f_y}}\overline{\lambda}_\theta + \overline{\lambda}_\theta^2\right] = 0,5\left[1+0,644\times0,65\sqrt{\frac{235}{275}}+0,644^2\right]$$

$$= 0,901$$

From Equation (8.31):

$$\chi_{fi,z} = \frac{1}{\phi_\theta + \sqrt{\phi_\theta^2 - \overline{\lambda}_\theta^2}} = \frac{1}{0,901+\sqrt{0,901^2-0,644^2}} = 0,653$$

$$\chi_{fi,min} = 0,653$$

From Equation (8.40):

$$\overline{\lambda}_{LT,\theta,com} = \overline{\lambda}_{LT}\left[\frac{k_{y,\theta,com}}{k_{E,\theta,com}}\right]^{0,5} = 0,276\left[\frac{0,374}{0,238}\right]^{0,5} = 0,346$$

From Equation (8.39):

$$\phi_{LT,\theta,com} = 0,5\left[1+0,65\sqrt{\frac{235}{f_y}}\overline{\lambda}_{LT,\theta,com} + \left(\overline{\lambda}_{LT,\theta,com}\right)^2\right]$$

$$= 0,5\left[1+0,346\times0,65\sqrt{\frac{235}{275}}+0.346^2\right] = 0,664$$

From Equation (8.38):

$$\chi_{LT,fi} = \cfrac{1}{\phi_{LT,\theta,com} + \sqrt{\left[\phi_{LT,\theta,com}\right]^2 - \left[\bar{\lambda}_{LT,\theta,com}\right]^2}} = \cfrac{1}{0,664 + \sqrt{0,664^2 - 0,346^2}}$$

$$= 0,813$$

From Equation (8.47):

$$\mu_{LT} = 0,15\,\bar{\lambda}_{z,\theta}\beta_{M,LT} - 0,15 = 0,15 \times 0,644 \times 2,5 - 0,15 = 0,092$$

This is less than 0,9 and therefore satisfactory.

From Equation (8.46):

$$k_{LT} = 1 - \cfrac{\mu_{LT}N_{fi,Ed}}{\chi_{z,fi}Ak_{y,\theta}\cfrac{f_y}{\gamma_{M,fi}}} = 1 - \cfrac{0,092 \times 600 \times 10^3}{0,653 \times 17400 \times 0,374 \times 275} = 0,953$$

This is less than 1,0 and therefore satisfactory.

From Equation (8.49):

$$\mu_y = (1,2\beta_{M,y} - 3)\,\bar{\lambda}_{y,\theta} + 0,44\beta_{M,y} - 0,29 = 0,368(1,2 \times 2,5 - 3)$$

$$+ 0,44 \times 2,5 - 0,29 = 0,81$$

As this is greater than the limiting value of 0,8, $\mu_y = 0,8$.

From Equation (8.48):

$$k_y = 1 - \cfrac{\mu_y N_{fi,Ed}}{\chi_{y,fi}Ak_{y,\theta}\cfrac{f_y}{\gamma_{M,fi}}} = 1 - \cfrac{0,8 \times 600 \times 10^3}{0,801 \times 17400 \times 0,374 \times 275} = 0,665$$

This is less than the limiting value of 3.

From Equation (8.51):

$$\mu_z = (2\beta_{M,z} - 5)\,\bar{\lambda}_{z,\theta} + 0,44\beta_{M,z} - 0,29$$

$$= 0,644(2 \times 2,5 - 5) + 0,44 \times 2,5 - 0,29 = 0,81$$

As this is greater than the limiting value of 0,8, $\mu_z = 0,8$.

From Equation (8.50):

$$k_z = 1 - \frac{\mu_z N_{fi,Ed}}{\chi_{z,fi} A k_{y,\theta} \dfrac{f_y}{\gamma_{M,fi}}} = 1 - \frac{0,8 \times 600 \times 10^3}{0,653 \times 17400 \times 0,374 \times 275} = 0,589$$

Check the interaction from Equations (8.42) and (8.43).

From Equation (8.42):

$$\frac{N_{fi,Ed}}{\chi_{min,fi} A k_{y,\theta} \dfrac{f_y}{\gamma_{M,fi}}} + \frac{k_y M_{y,fi,Ed}}{W_{pl,y} k_{y,\theta} \dfrac{f_y}{\gamma_{M,fi}}} + \frac{k_z M_{z,fi,Ed}}{W_{pl,z} k_{y,\theta} \dfrac{f_y}{\gamma_{M,fi}}}$$

$$= \frac{600 \times 10^3}{0,653 \times 17400 \times 0,374 \times 275} + \frac{0,665 \times 90 \times 10^3}{2297 \times 0,374 \times 275} + 0 = 0,767$$

From Equation (8.43):

$$\frac{N_{fi,Ed}}{\chi_{z,fi} A k_{y,\theta} \dfrac{f_y}{\gamma_{M,fi}}} + \frac{k_{LT} M_{y,fi,Ed}}{\chi_{LT,fi} W_{pl,y} k_{y,\theta} \dfrac{f_y}{\gamma_{M,fi}}} + \frac{k_z M_{z,fi,Ed}}{W_{pl,z} k_{y,\theta} \dfrac{f_y}{\gamma_{M,fi}}}$$

$$= \frac{600 \times 10^3}{0,653 \times 17400 \times 0,374 \times 275} + \frac{0,953 \times 90 \times 10^3}{0,813 \times 2297 \times 0,374 \times 275} + 0 = 0,96$$

It will be seen that in this case the check including lateral-torsional buckling is critical.

From Equation (8.22):

$$d_p = \frac{-\dfrac{1}{\lambda_p} \dfrac{V}{A_m} + \sqrt{\left(\dfrac{1}{\lambda_p} \dfrac{V}{A_m}\right)^2 + 4 \dfrac{\rho_p}{\rho_a} \dfrac{1}{\lambda_p} \left[\dfrac{t_{fi,d}}{40(\theta_{a,t} - 140)}\right]^{1,3}}}{2 \dfrac{\rho_p}{\rho_a} \dfrac{1}{\lambda_p}}$$

$$= \frac{-\dfrac{1}{70 \times 0,25} + \sqrt{\left(\dfrac{1}{70 \times 0,25}\right)^2 + 4 \dfrac{530}{7850 \times 0,25} \left[\dfrac{90}{40(640 - 140)}\right]^{1,3}}}{2 \dfrac{530}{7850 \times 0,25}} = 0,015 \text{ m}$$

From Equation (8.12):

$$d_p = 0,0152 \text{ m } (\theta_{a,t} = 641°C).$$

As the ASCE document does not appear to restrict its empirical formulae [Equations (8.72) to (8.74)] to columns with axial loads only, in this example $W = 137$ kg/m and $H_p = 2(h + b) = 2(320,5 + 309,2) = 1259$ mm, which satisfies the criterion $W/H_p = 0,109 < 0,215$. Hence

$$d_p = \frac{\sqrt{\left(\frac{W}{H_p}\right)^2 + 0,0064\left(\frac{t_{fi,requ}}{1,6}\right)^{\frac{4}{3}}} - \frac{W}{H_p}}{0,0016} = \frac{\sqrt{0,109^2 + 0,0064\left(\frac{1,5}{1,6}\right)^{\frac{4}{3}}} - 0,109}{0,0016}$$

$$= 15,2 \text{ mm}$$

This is again comparable to the exact solution of 15,0 mm.

8.3 OTHER STEELWORK CONSTRUCTIONS

8.3.1 External steelwork

It may be necessary in certain structures for the steel frame to be external to the cladding, i.e., outside the main envelope of the structure. Thus the design of the steelwork must consider the effects of a fire escaping from a compartment rather than retained within a compartment. Methods are therefore needed to calculate the temperatures in external steelwork. If the temperatures attained are low enough, it is possible that no protection need be applied to the steelwork. The problem therefore concerns temperature calculation rather than strength response. To carry out the calculations necessary, the reader is referred to Law (1978) or Law and O'Brien (1989) work that explains the theoretical basis behind the calculations and presents typical examples. The Annex in EN 1993-1-2 on bare external steelwork was adapted directly from Law and O'Brien.

8.3.2 Shelf angle floors

Only BS 5950 Part 8 gives an explicit method to cope with shelf angle floors, and it is concerned with the calculation of moment capacity. EN 1993-1-2 allows direct calculation of moment capacity but gives no guidance on the calculation of the temperature field within shelf angle floors. Further information including design charts to enable a simple check of the sufficiency of shelf angle floors is given in Newman (1993). BS 5950 Part 8 places the following limitations on shelf angle floors:

1. The connections at the end of the beam to any stanchions should be within the depth of the slab or protected to the same standard as the supporting member.
2. The supporting angles should be checked to ensure that their moment capacity based on the elastic section modulus of the leg is sufficient in a fire to resist the loads applied from precast units. A strength reduction factor corresponding to 1,5% proof strain should be used.

3. The weld on the upper face of the angle should be designed to resist both the applied vertical and longitudinal shears. The weld on the underside of the angle is to be neglected.

8.3.2.1 Calculation of temperature response

The beam is divided into a series of zones corresponding to the bottom flange, the exposed web, the exposed part of the shelf angles, and the part of the web and vertical legs of the shelf angles that attain temperatures of above 300°C. The temperature zones and alternative positions of the 300°C isotherm are given in Figure 8.8.

The temperature of the bottom flange θ_1 is determined from Table 10 of BS 5950 Part 8. The remaining temperatures are determined from θ_1 using Table C.1 of the code (Table 8.4). The position of the 300°C isotherm x_{300} above the top surface of the horizontal leg of the shelf angle is determined from a reference temperature θ_R by the following equation:

$$x_{300} = \frac{\theta_R - 300}{G} \tag{8.75}$$

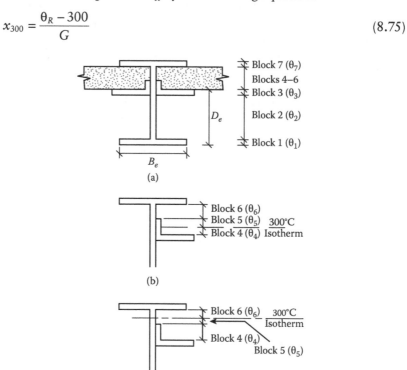

(a)

(b)

(c)

Figure 8.8 Temperature block definitions for shelf angle floor beams: (a) layout of blocks and corresponding temperatures (θ), (b) 300 isotherm in flange of angle, and (c) 300 isotherm in web of beam above angle (*Source*: BS 5950 Part 8. © British Standards Institute.)

Table 8.4 Block temperatures

	Block temperatures for given fire endurance period (min)								
	30			60			90		
Aspect ratio	θ_2 (°C)	θ_3 (°C)	θ_R (°C)	θ_2 (°C)	θ_3 (°C)	θ_R (°C)	θ_2 (°C)	θ_3 (°C)	θ_R (°C)
$\dfrac{D_e}{B} \le 0,6$	θ_1-140	475	350	θ_1-90	725	600	θ_1-60	900	775
$0,6 < \dfrac{D_e}{B} \le 0,8$	θ_1-90	510	385	θ_1-60	745	620	θ_1-30	910	785
$0,8 < \dfrac{D_e}{B} \le 1,1$	θ_1-45	550	425	θ_1-30	765	640	θ_1	925	800
$1,1 < \dfrac{D_e}{B} \le 1,5$	θ_1-25	500	425	θ_1	765	640	θ_1	925	800
$1,5 < \dfrac{D_e}{B}$	θ_1	550	425	θ_1	765	640	θ_1	925	800

Source: Table C.1 of BS 5950 Part 8. © British Standards Institution.

where G takes values of 2,3; 3,8; or 4,3°C/mm for fire resistance periods of 30, 60, or 90 min, respectively. The temperatures of blocks 4, 5 and 6 are calculated from the following equation:

$$\theta_x = \theta_R - Gx \ge 300°C \tag{8.76}$$

8.3.2.2 Calculation of moment capacity

With steel strengths reduced by factors determined for a 1,5% strain level, conventional plastic analysis is used to determine the moment capacity with the beam and angles replaced by rectangles and the fillets ignored.

Example 8.8: Design of shelf angle floor

Determine the moment capacity after a 30-min exposure to the standard furnace test for a shelf angle floor fabricated from a 406 × 178 × 54 Grade S275 UB with 125 × 75 × 12 Grade S355 angles with upper face of the long leg of the angles 282 mm above the soffit of the beam.

The example chosen is reported in Data Sheet No. 35 in Wainman and Kirby (1988). It should be noted that the fire endurance test was terminated at a limiting deflection of span/30. Thus the beam would still have had the ability to resist the applied moment for slightly longer than the time quoted in the test report.

Determination of steel temperatures — The flange thickness of the beam is 10,9 mm. Thus from Table 10 (BS 5950 Part 8), $\theta_1 = 760°C$

(flange thickness of 11 mm). The ratio $D_e/B_e = 282/177,7 = 1,59$. From Table C.1 (Table 8.4), $\theta_2 = \theta_1 = 760°C$; $\theta_3 = 550°C$; $\theta_R = 425°C$. From Equation (8.75) calculate x_{300},

$$x_{300} = \frac{\theta_R - 300}{G} = \frac{425 - 300}{2,3} = 54,3 \text{ mm}$$

This falls within the vertical leg of the angle. The upper portion of the angle and web are at temperatures below 300°C. Calculate the temperature θ_4 using Equation (8.76) at the mid height of the angle:

$$\theta_4 = \theta_R - Gx = 425 - 2,3 \times (54,3/2) = 363°C$$

The relevant strength reduction factors are found in Table 1 of BS 5950 Part 8. The relevant zones and dimensions for calculating temperatures are given in Figure 8.9.

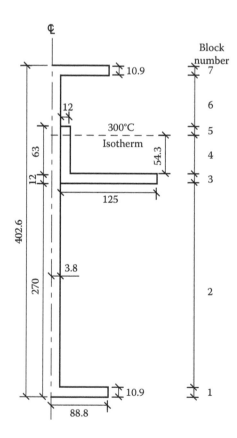

Figure 8.9   Temperature block data and section dimensions for Example 8.6.

Table 8.5 Calculation of moment capacity

Zone	f_{yk} (MPa)	θ (°C)	$k_{y,\theta}$	P_i (kN)		I_a (mm)	M_i (kNm)
1	275	760	0,143	37,94		5,45	0,207
2	275	760	0,143	39,23		140,45	5,510
3a	355	550	0,612	325,89		276,0	89,946
3b	275	550	0,612	7,78		276,0	2,147
4a	355	363	0,965	223,22	98,29 (t)	293,40	28,838
					124,93 (c)	321,11	−40,116
4b	275	363	0,965	55,48	24,43 (t)	293,40	7,168
					31,05 (c)	321,11	−9,970
5a	355	<300	1,0	37,06		340,65	−12,624
5b	275	<300	1,0	9,21		340,65	−3,137
6	275	<300	1,0	49,44		368,35	−18,211
7	275	<300	1,0	266,33		397,15	−105,773
			ΣP_i	1051,58		ΣM_i	−56,015

Note: c = compression, t = tension.

Calculation of moment capacity — The simplest method is to determine the total force on the section by summing the strengths of the individual zones. Then by calculating the tension or compression force resultant as half this value, the resultant position of the plastic neutral axis may be determined and hence moments may be taken about the soffit of the beam to determine the temperature reduced strength capacity.

The calculations are carried out for one half of the beam in Table 8.5. The total force on the section is 1051,58 kN, thus the compressive or tensile force is numerically equal to $0,5\Sigma P_i$ (= 525,79 kN); $P_1 + P_2 + P_3 = 403,06$ kN, $P_1 + P_2 + P_3 + P_4 = 681,76$ kN. Thus the neutral axis lies within Zone 4.

$$0,5\Sigma P_i - (P_1 + P_2 + P_3) = 525,79 - 403,06 = 122,73 \text{ kN}$$

$$x = \frac{122,73 \times 10^3}{0,965(12 \times 355 + 3,85 \times 275)} = 23,91 \text{ mm}$$

Zone 4 is partially in tension and partially in compression, thus dividing it into two sections as per column 6 of Table 8.9. Moments can then be taken about the soffit of the beam to determine ΣM_i. For half the beam, ΣM_i is numerically equal to 56,015 kNm; thus the total capacity of the beam is 112 kNm.

The test reported in Wainman and Kirby lasted 29 min with a deflection of L/30, with the beam carrying a total uniformly distributed load of 271,4 kN on a span of 4,5 m giving an applied moment of 152,7 kNm. Part of the discrepancy is due to the temperatures measured in the test and the assumed

Table 8.6 Comparison of assumed and measured temperatures (°C)

Zone	Position	Measured	Assumed
1	Bottom flange	733	760
2	Exposed web	715	760
3	Underside of shelf angle	571	550
4	Upstand of shelf angle	368	363
5	Unexposed web	167	<300
6	Top flange	97	<300

values in the calculations. These are summarized in Table 8.6. The test temperatures are lower in the bottom flange and web, thus increasing the values of P_i, but slightly higher in the underside and upstand of the shelf angle, thus decreasing the values of P_i. The latter will have a lesser effect on the moment capacity as they are closer to the centroidal axis.

Wainman and Kirby also reported that the actual strengths were 280 MPa for the Grade 275 beam and 381 MPa for the Grade 355 shelf angle. These strengths would also increase the capacity of the section. Thus the calculated moment capacity is acceptable.

8.4 STAINLESS STEEL

The principles of the design of stainless steel in fire are similar to those of normal carbon steel (Baddoo and Burgan, 2001). The major difference is in the determination of the strength reduction factors that depend upon the grade of stainless steel. If strength loss corresponding to 2% strain occurs, an additional factor $g_{2,\theta}$ is required as

$$p_{2,\theta} = p_{0,2proof,\theta} + g_{2,\theta}(U_{s,\theta} - p_{0,2proof,\theta}) \qquad (8.77)$$

where $p_{0,2proof,\theta}$ is 0,2% proof strain at temperature θ and $U_{s,\theta}$ is ultimate strength. The values of $k_{p0,2proof,\theta}$ $(= p_{0,2proof,\theta}/p_y)$ and $k_{s,\theta}$ $(= U_{s,\theta}/U_s)$ are tabulated. For the strength loss at 1% strain, the factor $g_{2,\theta}$ in Equation (8.77) is replaced by $0,5g_{2,\theta}$.

8.5 COLD-FORMED STEEL SECTIONS

The fire resistance of cold-formed steel sections depends on the strength of steel at elevated temperatures and the rate of heating of a relatively thin section. The loss of strength of cold-formed steel at elevated temperatures is generally worse than that of hot-rolled steel of the equivalent grade by 10 to 20% based on data obtained from tests (Sidey and Teague, 1988; Ranawaka and Mahendran, 2009).

In design guidance (BS 5950 Part 8), it was assumed that strains at least 1,5% can occur in beams, and strains at least 0,5% can occur in columns at the deflections corresponding to failure in a fire test. These strain limits are used in assessing the member strength at the fire limit state. The lower strain limit for columns is due to the effect of lateral buckling of columns.

The elastic modulus of cold-formed steel reduces more or less at the same rate as the strength of cold-formed steel at 0,5% strain. This effect leads to an increase in deflection due to loss of stiffness of the section.

Cold-formed steel members are often located within floors or walls and are heated from one side only. The compartment boundaries must provide the necessary barriers to prevent the passage of heat and smoke, and insulation criteria often determine their design. Typically, one layer of 12,5-mm fire resistance gypsum board is required beneath floor joists or on all sides of a load-bearing wall for 30-min fire resistance and two layers for 60-min fire resistance (SCI, 1993).

For individual beams and columns heated from all sides, the thicknesses of fire protection are based on rates of heating to reach a certain limiting temperature. The limiting temperatures of cold-formed steel members in fire conditions are functions of member type and load ratio. As a conservative approach, three limiting temperatures of 550, 500, and 450°C have been recommended for beams supporting concrete slabs, beams supporting timber floors or stocky columns in walls, and slender columns, respectively (SCI, 1993).

In recent years, effort has been made to develop performance-based design methods for cold-formed steel members at elevated temperatures. For example, Chen and Young (2007) analysed the behaviour of cold-formed steel-lipped channel columns at elevated temperatures using finite element methods. The analysis considers the reduced material properties of cold-formed steel at elevated temperatures and also the local and overall geometric imperfections.

The finite element analysis results were validated using test results both at room and elevated temperatures. The numerical results were also compared with the design strengths obtained using the effective width method specified in the North American specification and the direct strength method specified in a supplement to the North American specification by substituting reduced material properties. Both the effective width method and direct strength method using the reduced material properties provided conservative designs for assessing column strengths.

The distortional buckling strength of cold-formed steel members with uniform and nonuniform cross-section temperature distributions was studied by Shahbazian and Wang (2012) using ABAQUS software. The numerical results were also compared with the design strengths predicted using the direct strength method. It was found that the distortional buckling curve proposed in the direct strength method for ambient temperature can be still applied to heated columns with uniform temperature distributions in its cross section. To apply the direct strength method to heated columns with nonuniform temperature distributions, a new distortional buckling curve was proposed in their paper.

More recently, Kankanamge and Mahendran (2012) investigated the lateral–torsional buckling behaviour of simply supported cold-formed steel-lipped channel beams subjected to uniform bending at uniform elevated temperatures using nonlinear finite element analysis. The study showed that the lateral-torsional buckling capacity of a beam is influenced strongly by the level of nonlinearity of the stress–strain curve at elevated temperatures and that the level of nonlinearity can be represented by the ratio of the limit of proportionality to the yield strength of steel. The study also showed that it is inadequate to use a single buckling curve to obtain the ultimate moment capacities of cold-formed steel beams at different elevated temperatures. The conclusion was that design codes should include different buckling curves for the fire design of cold-formed steel members in different temperature ranges.

8.6 METHODS OF PROTECTION

8.6.1 Types of protection

The five common methods of fire protecting steelwork are board systems, spray systems, intumescent paints, brickwork/blockwork, and concrete encasement.

8.6.1.1 Board systems

These are now the most common, certainly within the UK, as they are quick and easy to fix and also produce relatively little mess because they use a dry process. Essentially the boards of mineral fibre or plaster board are secured together to form a box system for column protection or fixed to wood battens inserted between the flanges of beam systems. A further advantage of board systems is that building finishes may be applied directly, thus speeding construction.

8.6.1.2 Spray protection

The protection in the form of a gunite-type material is wet sprayed onto a member. The process can produce considerable mess. If used on exposed steelwork, it will require additional finishing by plastering or boarding to achieve the standard required for decorating. Generally spray systems are limited to beams used in areas with false ceilings.

8.6.1.3 Intumescent paints

When these paint systems are exposed to heat, they foam up and provide insulation in the form of aerated carbon similar to the char layer that forms on timber. The advantage is that these paints may be used to provide a final architectural finish on exposed steelwork. The drawbacks concern maintenance. Continual inspection is needed to examine the integrity of the paint layer. Replacement is required if a fire occurs. Intumescent paints also resist the effects of blast and hydrocarbon fires (Allen, 2006)

8.6.1.4 Brickwork and blockwork

This may be used as a convenient way of providing fire protection to free-standing stanchions or stanchions partially or totally built into masonry walls. The nonload-bearing masonry effectively acts as a heat sink and thus reduces the average temperatures within the steelwork. Freestanding stanchions with nonstructural infilling to the webs are capable of providing 30 to 60 min of fire resistance (Building Research Establishment, 1986). It is possible to provide structural infill to assist in load carrying, i.e., the steel column acts compositely with the infill. The design of such columns is covered in Chapter 9.

8.6.1.5 Concrete encasement

This method is now essentially obsolete partly owing to the large additional permanent loading imposed on the structure and partially because encasement is a long slow process. Formwork is needed for the concrete along with some reinforcement and curing time. Thus it is probably more economic to use fully reinforced concrete construction. Some of the problems can be mitigated by the use of precast concrete protection, but concrete will still need to be cast around the beam-to-column (or beam-to-beam) connections.

8.6.1.6 Manufacturer data

A compendium of manufacturer data is provided in the UK by the Loss Prevention Certification Board (2012). The thermal behaviour of insulation is conventionally determined using a failure temperature of 550°C for columns with normal protection materials and 620°C for beams supporting concrete floors over a limited range of section sizes. Thus manufacturers' design data imply such failure temperatures. This approach, is however, conservative generally and steel members can achieve higher temperatures than these at failure.

To allow test data to be extrapolated over a greater range of section sizes, an approach using linear regression was evolved (Barnfield, 1986). The standard fire resistance period t_{fi} is given by

$$t_{fi} = a_0 + a_1 \frac{V}{A_m} + a_2 d_p \tag{8.78}$$

where a_0, a_1 and a_2 are parameters relating to a given protection material. Equation (8.78) may be rewritten as:

$$d_p = \frac{t_{fi} - a_0 - a_1 \dfrac{V}{A_m}}{a_2} \tag{8.79}$$

These equations may then be used to predict values of d_p required on sections other than those tested. EN 13381-4 allows a similar but more complex approach as it also includes the temperature of the steel, and t_{fi} is now given by

$$t_{fi} = a_0 + a_1 d_p + a_2 \frac{d_p}{\dfrac{A_m}{V}} + a_3\theta + a_4 d_p\theta + a_5 d_p \frac{\theta}{\dfrac{A_m}{V}} + a_6 \frac{\theta}{\dfrac{A_m}{V}} + \frac{a_7}{\dfrac{A_m}{V}} \tag{8.80}$$

where a_0 to a_7 are regression coefficients and θ is the steel temperature. Equation (8.80) may also be rewritten as:

$$d_p = \frac{t_{fi} - a_0 - a_3\theta - a_6 \dfrac{\theta}{\dfrac{A_m}{V}} - \dfrac{a_7}{\dfrac{A_m}{V}}}{a_1 + a_4\theta + \dfrac{a_2}{\dfrac{A_m}{V}} + a_5 \dfrac{\theta}{\dfrac{A_m}{V}}} \tag{8.81}$$

Note that tests are done generally on I- or H-sections and modifications may be required under certain circumstances.

Structural hollow sections — The thicknesses d_p obtained from manufacturers' data need to be increased to $d_{p,mod}$ as follows:

$A/V < 250$:

$$d_{p,\text{mod}} = d_p \left(1 + 0,001 \frac{A}{V} \right) \tag{8.82}$$

$250 \leq A/V \leq 310$:

$$d_{p,\text{mod}} = 1,25 d_p \tag{8.83}$$

Castellated beams and beams with circular web openings — The issue with both these types of beams is that the web may heat up faster than the bottom flange and thus cause instability problems. Indeed the web temperature may exceed the bottom flange temperature.

For castellated beams, tabulated fire protection thicknesses of normal protection materials and intumescents should be increased by 20% with the A/V value obtained from the parent beam section (Loss Prevention Certification Board, 2012). For beams with circular web openings, the 20% rule applies only to conventional protection materials.

Newman et al. (2005) indicate that tabular data provided in SCI Advisory Desk note AD 269 (SCI, 2003) are limited to cases where the diameter of the circular openings is limited to 0,8 times the beam depth; the end post widths are at least 0,3 times the diameter of the openings; the spacing of openings to the opening diameter is greater than 0,4; and the beams are subject to typical office building loads. AD 269 states that the A_m/V ratio for three-sided heating should be based on the A_m/V ratio of the bottom tee-section at the centre line of the circular web opening. Table 1 of AD 269 gives multiplication factors for increase of protection thicknesses.

8.6.2 Connections

The resistance of connections should be checked by using temperature-reduced design values for bolts in shear or tension and for welds. The appropriate strength reduction factors are given in Annex D of EN 1993-1-2. The temperature of a connection may be assessed using the local A/V values of parts forming the connection. As an approximation, a uniform temperature may be assumed and calculated on the basis of the maximum A/V values of the members framing into the connection. For beam-to-beam or beam-to-column connections where the beams are supporting any type of concrete floor, the temperature of the bottom flange of the beam at mid-span may be used to determine the temperature in the connection.

The temperature distribution θ_h within the connection can be determined as follows:

For $D \leq 400$ mm:

$$\theta_h = 0{,}88\theta_0 \left[1 - 0{,}3\frac{h}{D} \right] \qquad (8.84)$$

For $D > 400$ mm and $h < D/2$:

$$\theta_h = 0{,}88\theta_0 \qquad (8.85)$$

For $D > 400$ mm and $h \geq D/2$:

$$\theta_h = 0{,}88\theta_0 \left[1 + 0{,}2\left(1 - \frac{2h}{D} \right) \right] \qquad (8.86)$$

where h is the height measured from the soffit, D is the depth of the beam, and θ_0 is the temperature of the bottom flange remote from the connection.

8.6.3 Aging and partial loss of protection

All the above calculation methods assume that insulating material remains in place, shows no deterioration in properties with age, and is effective for the whole of the fire resistance period required. Clearly there exists the requirement to ensure that the standard of workmanship for the execution of fire protection work is as high as the standard imposed for the remainder of the execution of the works.

8.6.3.1 Aging effects

A series of tests conducted over some 6 years by Kruppa (1992) indicated that the common methods of fire protection of structural steelwork revealed little evidence of deterioration in the insulation properties over time.

8.6.3.2 Partial loss of protection

The effect of loss of protection during a fire is to reduce substantially the fire performance of the structure. This has been observed in situations where a fire occurred in partially protected structures such as Broadgate Phase 8 (SCI, 1991) and also in simulation studies (Tomacek and Milke, 1993). It was found that although a reduction of standard fire test performance is observed for any loss of fire protection, the effect is far more marked for longer fire resistance periods. Tomacek and Milke determined that for a 3-hr design fire resistance period, an 18% loss of protection to the column flanges reduced the fire resistance of a column to around 40 to 60 min and that the reduction was sensibly independent of the column size. For a 1-hr design, the effect from an 18% loss was a reduction to a fire resistance of around 30 to 40 min.

A further effect of partial loss of protection is that any structural deformations during a fire will be increased as the steel attains higher temperatures with subsequent decreases in Young's modulus. The effect on cooling will be to maintain the increased deformations despite the little resultant effect on the residual strength of the steelwork (Section 13.3.1.2).

Having thus considered noncomposite steelwork elements, it is necessary to turn to composite steel–concrete construction in Chapter 9.

Chapter 9

Composite construction

This chapter is divided into four sections discussing composite slabs with profile sheet steel decking, composite beams where the composite action is achieved through shear studs welded to the top flanges of beams with or without sheet steel profiled decking to support the concrete slabs, concrete-filled steel I- and H-section columns, and concrete-filled steel hollow section columns.

9.1 COMPOSITE SLABS

It is not a general practice to apply fire protection to the soffit of a composite slab. This means that the profile sheet steel decking is exposed directly to a fire and therefore loses its strength very rapidly. Furnace tests revealed that simply supported composite slabs are inherently capable of providing 30-min fire resistance when exposed to the standard furnace test, and the decking may be assumed to retain about 5% of its original strength (Cooke et al., 1988). However, most calculation methods are conservative in ignoring this residual strength and determine the strength of the slab system purely on contributions of the concrete slab and any flexural reinforcement. As with reinforced concrete construction, the slab is required to satisfy both the load-bearing capacity and insulation limit states. The latter is generally satisfied by specifying overall slab depths.

9.1.1 Insulation requirement

Two approaches may be used (Annex D, EN 1994-1-2). The first is a calculation method and the second is a simple approach based on effective thickness.

9.1.1.1 Calculation approach

The time to failure of the insulation limit state t_i is given by

$$t_1 = a_0 + a_1 h_1 + a_2 \Phi + a_3 \frac{A}{L_r} + a_4 \frac{1}{l_3} + a_5 \frac{A}{L_r} \frac{1}{l_3} \qquad (9.1)$$

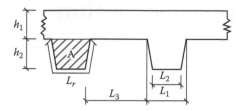

Figure 9.1 Rib geometry for calculation of A/Lr. (Source: EN 1994-1-2. © British Standards Institute.)

where a_0 to a_5 are coefficients given in Table D.1 (EN 1994-1-2), A/L_r is a rib geometry factor, Φ is the view factor for the upper flange, and l_3 is the width of the upper flange (Figure 9.1). A/L_r and Φ are given by

$$\frac{A}{L_r} = \frac{h_2 \frac{l_1 + l_2}{2}}{l_2 + 2\sqrt{h_2^2 + \left(\frac{l_1 - l_2}{2}\right)^2}}$$

(9.2)

$$\Phi = \frac{\sqrt{h_2^2 + \left(l_3 + \frac{l_1 - l_2}{2}\right)^2} - \sqrt{h_2^2 + \left(\frac{l_1 - l_2}{2}\right)^2}}{l_3}$$

(9.3)

9.1.1.2 Effective thickness

This is satisfied by the determination of a minimum effective thickness h_{eff} which is given as:

For $h_2/h_1 \leq 1{,}5$ and $h_1 > 40$ mm:

$$h_{eff} = h_1 + 0{,}5h_2 \frac{l_1 + l_2}{l_1 + l_3}$$

(9.4)

For $h_2/h_1 > 1{,}5$ and $h_1 > 40$ mm:

$$h_{eff} = h_1 \left(1 + 0{,}75 \frac{l_1 + l_2}{l_1 + l_3}\right)$$

(9.5)

where h_1, h_2, l_1, l_2, and l_3 are defined in Figure 9.2. If $l_3 > 2l_1$, $h_{eff} = h_1$. The required values of h_{eff} for normal weight concrete are given in Table 9.1. For lightweight concrete, these values should be reduced by 10%.

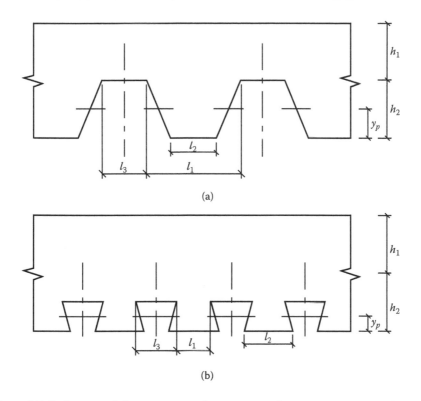

(a)

(b)

Figure 9.2 Definition of dimensions to determine insulation requirements of composite deck for (a) trapezoidal section and (b) dovetail section. (*Source:* Adapted from Figure 4.1 of EN 1994-1-2. © British Standards Institute.)

Table 9.1 Values of effective thickness

Standard fire resistance (min)	Minimum effective thickness (mm)
R30	$60-h_3$
R60	$80-h_3$
R90	$100-h_3$
R120	$120-h_3$
R180	$150-h_3$
R240	$175-h_3$

Note: h_3 is thickness of the screed layer if present.

Source: Table D.6 of EN 1994-1-2.

9.1.2 Load-bearing capacity

A slab should be analyzed assuming the formation of plastic hinges at mid-span and the supports where appropriate. For end spans where continuity exists at one support only, it is sufficiently accurate to assume the hinge in the sagging régime occurs at mid-span.

9.1.2.1 Calculation of moment capacity

9.1.2.1.1 Mids-span (sagging moments)

The strength of concrete should be taken as its ambient strength. This is not unreasonable as the depth of the concrete stress block will be small because the tensile force of the reinforcement that the concrete in compression is required to balance is low. This means that it is very unlikely that the concrete strength will be affected by excessive temperature rise. The temperature θ_a in the bottom flange, web, or top flange of the decking is given by

$$\theta_a = b_0 + b_1 \frac{1}{l_3} + b_2 \frac{A}{L_r} + b_3 \Phi + b_4 \Phi^2 \tag{9.6}$$

where b_0 to b_4 are coefficients dependent upon the type of concrete and the fire resistance period and are given in Table D.2 of EN 1994-1-2. The temperature of the reinforcement θ_s is given by

$$\theta_s = c_0 + c_1 \frac{u_3}{h_1} + c_2 z + c_3 \frac{A}{L_r} + c_4 \alpha + c_5 \frac{1}{l_3} \tag{9.7}$$

where c_0 to c_5 are coefficients dependent upon the fire period and concrete type and are given in Table D.3 of EN 1994-1-2; u_3 and z characterize the position of the reinforcing bar; and α is the angle of the web (in degrees). The parameter z is given by

$$\frac{1}{z} = \frac{1}{\sqrt{u_1}} + \frac{1}{\sqrt{u_2}} + \frac{1}{\sqrt{u_3}} \tag{9.8}$$

where u_1, u_2 and u_3, measured in millimeters, are defined in Figure 9.3. For a re-entrant profile where perpendicular distances u_2 and u_3 are undefined owing to the position of the rebar, u_2 and u_3 are taken to the nearest corner of the dovetail. The moment capacity is then determined using conventional reinforced concrete theory.

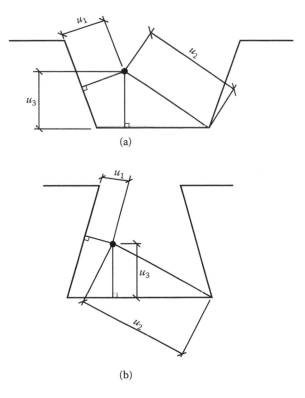

Figure 9.3 Definition of parameters required to determine reinforcement temperatures within profiled deck for (a) trapezoidal section and (b) dovetail section.
(*Source:* EN 1994-1-2. © British Standards Institute.)

9.1.2.1.2 Supports (hogging moments): method I

The contribution of the steel decking may be ignored (as it is small). Only
the concrete cross section with temperatures less than a limiting value of
θ_{lim} is considered as contributing to the moment capacity (see Figure 9.4).

$$\theta_{\lim} = d_0 + d_1 N_s + d_2 \frac{A}{L_r} + d_3 \Phi + d_4 \frac{1}{l_3} \tag{9.9}$$

where d_0 to d_4 are coefficients given in Table D.3.3 of EN 1994-1-2 and N_s
is the force in the hogging reinforcement. The positions I to VI are defined
by the following coordinates:

$$X_I = 0 \tag{9.10}$$

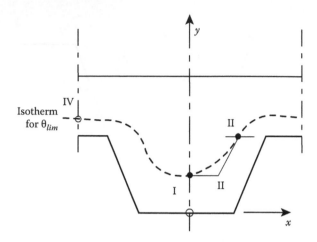

Figure 9.4 Idealization of isotherm for θ_{lim}. (*Source:* EN 1994-1-2. © British Standards Institute.)

$$Y_I = Y_{II} = \cfrac{1}{\left(\cfrac{1}{z} - \cfrac{4}{\sqrt{l_1 + l_3}}\right)^2} \tag{9.11}$$

$$X_{II} = \frac{l_2}{2} + \frac{Y_I}{\sin\alpha}(\cos\alpha - 1) \tag{9.12}$$

$$X_{III} = \frac{l_1}{2} - \frac{b}{\sin\alpha} \tag{9.13}$$

$$Y_{III} = b_2 \tag{9.14}$$

$$X_{IV} = \frac{l_1}{2} \tag{9.15}$$

$$Y_{IV} = b_2 + b \tag{9.16}$$

where

$$\alpha = \arctan\left(\frac{2b_2}{l_1 - l_2}\right) \tag{9.17}$$

$$a = \left(\frac{1}{z} - \frac{1}{\sqrt{b_2}}\right)l_1 \sin\alpha \tag{9.18}$$

$$b = \frac{l_1}{2}\sin\alpha\left(1 - \frac{\sqrt{a^2 - 4ac}}{a}\right) \tag{9.19}$$

For $a \geq 8$:

$$c = -8(1 + \sqrt{1+a}) \tag{9.20}$$

For $a < 8$:

$$c = 8(1 + \sqrt{1+a}) \tag{9.21}$$

The value of z should be determined using Equation (9.7) with $\theta_R = \theta_{lim}$ and $u_3/h_2 = 0{,}75$. If $Y_I > h_2$, the concrete in the ribs may be ignored. A conservative approach to determine the positions of isotherms in a 100-mm thick normal weight concrete is given in Table 9.2 (from Table D.5 of EN 1994-1-2). For lightweight concrete, the temperatures in Table D.5 may be reduced by 10%.

9.1.2.1.3 Supports (hogging moments): method 2.

The profiled concrete slab is replaced by a flat slab having a depth equal to h_{eff} measured from the top of the slab excluding any nonstructural screed.

Table 9.2 Concrete temperature distribution in slab

Depth	Temperature θ_c (°C) for fire duration					
(mm)	30 min	60 min	90 min	120 min	180 min	240 min
5	535	705				
10	470	642	738			
15	415	581	681	754		
20	350	525	627	697		
25	300	469	571	642	738	
30	250	421	519	591	689	740
35	210	374	473	542	635	700
40	180	327	428	493	590	670
45	160	289	387	454	549	645
50	140	250	345	415	508	550
55	125	200	294	369	469	520
60	110	175	271	342	430	495
80	80	140	220	270	330	395
100	60	100	160	210	260	305

Source: Table D.5 of EN 1994-1-2.

The temperature profile in the concrete is given by the isotherm values in Table 9.2, where the depth is measured from the bottom of the equivalent deck. The strength reduction factors for normal weight concrete are given in Table 5.8.

Where appropriate, the reinforcement strength should be reduced using the concrete temperature at the level of the reinforcement. The moment capacity must then be calculated using basic theory, but with an iterative procedure as the depth of the neutral axis is not known *a priori*.

Example 9.1: Fire engineering design of composite slab

The composite slab detailed in Figure 9.5 is to be checked under a 90-min fire resistance period. The concrete is Grade 25/30 normal weight with structural mesh over the supports to resist hogging moments. The pro-filed deck is Richard Lee's Super Holorib deck of dovetail profile.

Insulation — EN 1994-1-2 allows two approaches. In this example both will be used, although clearly in practice only one is necessary. The dimensions of the deck and sheeting are given in Figure 9.5.

(1) **Effective depth approach (Clause D.4, EN 1994-1-2).** First deter-mine h_2/h_1: $h_2/h_1 = 51/59 < 1,5$ and $h_1 > 40$ mm, so use Equation (9.4):

$$h_{eff} = h_1 + 0,5h_2 \frac{l_1 + l_2}{l_1 + l_3} = 59 + 0,5 \times 51 \frac{112 + 138}{112 + 38} = 101 \text{ mm}$$

From Table 9.1 with no screed (i.e., $h_3 = 0$), the minimum value of h_{eff} is 100 mm, i.e., the slab just satisfies the insulation requirement.

(2) **Calculation approach (Clause D.1, EN 1994-1-2).** First determine parameter A/L_r from Equation (9.2):

$$\frac{A}{L_r} = \frac{h_2 \frac{l_1 + l_2}{2}}{l_2 + 2\sqrt{h_2^2 + \left(\frac{l_1 - l_2}{2}\right)^2}} = \frac{51 \frac{112 + 138}{2}}{138 + 2\sqrt{51^2 + \left(\frac{112 - 138}{2}\right)^2}} = 26,2 \text{mm}$$

Determine Φ from Equation (9.3):

$$\Phi = \frac{\sqrt{h_2^2 + \left(l_3 + \frac{l_1 - l_2}{2}\right)^2} - \sqrt{h_2^2 + \left(\frac{l_1 - l_2}{2}\right)^2}}{l_3}$$

$$= \frac{\sqrt{51^2 + \left(38 + \frac{112 - 138}{2}\right)^2} - \sqrt{51^2 + \left(\frac{112 - 138}{2}\right)^2}}{38} = 0,11$$

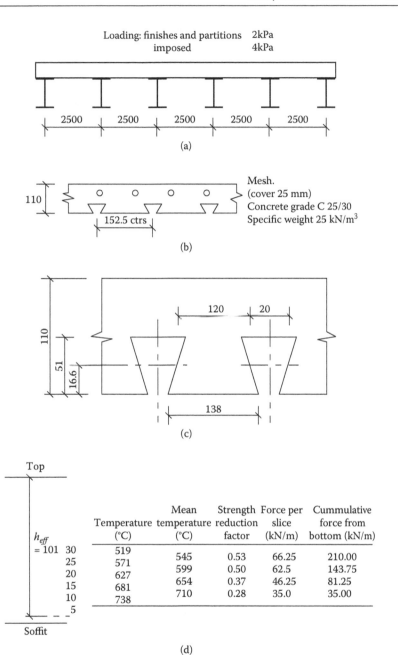

Figure 9.5 Design data for determination of fire performance of composite deck (Example 9.1): (a) longitudinal view, (b) cross section, (c) calculation of h_{eff} and (d) calculation of compressive force.

With the values of a_0 to a_5 taken from Table D.1 of EN 1994-1-2, Equation (9.1) is evaluated as

$$t_1 = a_0 + a_1 h_1 + a_2 \Phi + a_3 \frac{A}{L_r} + a_4 \frac{1}{l_3} + a_5 \frac{A}{L_r} \frac{1}{l_3}$$

$$= -28,8 + 1,55 \times 59 - 12,6 \times 0,11 + 0,33 \times 26,2 - 735 \times \frac{1}{38} + 48 \times \frac{26,2}{38} = 84 \, \text{min}$$

This is less than the required 90 min (albeit by a small margin). It would possibly be expected that the effective thickness approach would have been more conservative!

Strength — Using ψ values from EN 1991-1-2 applied to the variable load, $M_{fi,Ed} = ((0,11 \times 25 + 2,0) + 0,3 \times 4,0) \times 2,5^2/8 = 4,65$ kNm/m.

Interior span — Replace the dovetail profile slab by a slab of depth h_{eff} measured from the top of the slab (Figure 9.5). The temperature at h_{eff} minus centroidal distance to top steel, i.e., at 102 – 30 = 72 mm is around 240°C (Table 9.2). From Equation (7.26), the reinforcement suffers no strength reduction.

Two options can be considered: (1) design and detail the mesh such that no bottom reinforcement is required to carry any sagging moments, or (2) design and detail the mesh such that bottom reinforcement is only required in the end span. Although the former is more economical and more practical on site, both solutions will be investigated here to demonstrate the principles involved.

Method A (no bottom reinforcement) — Supply B283 mesh. Force in reinforcement $F_t = 283 \times 500 = 141,5$ kN. This must be balanced by the force in the strength-reduced concrete. Using a series of 5-mm deep strips in the concrete, determine the concrete temperatures at 0, 10, 15 mm etc. up from the soffit and calculate the temperatures at the mid-points of the strips together with the strength reduction factors to enable the concrete force in each strip to be determined. Then take sufficient 5-mm strips to balance the concrete and steel force.

This calculation has been carried out in Figure 9.5, where the bottom 10 mm of concrete has been ignored since there are no data for 0 and 5 mm in Table 9.2 for concrete temperatures at 90-min exposure.

Assuming that where a partial depth strip occurs, the concrete strength reduction factor is taken as that for the whole 5-mm strip; the total depth of the concrete x required to balance the tensile force in the steel is given by

$$x = 25 + \frac{(141,5 - 81,25) \times 10^3}{1000 \times 0,50 \times 25} = 24,82 \, \text{mm}$$

Taking moments about the reinforcement, $M_{fi,d}$ is given as

$$M_{fi,Rd^-} = 35,0(72 - 12,5) + 81,25(72 - 17,5) + 60,25\left(72 - 20 - \frac{4,82}{2}\right)$$

$$= 9,5 \, \text{kNm/m}$$

This is around double the free bending moment of 4,65 kNm/m, and thus will be adequate over all the spans, including the end span.

Method B (bottom reinforcement in end span) — Supply B196 mesh. Force in reinforcement $F_t = 196 \times 500 = 98$ kN. This must be balanced by the force in the strength reduced concrete. Using a series of 5-mm deep strips in the concrete, determine the concrete temperatures at 0, 10, 15 mm etc. up from the soffit and calculate the temperatures at the mid-points of the strips together with the strength reduction factors to enable the concrete force in each strip to be determined. Then take sufficient 5-mm strips to balance the concrete and steel force.

This calculation has been carried out in Figure 9.5, where the bottom 10 mm of concrete has been ignored since there are no data for 0 and 5 mm in Table 9.2 for concrete temperatures at 90 min exposure.

Assuming that where a partial depth strip occurs, the concrete strength reduction factor is taken as that for the whole 5 mm strip, the total depth of the concrete x required to balance the tensile force in the steel is given by:

$$x = 20 + \frac{(98 - 81,25) \times 10^3}{1000 \times 0,50 \times 25} = 21,34 \text{ mm}$$

Taking moments about the reinforcement, M_{fi,d^-} is given as

$$M_{fi,Rd^-} = 35,0(72 - 12,5) + 81,25(72 - 17,5) + 16,75\left(72 - 20 - \frac{1,34}{2}\right)$$

$$= 7,37 \text{ kNm/m}$$

This is greater than the free bending moment of 4,65 kNm/m, and thus will be adequate for internal spans. If the hinge occurs in the centre of the end span, the required moment is approximately $4,65 - 7,41/2 \approx 1$ kNm/m. It will be conservative to ignore any contribution of the decking; then the sagging reinforcement can be designed as if for a slab, but with the steel strength taken as $k_y(\theta)f_y$. Place the required reinforcement level with the top of the dovetail, i.e., the effective depth is h_1 or 59 mm.

Determination of θ_s using Equation (9.7) — Determine z first. $u_1 = 51$ mm. $u_2 = u_3 = 0,5(150 - 38) = 56$ mm. From Equation (9.8) $1/z$ is given by

$$\frac{1}{z} = \frac{1}{\sqrt{u_1}} + \frac{1}{\sqrt{u_2}} + \frac{1}{\sqrt{u_3}} = \frac{1}{\sqrt{51}} + \frac{1}{\sqrt{56}} + \frac{1}{\sqrt{56}} = 0,407$$

Thus $z = 2,46$ mm.

$$\theta_s = c_0 + c_1 \frac{u_3}{h_1} + c_2 z + c_3 \frac{A}{L_r} + c_4 \alpha + c_5 \frac{1}{l_3}$$

$$= 1342 - 256 \frac{51}{51} - 235 \times 2,56 - 5,3 \times 26,2 + 1,39 \times 104 - \frac{1267}{38} = 483 °C$$

From Equation (7.27), $k_y(\theta) = 0{,}80$. Noting that for reinforced concrete design in the fire limit state the values of γ_c, γ_s, and α_{cc} are all 1,0, and that for $f_{ck} = 25$ MPa, $\eta = 1{,}0$ and $\lambda = 0{,}8$, Equation (6.15) of Martin and Purkiss (2006) reduces to:

$$\frac{A_s k_y(\theta) f_y}{bd f_{ck}} = 1{,}0 - \sqrt{1 - 2\frac{M}{bd^2 f_{ck}}}$$

$M/bd^2 f_{ck} = 1{,}0 \times 10^6/(1000 \times 59^2 \times 25) = 0{,}0115$ and $A_s k_y(\theta)/bd f_{ck} = 0{,}0116$ or $A_s = 0{,}0116 \times 1000 \times 59 \times 25/(0{,}80 \times 500) = 43$ mm²/m. Supply H6 bars in alternate ribs, i.e., $A_s = 95$ mm²/m.
$A_s k_y(\theta)/bd f_{ck} = 95 \times 0{,}80 \times 500/(1000 \times 59 \times 25) = 0{,}0258$. From Equation (6.11) of Martin and Purkiss (2006):

$$\frac{x}{d} = 1{,}25\frac{A_s k_y(\theta) f_y}{bd f_{ck}} = 1{,}25 \times 0{,}0258 = 0{,}0323$$

From Equation (6.12) of Martin and Purkiss (2006):

$$M_{Rd} = A_s k_y(\theta)(d - 0{,}4x) = 95 \times 0{,}80 \times 500(59 - 0{,}4 \times 0{,}0323 \times 59)$$

$$= 2{,}21 \text{ kNm/m}$$

Check the capacity of the end span using the equation proposed by Cooke et al. (1988):

$$M_{fi,Rd^+} + 0{,}5M_{fi,Rd^-}\left(1 - \frac{M_{fi,Rd^-}}{8M_{fi,Ed}}\right) \geq M_{fi,Ed} \qquad (9.22)$$

where M_{fi,Rd^+} and M_{fi,Rd^-} are the sagging and hogging moments and $M_{fi,Ed}$ is the free bending moment.

$$M_{fi,Rd^+} + 0{,}5M_{fi,Rd^-}\left(1 - \frac{M_{fi,Rd^-}}{8M_{fi,Ed}}\right) = 2{,}21 + 0{,}5 \times 7{,}37\left(1 - \frac{7{,}37}{8 \times 4{,}65}\right)$$

$$= 5{,}16 \text{ kNm/m}$$

This is greater than the free bending moment of 4,65 kNm/m and is therefore satisfactory.

9.2 COMPOSITE BEAMS

Figure 9.6 shows the basic configuration of the composite beam that will be considered. The two types of deck can generically be described as trapezoidal and dovetail. It is also important to distinguish the relative directions of the beam span and deck span as they may affect the temperatures within the steelwork (Newman and Lawson, 1991).

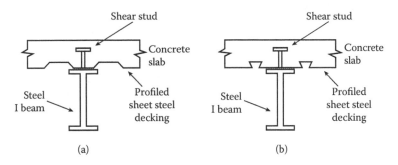

Figure 9.6 Typical steel–concrete composite beam details. Composite construction with (a) trapezoidal profile steel decking and (b) dovetail profile steel decking.

The case where beams are infilled or encased in concrete will not be considered as these are covered by a tabular approach. For the determination of structural behavior, EN 1994-1-2 adopts one of two approaches concerned with the calculation of either critical temperature or full moment capacity. In both cases, the calculation of temperatures uses the equations for non-composite steelwork that assume no thermal gradients.

9.2.1 Critical temperature approach

This method may only be used where the beam depth is less than 500 mm, the slab depth h_c exceeds 120 mm, and the composite beam system is simply supported under sagging moments only. The critical temperature θ_{cr}, for a fire resistance period R30 is determined corresponding to a critical steel strength $f_{ay,\theta cr}$ given by:

$$0,9\eta_{fi,t} = f_{ay,\theta cr}/f_{ay} \tag{9.23}$$

and for all other cases

$$1,0\eta_{fi,t} = f_{ay,\theta cr}/f_{ay} \tag{9.24}$$

where f_{ay} is the ambient yield or characteristic strength, and the load level $\eta_{fi,t}$ is given by

$$\eta_{fi,t} = \frac{E_{fi,d,t}}{R_d} \tag{9.25}$$

The temperature rise in the steel section may be determined using the A_m/V factor for the lower flange.

9.2.2 Full moment calculation

Plastic theory is used to determine the moment capacity for all but Class 4 sections. If shear connectors are provided, the compression flange may be taken as Class 1, but this may be used only to determine the sagging moment capacity. Calculations are made of the steel temperatures, assuming the web and both flanges are at uniform temperatures and the compression zone of the concrete slab is unaffected by temperature. The notation used is given in Figure 9.7.

To determine the temperatures in the flanges for unprotected members or members with contour protection, the section factor A/V_i is calculated as follows:

(a) Top flange when at least 85% of the concrete slab is in contact with the upper flange and any voids are filled with non-combustible materials:

$$\frac{A}{V} = \frac{b_2 + 2e_2}{b_2 e_2} \tag{9.26}$$

(b) Top flange when less than 85% of the deck is in contact:

$$\frac{A}{V} = 2\frac{b_2 + e_2}{b_2 e_2} \tag{9.27}$$

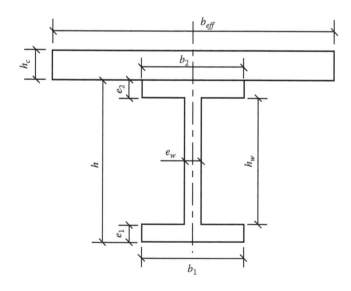

Figure 9.7 Definition of symbols for moment capacity method of EN 1994-1-2.

(c) Bottom flange

$$\frac{A}{V} = 2\frac{b_1 + e_1}{b_1 e_1} \tag{9.28}$$

where b_1, e_1, b_2, and e_2 are the widths and thicknesses of the bottom and top flanges, respectively. Note that where the overall beam depth is less than 500 mm, the web temperature may be taken equal to that in the bottom flange. For box protection, a uniform temperature over the whole section may be assumed with the A/V value taken as that for the box. For unprotected composite steel beams, Equation (8.7) is used but with a shadow factor k_{shadow} introduced in the equation to give

$$\Delta\theta_{a,t} = k_{shadow}\frac{\alpha}{c_a\rho_a}\frac{A}{V}(\theta_t - \theta_{a,t})\Delta t \tag{9.29}$$

where k_{shadow} is given by

$$k_{shadow} = [0,9]\frac{e_1 + e_2 + \frac{b_1}{2} + \sqrt{h_w^2 + \frac{(b_1 - b_2)^2}{4}}}{h_w + b_1 + \frac{b_2}{2} + e_1 + e_2 - e_w} \tag{9.30}$$

The 0,9 factor is a nationally determined parameter.

Annex E of EN 1994-1-2 details the model that may be used to determine both the sagging and hogging moment resistances of composite beams. As only the former will be considered here, the plus (+) superscript will be omitted.

The tensile capacity of the steel beam T calculated using temperature-reduced steel strengths (Table 5.6) is given by

$$T = \frac{f_{ay,\theta1}b_1e_1 + f_{ay,\theta w}b_w e_w + f_{ay,\theta2}b_2 e_2}{\gamma_{M,fi,a}} \tag{9.31}$$

where $f_{ay,\theta2}$, $f_{ay,\theta w}$, and $f_{ay,\theta1}$ are the temperature-reduced stresses in the top flange, web, and bottom flange respectively, $\gamma_{M,fi,a}$ is the materials partial safety factor applied to the steel section, and h_w and e_w are the height and thickness of the web, respectively. The tensile force T acts at a distance y_T from the bottom of the beam which is calculated from

$$y_T = \frac{f_{ay,\theta1}\frac{b_1e_1^2}{2} + f_{ay,\theta w}b_w e_w\left(e_1 + \frac{h_w}{2}\right) + f_{ay,\theta2}b_2 e_2\left(h - \frac{e_2}{2}\right)}{T\gamma_{M,fi,a}} \tag{9.32}$$

where h is the overall depth of the beam. The tensile force T is limited to

$$T \leq NP_{fi,Rd} \tag{9.33}$$

where N is the number of shear connectors in any critical length of the beam and $P_{fi,Rd}$ is temperature-reduced capacity of the shear connectors. The tensile force T is resisted by the force from the compression block in the concrete where the depth of the concrete block h_u is given by

$$h_u = \frac{T}{\dfrac{b_{eff} f_{ck}}{\gamma_{M,fi,c}}} \tag{9.34}$$

where b_{eff} is the effective width of the slab taken as the ambient design condition and f_{ck} is the ambient strength of the concrete.

If $h_c - h_u > h_{cr}$, the temperature of the concrete below 250°C and a full strength may be taken.

If $h_c - h_u \leq h_{cr}$, some layers of the concrete are at temperatures exceeding 250°C. The calculation of T is then iterative and based on using 10-mm thick layers in the temperature-affected zone, and T is then given by

$$T = \frac{b_{eff}(h_c - h_{cr}) f_{ck} + \displaystyle\sum_{i=2}^{n-1} 10 b_{eff} f_{cf,\theta i} + h_{u,n} b_{eff} f_{ck,\theta n}}{\gamma_{M,fi,c}} \tag{9.35}$$

where

$$h_u = (h_c - h_{cr}) + 10(n-2) + h_{u,n} \tag{9.36}$$

where n is the total number of layers of concrete including the top layer with a temperature less than 250°C. The position of h_{cr} and the elemental temperatures should be taken from Table D.5.

The point of application of the compression force y_F is given by

$$y_F = h + h_c - \frac{h_u}{2} \tag{9.37}$$

and the moment capacity of the section $M_{fi,Rd}$ is given by

$$M_{fi,Rd} = T\left(h + h_c - y_T - \frac{h_u}{2}\right) \tag{9.38}$$

The shear stud capacity must also be checked in that the stud capacity $P_{fi,Rd}$ is given as the lesser of

$$P_{fi,Rd} = 0,8k_{u,\theta}P_{Rd} \qquad (9.39)$$

or

$$P_{fi,Rd} = k_{c,\theta}P_{Rd} \qquad (9.40)$$

where $k_{u,\theta}$ is the temperature reduction on the ultimate strength of steel and $k_{c,\theta}$ is the concrete strength reduction factor. The values of P_{Rd} should be calculated in accordance with EN 1994-1-1 except that the partial safety factor γ_v should be replaced by $\gamma_{M,fi,v}$. The temperature θ_v in the shear studs may be taken as 0,8 of the temperature in the top flange of the steel beam, and the temperature in the concrete θ_c as 0,4 of that in the top flange.

Example 9.2: Determination of required fire protection for composite beam

Determine the thickness of sprayed gypsum plaster to give a beam 90-min fire resistance.

The design data for the composite beam are given in Figure 9.8. The dovetail decking (Richard Lee's Super Holorib) runs normal to the

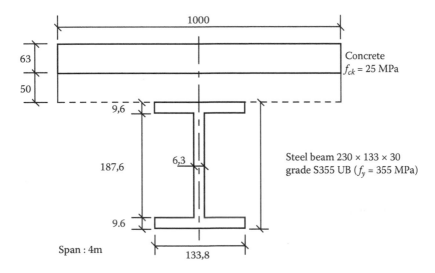

Figure 9.8 Design data for determination of fire performance of composite steel beam (Example 9.2).

span of the beam. Although no specific recommendations are made in EN 1994-1-2 for the effects of the loss of concrete in the deck, the effective thickness will be used for structural calculations. Applied loadings (actions) are:

> Imposed (variable): 4 kPa
> Dead (permanent): 2,75 kPa
> Finishes (permanent): 2,0 kPa

The beam is at an internal support of a composite deck system. From the ambient structural analysis of the decking, the application of a uniformly distributed load of q to the deck gives a beam reaction per unit run of beam of 2,98 q. Actions in the fire limit state are:

Per unit area:

$$q_{fi} = (25 \times 0,11 + 2,0) + 0,3 \times 4 = 5,95 \text{ kPa.}$$

Per unit run:

$$q_{fi} = 2,98 \times 5,95 = 17,73 \text{ kN/m.}$$

$$M_{fi,Ed} = 17,73 \times 4^2/8 = 35,46 \text{ kNm.}$$

Moment capacity approach — This solution is also iterative for both the temperature determination and the resultant moment capacity calculations. Only the final result is given here and not previous invalid determinations. Note: To comply with the units in the heat transfer equations, all dimensions are in metres. To calculate A_p/V_i values:

> Bottom flange from Equation (9.28): $A/V = 2(b_1 + e_1)/b_1e_1 = 223 \text{ m}^{-1}$.
> Top flange from Equation (9.26), when at least 85% of profile deck is in contact with beam: $A/V = (2e_2 + b_2)/b_2e_2 = 119 \text{ m}^{-1}$.
> Beam depth is less than 500 mm, so the web temperature can be taken as the lower (bottom) flange temperatures.

It was found that an insulation thickness of 15,4 mm produced the following temperatures and strength reduction factors:

> Bottom flange (and web): 834°C and 0,093
> Top flange: 691°C and 0,253

From Equation (9.31):

$$T = \frac{f_{ay,\theta 1}b_1e_1 + f_{ay,\theta w}b_we_w + f_{ay,\theta 2}b_2e_2}{\gamma_{M,fi,a}}$$

$$= 355\frac{0,093\times 9,6\times 133,8 + 0,093\times 187,6\times 6,3 + 0,253\times 9,6\times 133,8}{1,0} = 196,8\text{kN}$$

Determine y_T from Equation (9.32):

$$y_T = \frac{f_{ay,\theta 1}\dfrac{b_1 e_1^2}{2} + f_{ay,\theta w}b_w e_w\left(e_1 + \dfrac{b_w}{2}\right) + f_{ay,\theta 2}b_2 e_2\left(h - \dfrac{e_2}{2}\right)}{T\gamma_{M,fi,a}}$$

$$= \frac{355\left(0,093\dfrac{133,8 \times 9,6^2}{2} + 0,093 \times 6,3 \times 187,6\left(9,6 + \dfrac{187,6}{2}\right)\right)}{196,8 \times 10^3 \times 1,0}$$

$$+ \frac{355\left(+0,253 \times 133,8 \times 9,6\left(9,6 + 187,6 + \dfrac{9,6}{2}\right)\right)}{196,8 \times 10^3 \times 1,0}$$

$$= 140\ mm$$

As in the ambient design, the effective width of the concrete flange b_{eff} is 1000 mm. Thus from Equation (9.33), the depth of the compression block h_u (assuming concrete temperatures are below 250°C) is given by

$$h_u = \frac{T}{\dfrac{b_{eff}f_{ck}}{\gamma_{M,fi,c}}} = \frac{196,8 \times 10^3}{\dfrac{1000 \times 25}{1,0}} = 7,87\ \text{mm}$$

Interpolating the values in Table 9.2 gives $h_{cr} = 68$ mm for a temperature of 250°C. The value of h_{eff} is 101 mm (from Example 9.1). Thus the allowable depth of the compression block before concrete strength loss needs to be considered is $h_{eff} - h_{cr} = 101 - 68 = 37$ mm; h_u is less than this critical value. The moment capacity $M_{fi,Rd}$ is given by Equation (9.38),

$$M_{fi,Rd} = T\left(h + h_c - y_T - \frac{h_u}{2}\right) = 196,8\left(187,6 + 2 \times 9,6 + 110 - 140 - \frac{7,87}{2}\right)$$

$$= 34\ \text{kNm}$$

This is slightly smaller than the applied moment of 35,46 kNm.

Calculation of shear stud capacity — In the ambient design, 19-mm studs were placed in each trough giving 13,3 (or 14 in practice) studs per half span of the beam at a spacing of 150 mm. The temperature in the stud is taken as 80% of the top flange temperature, i.e., $0,8 \times 691 = 553$°C and the concrete temperature as 40%, i.e., $0,40 \times 691 = 276$°C. From Table 5.8, $k_{c,\theta} = 0,874$; and from Table 3.2 of EN 1994-1-2, $k_{u,\theta} = 0,616$.

Temperature-reduced ultimate shear capacity, $P_{fi,Rd}$ from Equation (9.39)

$$P_{fi,Rd} = 0,8k_{u,\theta}\left[\frac{0,8f_u\dfrac{\pi d^2}{4}}{\gamma_{M,fi,v}}\right] = 0,8 \times 0,616\left[\frac{0,8 \times 450\dfrac{\pi 19^2}{4}}{1,0}\right] = 50,3\ \text{kN}$$

Temperature-reduced crushing capacity — From Table 3.1 of EN 1992-1-1, $E_{cm} = 33$ GPa for concrete grade C30 and $\alpha = 1,0$ for the stud layout. From Equation (9.40)

$$P_{fi,Rd} = k_{c,\theta} \frac{0,29\alpha d^2 \sqrt{f_{ck} E_{cm}}}{\gamma_{M,fi,v}} = 0,874 \frac{0,29 \times 1,0 \times 19^2 \sqrt{25 \times 33 \times 10^3}}{1,0} = 83,1 \text{kN}$$

Minimum fire capacity of the shear studs is 50,3 kN. The value of T is limited to a value of T_{lim} given by

$$T_{lim} = NP_{fi,Rd} = \frac{13,3}{2} 50,3 = 334 \text{kN}$$

The actual value of T is 196,8 kN, and therefore the section capacity is not limited by the shear stud capacity.

Limiting temperature approach — Note: The example is strictly outside the limits where the critical temperature approach may be used since the slab is *only* 110 mm thick and is therefore 10 mm below the lower limit of 120 mm. However, rather than use another example, the calculations that follow are *illustrative* only. The design moment of resistance of the composite beam is 200 kNm, so from Equation (9.25):

$$\eta_{fi,t} = \frac{E_{fi,t,d}}{R_d} = \frac{35,46}{200} = 0,177$$

$k_{y,\theta} = 0,177$ or $\theta_{cr} = 753°$C (from Table 5.6). The heat transfer calculations were carried out on a spreadsheet and gave a protection thickness of 18,3 mm with a temperature of 752°C.

9.3 CONCRETE-FILLED STEEL I AND H SECTION COLUMNS

In these columns, the areas contained between the webs and the flanges of a normal I or H section are infilled with concrete so that the concrete acts compositely with the steel. Composite action is provided using proprietary shot fired shear connectors.

Data reported by Newman (1992) indicate that such columns are capable of attaining a 60-min fire test rating, provided there is no possibility of premature failure at the ends of the columns and that the original section is designed as totally noncomposite. Newman also provides design tables for UK sections. EN 1994-1-2 gives design charts for European sections to Grade 355 and a concrete grade of 40/50.

9.4 CONCRETE-FILLED STEEL TUBE COLUMNS

Concrete-filled steel tube (CFST) columns are widely used in building construction. They have many advantages compared with ordinary steel or reinforced concrete columns including high load carrying capacity, good cross-sectional properties, fast speed of construction, and high fire resistance. CFST columns can be divided into those that are externally protected against fire by fire-rated boards, lightweight sprayed protection or intumescent coatings, and those that have no such protection. A further division can be made by distinguishing those that are filled with plain concrete mixes from those containing steel reinforcement within the mix.

Externally protected CFST columns are designed compositely at ambient temperature, and external fire protection is applied to achieve the required fire rating. In general, such columns will not need to contain reinforcement in the mix. Composite action is maintained in the fire limit state. The external protection serves to limit the rise in steel temperature such that the column capacity is always in excess of the fire limit state design load over the required fire resistance period. CFST columns without external fire protection are normally designed using the concrete core alone to meet the fire limit state load requirements but the capacity of the composite section is checked for the ambient temperature design case. In general, such columns need to contain reinforcement in the mix to minimise section dimensions and sustain the required fire limit state design loads for practical fire resistance periods of 60 min or more (Hicks and Newman, 2002).

During a fire, heat is transferred from outer to inner layers. As the concrete temperature increases, water vaporizing takes place gradually from outer to inner layers. To allow steam to escape, venting is needed in steel tubes. BS5950-8 and EN1994-1-2 recommend that steel tubes contain one vent hole, with a minimum diameter of 20 mm, at the top and bottom of each storey. The longitudinal spacing of these holes should never exceed 5 m. Care must also be taken to ensure that these vent holes are positioned such that they are not within the depth of the floor construction.

The most important parameters that can affect the fire performance of CFST columns include material strength, column size, effective length, applied load, eccentricity, reinforcement, and external protection (if applied). A detailed discussion on how each of these parameters affects the fire performance of CFST columns is given in Bailey (2000), and Hicks and Newman (2002).

9.4.1 Design of unprotected CFST columns using SCI–corus guides

EN 1994-1-2 gives general design methods for composite columns and, in Annex H, specific rules for CFST columns. Following a calibration exercise, SCI and Corus concluded that a combination of the two methods

(Eurocode and SCI) provided a best approach that is fully in line with the principles of the Eurocode. The detailed design guides recommended by SCI and Corus were published by Hicks and Newman (2002). Discussion of the actual difference between EN 1994-1-2 and SCI–Corus guides can be found in Wang and Orton (2008). In the SCI–Corus design guides, the calculation is carried out in five steps:

1. Carry out a thermal analysis to establish the temperature distribution throughout the cross section.
2. Calculate the properties of the cross section.
3. Calculate the buckling resistance of the column.
4. Calculate the bending resistance of the column.
5. Calculate the effect of interaction between axial load and bending.

The thermal analysis can be performed numerically by solving the heat transfer equation defined by Equation (6.3) with the fire-exposed surface boundary condition defined by Equation (6.4a). Note that EN 1991-1-2 includes partial factors in the convective and radiative heat flux terms to allow heating rates in different furnaces to be modelled. However, according to the SCI analysis of test data, these two partial factors are not necessary and can be taken as 1,0 for CFST columns.

The properties of the cross section can be calculated by dividing the cross section into a number of elements, for instance, as used in thermal analysis. Within each element, temperature and material properties are assumed to be constants. Hence, the structural section properties can be obtained by summing the properties of all the elements, with due consideration paid to the effect of the temperature on each element. In this case, the plastic resistance $N_{fi,pl,Rd}$ and effective flexural stiffness $(EI)_{fi,eff}$ of a CFST column can be expressed as:

$$N_{fi,pl,Rd} = \sum_{j=1} \frac{A_{a,\theta}f_{a,max,\theta}}{\gamma_{M,fi,a}} + \sum_{k=1} \frac{A_{s,\theta}f_{s,max,\theta}}{\gamma_{M,fi,s}} + \sum_{m=1} \frac{A_{c,\theta}f_{c,\theta}}{\gamma_{M,fi,c}} \qquad (9.41)$$

$$(EI)_{fi,eff} = \sum_{j=1} E_{a,\theta,\sigma}I_{a,\theta} + \sum_{k=1} E_{s,\theta,\sigma}I_{s,\theta} + \sum_{m=1} E_{c,\theta,\sigma}I_{c,\theta} \qquad (9.42)$$

where $A_{a,\theta}$, $A_{s,\theta}$, and $A_{c,\theta}$ are the element areas related to steel tube, steel reinforcement and concrete; $f_{a,max,\theta}$, $f_{s,max,\theta}$, and $f_{c,\theta}$ are the strengths of steel, steel reinforcement, and concrete elements at temperature θ; $\gamma_{M,fi,a}$, $\gamma_{M,fi,s}$, and $\gamma_{M,fi,c}$ are the partial material safety factors for steel tube, steel reinforcement and concrete; $E_{a,\theta,\sigma}$, $E_{s,\theta,\sigma}$, and $E_{c,\theta,\sigma}$ are the tangent moduli of the stress–strain relationships of steel tube, steel reinforcement, and concrete at temperature θ and stress σ; and $I_{a,\theta}$, $I_{s,\theta}$, and $I_{c,\theta}$ are the second moments of area of steel tube, steel reinforcement, and concrete elements, respectively. The reduced strengths and Young's moduli of structural steel, steel reinforcement bars, and structural concrete due to elevated temperatures are provided in Annex H of EN 1994-1-2 and take into account the confinement effect of stresses perpendicular to the longitudinal axis of the column.

The critical buckling load of the CFST column is calculated using the Euler formula:

$$N_{fi,cr} = \frac{\pi^2 (EI)_{fi,eff}}{l_\theta^2} \qquad (9.43)$$

where l_θ is the effective length of the CFST column. The nondimensional slenderness is obtained in terms of the plastic resistance and critical buckling load calculated in Equations (9.41) and (9.43):

$$\bar{\lambda}_\theta = \sqrt{\frac{N_{fi,pl,Rd}}{N_{fi,cr}}} \qquad (9.44)$$

The buckling resistance $N_{fi,Rd}$ of the CFST column is obtained by applying a reduction factor to the plastic resistance:

$$N_{fi,Rd} = \chi N_{fi,pl,Rd} \qquad (9.45)$$

where χ is the reduction coefficient for buckling obtained from the Section 6.3.1.2 of EN 1993-1-1 as follows:

$$\chi = \frac{1}{\phi + \sqrt{\phi^2 - \bar{\lambda}_\theta^2}} \qquad \text{but } \chi \leq 1,0 \qquad (9.46)$$

where $\phi = 0,5[1 + \alpha(\bar{\lambda}_\theta - 0,2) + \bar{\lambda}_\theta^2]$ and $\alpha = 0,49$ is the imperfection factor. If the column is subjected to combined compression and bending, the position of the plastic neutral axis is calculated by:

$$\sum_{j=1} \frac{A_{a,\theta} f_{a,max,\theta}}{\gamma_{M,fi,a}} + \sum_{k=1} \frac{A_{s,\theta} f_{s,max,\theta}}{\gamma_{M,fi,s}} + \sum_{m=1} \frac{A_{c,\theta} f_{c,\theta}}{\gamma_{M,fi,c}} = 0 \qquad (9.47)$$

where the stresses are considered as positive in tension and negative in compression. Figure 9.9 shows how Equation (9.47) is executed in a section. Based on the new plastic neutral axis, the plastic moment resistance is calculated as:

$$M_{fi,Rd} = \sum_{j=1} \frac{y A_{a,\theta} f_{a,max,\theta}}{\gamma_{M,fi,a}} + \sum_{k=1} \frac{y A_{s,\theta} f_{s,max,\theta}}{\gamma_{M,fi,s}} + \sum_{m=1} \frac{y A_{c,\theta} f_{c,\theta}}{\gamma_{M,fi,c}} \qquad (9.48)$$

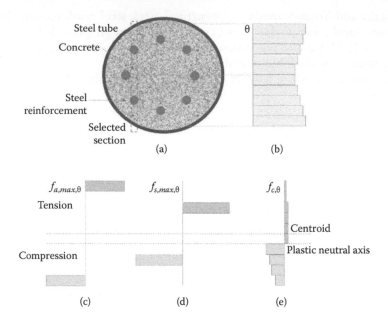

Figure 9.9 Calculation of plastic neutral axis: (a) a selected section of CFST column, (b) temperature distribution in section, (c) strength distribution of steel tube in section, (d) strength distribution of steel reinforcement in section, and (e) strength distribution of concrete in section.

where y is the distance from the centroid of each element to the plastic neutral axis. The interaction of axial compression and bending is checked using the following expression:

$$\frac{N_f}{N_{fi,Rd}} + \frac{k_y M_{fy}}{M_{fi,Rd}} + \frac{k_z M_{fz}}{M_{fi,Rd}} \le 1,0 \tag{9.49}$$

where N_f is the applied axial load, M_{fy} and M_{fz} are the applied moments about the y and z axis, respectively, and k_y and k_z are the modification factors to take account of any possible second order effects in slender columns in accordance with the composite column design at ambient temperature specified in EN 1994-1-1. If the second order effects about an axis or both axes are not significant, the modification factor about that axis or both axes can be taken as 1,0.

SCI–Corus recommended that the second order effects need be considered only when both of the following conditions are satisfied:

$$\frac{N_f}{N_{fi,Rd}} \ge 0,1 \quad \text{and} \quad \bar{\lambda}_\theta > 0,2(2-r) \tag{9.50}$$

where r is the ratio of the smaller to the larger end moment. If there is any transverse loading, r should be taken as $1,0$. In this case the modification factor about an axis can be calculated as:

$$k = \frac{\beta}{1 - \dfrac{N_f}{N_{cr,f}}} \geq 1,0 \qquad\qquad (9.51)$$

where k is the modification factor about the axis where the critical buckling load is $N_{cr,f}$ and β is the equivalent moment factor. For a column with transverse loading within the column length, $\beta = 1,0$. For a column with pure end moments, β is taken as:

$$\beta = 0,66 + 0,44r \qquad \text{but} \qquad \beta \geq 0,44 \qquad\qquad (9.52)$$

Programming is normally required for the above calculation. To facilitate the use of the design procedure, SCI developed a special computer code that covers the design of unprotected CFST columns for both normal conditions and fire and is available on a Corus CD for general use.

It should be noted that the design method is recommended only for CFST columns with circular or rectangular steel tubes, carrying predominantly axial loads with ratios of applied moment-to-moment capacity not greater than $0,67$.

9.4.2 Empirical design methods of CFST columns

Following tests and a parametric investigation Kodur and Lie (1985) and Kodur (1999) suggested that the fire endurance $t_{fi,Rd}$ of a square or circular concrete-filled hollow section could be expressed as:

$$t_{fi,Rd} = f_1 \frac{f_{ck} + 20}{l_\theta - 1000} \frac{D^{2,5}}{\sqrt{N_{fi,Ed}}} \qquad\qquad (9.53)$$

where f_{ck} is the characteristic concrete strength (MPa), l_θ is the buckling length (mm), D is the diameter of a circular column or side length of a square column (mm), $N_{fi,Ed}$ is the applied load in the fire limit state (kN), and f_1 is a factor (see Table 9.3) that allows for type of concrete (plain or non-fibre) and level of reinforcement.

Wang (2000a) also provides an empirical approach whereby the squash load $N_{fi,Rd}$ at time t_{fi} (min) for a protected circular column of diameter D (mm) is given by:

For $t_{fi} \leq D$:

$$N_{fi,Rd} = \frac{t_{fi}}{D}\left(N_{fi,\theta,const} - N_{fi,\theta,a}\right) + N_{fi,\theta,a} \qquad\qquad (9.54)$$

Table 9.3 Values of f_i

Aggregate	Column type	Plain concrete	Fibre concrete	Reinforcement ratio and cover			
				<3%		>3%	
				< 25	>25	<25	>25
Siliceous:	CHS	0,07	0,075	0,075	0,08	0,08	0,085
	SHS	0,06	0,065	0,065	0,07	0,07	0,075
Carbonate :	CHS	0,08	0,07	0,085	0,09	0,09	0,095
	SHS	0,06	0,065	0,075	0,08	0,08	0,085

Source: Kodur,V.K.R. (1999) Journal of Constructional Steel Research, 51, 21–36. With permission.

For $t_{fi} > D$:

$$N_{fi,Rd} = N_{fi,\theta,const} \tag{9.55}$$

where $N_{fi,\theta,const}$ is the column squash load with the steelwork and concrete at the same constant temperature and $N_{fi,\theta,a}$ is the squash load when the steelwork is at maximum temperature and the concrete is considered cold (20°C). For unprotected columns, the contribution of the concrete needs to be determined incrementally. An alternative form of composite column is to use a standard column or beam section with the web infilled.

9.4.3 Finite element analysis of fire performance of CFST columns

Since the publication of EN 1994-1-2 in 2005, extensive experimental and numerical investigations of the fire performances of CFST columns have been carried out. For example, Lua et al. (2009) reported their investigations on the behaviour of high strength self-compacting CFST stub columns exposed to standard fire by using both experimental and numerical methods. All the tested stub columns were found to fail in a ductile way. Specimens retained their integrity after testing despite local bulge of the steel hollow section and local crush of the concrete. This demonstrates the interaction of the concrete and steel in a CFST column during fire exposure.

Song et al. (2010) developed a finite element analysis model to predict the load versus deformation relationships of CFST stub columns subjected to a combination of temperature and axial compression. This model was used to simulate a set of CFST stub column experiments under various thermal and mechanical loading conditions including tests at high temperature, tests on the residual strength of specimens subjected to uniform heating, and tests on specimens exposed to the ISO-834 standard fire without initial loads. Comparisons between the predicted and test results showed that the model can predict the load versus deformation relationships with reasonable accuracy.

The finite element analysis model was then used to investigate the behaviour of CFST stub columns in a complete loading history including initial loading, heating, and cooling by examining the cross-sectional stress distribution and confinement stress development at different loading phases. All specimens were loaded to ultimate strength after cooling, and the residual stress indices of a group of parameters were studied. The ultimate strength when considering the mutual actions of temperature and loading was slightly lower than that after exposure to fire without initial load, but the peak strain corresponding to the ultimate strength increased significantly.

The confinement stress and the longitudinal stress distribution of CFST stub columns under combined thermal and mechanical loading were analysed using the finite element analysis model, and the results at various phases were presented. The columns of circular hollow steel sections developed significant confinement stress during the cooling phase and thus enhanced the post-fire behaviour of the column. However, the columns of square hollow steel sections are not so effective, and confinement stresses can be generated only at the four corners.

In addition to the concrete-filled circular and rectangular steel hollow section columns, concrete-filled polygonal steel hollow section columns and concrete-filled double skin steel tubular columns have been proposed in recent years (Yu et al., 2013). The fire performance data for these new types of CFST columns have yet to be developed.

Chapter 10

Design of timber elements

Unlike the situation where the design of concrete, steel, or steel–concrete composite construction is concerned, the calculation procedure for timber is simplified. There is no explicit requirement to calculate the temperature distribution within the element as the strength calculation is carried out on a residual section after the depth of charring is removed from the original section. There is also generally no need to consider strength reduction in the residual section as any temperature rise can be considered small and therefore ignored. This chapter considers fire design to EN 1995-1-2 and various empirical methods.

10.1 DESIGN TO EN 1995-1-2

The design process involves two stages. The first is to calculate the depth of charring and the second is to determine the strength of the residual section.

10.1.1 Depth of charring

Two cases need to be considered, namely exposure to the standard furnace curve or to a real or parametric fire.

10.1.1.1 Exposure to standard furnace curve

EN 1995-1-2 gives two values of charring rates: β_0 for single face exposure and β_n for multiface exposure. The values of β_n include an allowance for arris rounding. Values of β_0 and β_n for timber are given in Table 10.1. To calculate the depth of charring, two cases are to be considered:

Single face exposure — The depth of charring $d_{char,0}$ at time t is given by

$$d_{char.0} = \beta_0 t \tag{10.1}$$

Table 10.1 Charring rates from EN 1995-1-2

Timber type	Charring rate (mm/min)	
	β_0	β_n
Softwood and Beech		
Glulam ($\rho \geq 290$ kg/m3)	0,65	0,70
Solid ($\rho \geq 290$ kg/m3)	0,65	0,80
Hardwood		
Solid or glulam ($\rho = 290$ kg/m³)	0,65	0,70
Solid or glulam ($\rho > 450$ kg/m³)	0,50	0,55

Note: For solid hardwoods except beech, charring rates for densities between 240 and 450 kg/m³ may be obtained using linear interpolation. Beech should be treated as a softwood.

Source: Table 3.1 of EN 1995-1-2 (abridged).

Multiface exposure — Two ways to determine this factor are (1) by the use of β_n or by using β_0 and modifying values:

$$d_{char,n} = \beta_n t \tag{10.2}$$

or (2) if the value of the minimum width of the section b_{min} satisfies the following relationships, β_0 may be used, but then arris rounding with a radius of $d_{char,0}$ must be taken into account.

For $d_{char,0} \geq 13$ mm:

$$b_{min} = 2d_{char,0} + 80 \tag{10.3}$$

For $d_{char,0} < 13$ mm:

$$b_{min} = 8,15 d_{char,0} \tag{10.4}$$

10.1.1.2 Charring to natural fire exposure

Hadvig (1981) gave the following equations with the time to maximum charring rate t_0 given by

$$t_0 = \frac{0,0175 q_{t,d}}{O} \tag{10.5}$$

For $t \leq 0,33 t_0$:

$$d_{char} = \beta_0 t \tag{10.6}$$

For $0,33t_0 < t < t_0$:

$$d_{char} = \beta_0 \left(1,5t - \frac{t_0}{12} - \frac{0,75t^2}{t_0} \right) \qquad (10.7)$$

where β_0 is given by

$$\beta_0 = 1,25 - \frac{0,035}{O + 0,021} \qquad (10.8)$$

The equations in Annex A of EN 1995-1-2 to calculate the charring depth d_{char} for timber exposed to parametric fires are given as:
For $0 \leq t \leq t_0$:

$$d_{char} = \beta_{par} t \qquad (10.9)$$

For $t_0 \leq t \leq 3t_0$:

$$d_{char} = \beta_{par} \left(1,5t_0 - \frac{t^2}{4t_0} - \frac{t_0}{4} \right) \qquad (10.10)$$

For $3t_0 \leq t \leq 5t_0$:

$$d_{char} = 2\beta_{par} t_0 \qquad (10.11)$$

where t is the time in min, t_0 is a parameter that determines the time to maximum charring and is dependent upon the characteristics of the fire and the compartment and is defined by

$$t_0 = 0,009 \frac{q_{t,d}}{O} \qquad (10.12)$$

where $q_{t,d}$ is the design fire load density with respect to the total compartment area (MJ/m²) and O is the ventilation factor defined as $A_v \sqrt{h_{eq}}/A_t$ where A_t is the total area of the compartment (m²), A_v and h_{eq} are the area (m²) and height (m) of the openings, respectively, and β_{par} (mm/min) is defined by

$$\beta_{par} = 1,5\beta_n \frac{0,2\sqrt{\Gamma} - 0,04}{0,16\sqrt{\Gamma} + 0,08} \qquad (10.13)$$

where Γ accounts for the thermal properties of the compartment and is given by

$$\Gamma = \frac{\left(\dfrac{O}{\sqrt{\rho c\lambda}}\right)^2}{\left(\dfrac{0,04}{1160}\right)^2} \qquad (10.14)$$

where $\sqrt{\rho c\lambda}$ is the thermal inertia of the compartment. This method may be used only if $t \leq 40$ min and d_{char} is less than both $b/4$ and $d/4$. There is a clear anomaly between the two sets of equations as Hadvig gives a value of t_0 approximately twice that of the equation in Annex A of EN 1995-1-2. Note that the UK National Annex indicates the equations in EN 1995-1-2 Annex A should not, in fact, be used.

10.1.2 Calculation of structural capacity

The strength and stiffness properties for the fire limit are based not on the usual 5% fractile values used at ambient limit state but on 20% fractiles. The design values of strength f_k and stiffness S_{05} are multiplied by a factor k_{fi} that has values of 1,25 for solid timber and 1,15 for glulam.

The background to the determination of cross-sectional resistance determination is given in Kersken-Bradley (1993). The first design method uses an increased charring depth to allow for potential strength loss in the core and is known as the effective section method. The second uses a lower char depth together with factors to allow for the reduction of properties with temperature.

10.1.2.1 Effective section method (standard furnace exposure)

The total depth of section reduction d_{ef} is given as the sum of two components $d_{char,n}$ and $k_0 d_0$ such that

$$d_{ef} = d_{char,n} + k_0 d_0 \qquad (10.15)$$

where d_0 is taken as 7 mm, and k_0 is given by
  For $t \leq 20$ min:

$$k_0 = \frac{t}{20} \qquad (10.16)$$

For $t > 20$ minutes:

$$k_0 = 1,0 \tag{10.17}$$

The values of $\gamma_{M,fi}$ are taken as 1,0.

10.1.2.2 Reduced strength and stiffness method (standard furnace exposure)

The reduction in section dimensions is taken as due only to the charring. For $t \geq 20$ min, the strengths and Young's modulus are reduced by modification factors $k_{mod,fi}$:

For bending strength:

$$k_{\mathrm{mod},fi} = 1,0 - \frac{1}{200}\frac{p}{Ar} \tag{10.18}$$

For compressive strength:

$$k_{\mathrm{mod},fi} = 1,0 - \frac{1}{125}\frac{p}{A_r} \tag{10.19}$$

For tension strength and Young's modulus:

$$k_{\mathrm{mod},fi} = 1,0 - \frac{1}{330}\frac{p}{A_r} \tag{10.20}$$

where p/A_r is the section factor (m^{-1}) defined as a heated perimeter/cross sectional area for the reduced section. For $t < 20$ min, the modification factors are obtained using linear interpolation between those for Equations (10.18) to (10.20) and a value of $k_{mod,fi}$ of 1,0 at $t = 0$.

10.1.2.3 Residual strength and stiffness method (parametric exposure)

For flexure about the major axis, the residual cross section should be determined using the charring depths calculated from Equations (10.9) to (10.11) as appropriate. For softwoods, the modification factor $k_{mod,fi}$ should be determined as shown below. For $t < 3t_0$, Equation (10.18) should be used, and for $t = 5t_0$

$$k_{\mathrm{mod},fi} = 1,0 - 3,2\frac{d_{char,n}}{b} \tag{10.21}$$

where $d_{char,n}$ is the notional charring depth.
  For $3t_0 \leq t < 5t_0$, linear interpolation should be used.

All design examples will be done using Grade C22 timber for which $f_{m,k}$ = 22 MPa, E_{mean} = 10 kPa, $E_{0,05}$ = 6,7 kPa, and $f_{v,k}$ = 2,4 MPa with exposure to the standard furnace curve. Determine values of 20% fractile strength and elasticity values as follows:

$f_{m,20} = k_{fi} f_{m,k} = 1,25 \times 22 = 27,5$ MPa
$f_{v,20} = k_{fi} f_{v,f} = 1,25 \times 2,4 = 3,0$ MPa
$E_{20} = k_{fi} E_{0,05} = 1,25 \times 6,7 = 8,5$ kPa.
$f_{c,20} = k_{fi} f_{c,0,k} = 1,25 \times 20 = 25$ MPa

Example 10.1 Determination of fire performance of beam

A timber beam is 250 mm deep × 75 mm wide and is simply supported over a span of 4,5 m. The beam system is at 600 mm spacing and carries a permanent action on the beam of 0,2 kN/m and a total variable load of 2,5 kPa. The required fire endurance period is 30 min.

With a ψ factor of 0,3, the load in the fire limit state is 0,3 × (0,6 × 2,5) + 0,2 = 0,65 kN/m. Maximum bending moment $M_{Ed,fi}$ = 0,65 × $4,5^2/8$ = 1,65 kNm.

Effective cross section method — From Table 10.1, the charring rate β_n is 0,80 mm/min for a softwood of density greater than 290 kg/m³.

> Depth of charring $d_{char,n}$ = 30 × 0,80 = 24 mm.
> As t > 20 min, k_0 = 1,0, $k_0 d_0$ = 7 mm, so d_{ef} = 24 + 7 = 31 mm.
> Reduced width $b = B - 2d_{ef}$ = 75 – 2 × 31 = 13 mm.
> Reduced depth $d = D - d_{ef}$ = 250 – 31 = 219 mm.
> (i) Flexure:
> > Elastic section modulus $W_{el} = bd^2/6 = 13 \times 219^2/6 = 104 \times 10^3$ mm³.
> > $f_{m,20}$ = 27,5 MPa, so $M_{Rd,fi}$ = 27,5 × 104 × 10⁻³ = 2,86 kNm.
> > $M_{Ed,fi}$ = 1,65 kNm, so the beam is satisfactory.
> (ii) Shear:
> > Maximum shear force is 0,65 × 4,5/2 = 1,46 kN.
> $V_{Rd,fi} = 1,5bdf_{v,20} = 1,5 \times 13 \times 219 \times 3,0 \times 10^{-3}$ = 12,8 kN, which is satisfactory.

Although there appears no specific requirement to check deflection, it is worthwhile to carry out the check.

> (iii) Deflection:
> > $I = bd^3/12 = 13 \times 219^3/12 = 11,4 \times 10^6$ mm⁴.
> > $\delta_b = 5ML^2/48EI$
> > $= 5 \times 1,65 \times 10^3 \times 4,5^2/(48 \times 8,5 \times 10^{-6} \times 11,4 \times 10^6) = 0,035$ m.

This represents a span/deflection ratio of 4,5/0,035 = span/130 which is satisfactory.

Reduced strength and stiffness method — From Table 10.1, the charring rate is 0,80 mm/min.

Depth of charring $d_{char} = 30 \times 0,80 = 24$ mm.
Reduced width $b = B - 2d_{char} = 75 - 2 \times 24 = 27$ mm.
Reduced depth $d = D - d_{char} = 250 - 24 = 226$ mm.
Residual area $A_r = 226 \times 27 = 6102$ mm^2.
Residual perimeter calculated for fire exposed faces only: $p_r = 2 \times 226 + 27 = 479$ mm.
$p/A_r = 479/6102 = 0,078$ mm^{-1} = 78 m^{-1}.
  (i) Flexure:
      Elastic section modulus $W_{el} = bd^2/6 = 230 \times 10^3$ mm^3.
      Flexure $k_{mod,fi}$ is given by Equation (10.18):
      $k_{mod,fi} = 1 - (1/200)p/A_r = 1 - 78/200 = 0,61$.
      $M_{Rd,fi} = k_{mod,fi}W_{el}f_{20} = 0,61 \times 0,230 \times 27,5 = 3,86$ kNm.
$M_{Ed,fi} = 1,65$ kNm.

The beam is therefore satisfactory.

  (iii) Reduction factors:

These factors are not given for shear, but based on the large margin in (a), this should not be a problem.

  (iv) Deflection:
      $I = bd^3/12 = 27 \times 226^3/12 = 26,0 \times 10^6$ mm^4.
      Modification factor for Young's modulus calculated from Equation (10.20):

      $k_{mod,fi} = 1 - (1/330)p/A_r = 1 - 78/330 = 0,766$.
      $E_{f,d} = k_{mod,fi}E_{mean}/\Gamma_{m,s} = 0,766 \times 8500 = 6511$ MPa.
      $\delta_b = 5ML^2/48E_{f,d}I = 5 \times 1,65 \times 10^3 \times 4,5^2/(48 \times 6,511 \times 10^{-6} \times 26,0 \times 10^6) = 0,021$ m.

This represents a span/deflection ratio of $4,5/0,021 = 214$ which is satisfactory.

Example 10.2: Fire performance of column

Check the fire performance of a 150 mm square 3 m high column to last 30 min carrying an axial load in the fire limit state of 45 kN. Using both the effective section method and reduced strength and stiffness method, assume the column is isolated and that the effective length may be taken as 3 m.

The buckling strength is then determined using the methods of EN 1995-1-1 but with values of strength and stiffness appropriate to the fire limit state. As the column is square, suffixes relating to axes will be omitted from symbols where appropriate.

Effective section method — The charring rate β_n is taken as 0,80 mm/min.

$d_{char,n} = 0,80 \times 30 = 24$ mm.
$k_0 = 1,0$ ($t > 20$ min), so $k_0 d_0 = 7$ mm.
$d_{ef} = 24 + 7 = 31$ mm.
$d = D - 2d_{ef} = 150 - 2 \times 31 = 88$ mm.

$$\lambda = \frac{L}{i} = \frac{L}{\dfrac{d}{2\sqrt{3}}} = \frac{3000}{\dfrac{88}{2\sqrt{3}}} = 118$$

$$\lambda_{rel} = \frac{\lambda}{\pi}\sqrt{\frac{f_{c,20}}{E_{20}}} = \frac{118}{\pi}\sqrt{\frac{25}{8,5\times10^3}} = 2,04$$

$$k = 0,5\left[1 + \beta_c\left(\lambda_{rel} - 0,3\right) + \lambda_{rel}^2\right] = 0,5\left[1 + 0,2\left(2,04 - 0,3\right) + 2,04^2\right] = 2,75$$

($\beta_c = 0,2$ for solid timber).

$$k_c = \frac{1}{k + \sqrt{k^2 - \lambda_{rel}^2}} = \frac{1}{2,75 + \sqrt{2,75^2 - 2,04^2}} = 0,218$$

$$N_{Rd,fi} = k_c f_{c,20} A_c = 0,218 \times 25 \times 88^2 \times 10^{-3} = 42 \text{ kN}.$$

This is less than the applied load; thus the load-carrying capacity at ultimate limit state should be reduced to $42/0,7 = 60$ kN. Thus the carrying capacity at the fire limit state controls the performance of the column.

Reduced strength and stiffness method —

$d_{char,n} = 0,80 \times 30 = 24$ mm.
$d = D - 2d_{char,n} = 150 - 2 \times 24 = 102$ mm.
For a column heated on four sides, $p/A_r = 4/d$, so $p/A_r = 4/0,102 = 39,2$ m^{-1}.

Assume the column is isolated and that the effective length may be taken as 3 m. The buckling strength is then determined using the methods of EN 1995-1-1 but with values of strength and stiffness appropriate to the fire limit state. As the column is square, suffixes relating to axes will be omitted from symbols where appropriate.

$$\lambda = \frac{L}{i} = \frac{L}{\dfrac{d}{2\sqrt{3}}} = \frac{3000}{\dfrac{102}{2\sqrt{3}}} = 102$$

$$\lambda_{rel} = \frac{\lambda}{\pi}\sqrt{\frac{f_{c,20}}{E_{20}}} = \frac{102}{\pi}\sqrt{\frac{25}{8,5\times10^3}} = 1,76$$

$$k = 0,5\left[1 + \beta_c\left(\lambda_{rel} - 0,3\right) + \lambda_{rel}^2\right] = 0,5\left[1 + 0,2\left(1,76 - 0,3\right) + 1,76^2\right] = 2,19$$

($\beta_c = 0,2$ for solid timber)

$$k_c = \frac{1}{k + \sqrt{k^2 - \lambda_{rel}^2}} = \frac{1}{2,19 + \sqrt{2,19^2 - 1,76^2}} = 0,286$$

The modification factor for compressive strength is given by

$$k_{mod,fi} = 1,0 - \frac{1}{125}\frac{p}{A_r} = 1,0 - \frac{39,2}{125} = 0,686$$

$$N_{Rd,fi} = k_{mod,fi}k_c f_{c,20} A_c = 0,686 \times 0,286 \times 25 \times 102^2 \times 10^{-3} = 51kN$$

This is less than the applied load. Thus the load-carrying capacity at ultimate limit state should be reduced to $51/0,7 = 73$ kN. The carrying capacity at the fire limit state controls the performance of the column.

For both the beam and column examples, the reduced properties method is less conservative than the reduced cross-section method.

10.2 EMPIRICAL APPROACHES

A number of empirical approaches for the assessment of the fire performance of timber elements were developed by Ödeen (1969), Lie (1977), ASCE (2007), and Stiller (1983).

10.2.1 Approach of Ödeen

For members heated on four sides, Ödeen (1967) proposed the following empirical equations to determine the fire endurance $t_{fi,d}$ of timber members exposed to the standard furnace curve at a constant charring rate:

$$t_{fi,d} = \left(1 - \frac{d}{D}\right)\frac{D}{2\beta} \tag{10.22}$$

where the depth of the residual section d is determined from

$$\left(\frac{k}{\alpha}\right)\left(\frac{\dfrac{B}{D}}{\dfrac{d}{D} - \left(1 - \dfrac{B}{D}\right)}\right) = \left(\frac{d}{D}\right)^2 \tag{10.23}$$

where k is the ratio between the maximum stress induced by the loading before the fire to the ultimate strength of the timber, α is the ratio between the ultimate strength of the timber in its temperature-affected state to its ambient strength, B and D are the original width and depth of the section, d is the fire affected depth, $t_{fi,d}$ is the required fire resistance given by standard ratings, and β is the appropriate rate of charring.

For a beam heated on three sides, the equations are modified to

$$t_{fi,d} = \left(1 - \frac{d}{D}\right)\frac{D}{\beta} \tag{10.24}$$

$$\left(\frac{k}{\alpha}\right)\left(\frac{\dfrac{B}{D}}{\dfrac{B}{D} - 2\left(1 - \dfrac{d}{D}\right)}\right) = \left(\frac{d}{D}\right)^2 \tag{10.25}$$

Ödeen suggests that the value of the strength reduction factor with temperature α should lie between 0,85 and 0,9. Following Malhotra (1982a), the value of α is taken as 0,87 in all the ensuing examples.

In the original work, the value of k was defined as the ratio between the maximum bending stress before the fire to the ultimate bending strength before the fire. At that point, timber was generally designed using working (or serviceability) loads, and it would have been assumed that the structural loading in the fire limit state was the normal design loading. Given that the variable portion of the fire loading may be reduced by ψ factors, it is more appropriate to use the structural fire load to determine the value of k. Also, the determination of the load level to calculate over-design parameter f should be based on structural fire loading (not service loading).

Example 10.3: Fire performance of beam in example 10.1

Check the fire performance of the beam described in Example 10.1 using Ödeen's approach. Using the modified definition of k:

From Example 10.1, the design fire moment $M_{Ed,fi} = 1,65$ kNm.
$F_{Ed,fi} = 1,65 \times 10^6/(75 \times 250^2/6) = 2,11$ MPa.
Design strength (assuming medium-term loading) is $0,8 \times 22/1,3 = 13,54$ MPa.
$k = 2,11/13,54 = 0,156$.

From Equation (10.25)

$$\left(\frac{k}{\alpha}\right)\left(\frac{\dfrac{B}{D}}{\dfrac{B}{D} - 2\left(1 - \dfrac{d}{D}\right)}\right) - \left(\frac{d}{D}\right)^2 = 0 = \left(\frac{0,156}{0,87}\right)\left(\frac{\dfrac{75}{250}}{\dfrac{75}{250} - 2\left(1 - \dfrac{d}{D}\right)}\right) - \left(\frac{d}{D}\right)^2$$

or

$$\left(\frac{d}{D}\right)^2\left(0,3 - 2\left(1 - \frac{d}{D}\right)\right) = 0,052$$

or

$$d/D = 0,883$$

Take the charring rate as β_n as arris rounding has not been taken into account in the formulae. So from Equation (10.24)

$$t_{fi,d} = \frac{D}{\beta}\left(1 - \frac{d}{D}\right) = \frac{250}{0,8}(1 - 0,883) = 37\,\text{min}$$

Thus the beam would last in excess of 30 min.

10.2.2 Lie's approach for beams

For beams, Lie (1977) proposed simpler equations that give a direct evaluation of fire resistance:

Beams heated on four sides:

$$t_{fi,d} = 0,1fB\left(4 - \frac{2B}{D}\right) \tag{10.26}$$

Beams heated on three sides:

$$t_{fi,d} = 0,1fB\left(4 - \frac{B}{D}\right) \tag{10.27}$$

where f is a factor allowing for effective over-design due to the availability of timber sizes and all dimensions are in millimetres. Equations (10.26) and (10.27) imply a charring rate of 0,6 mm/min. Values of the parameter f are given in Table 10.2.

Table 10.2 Values of over-design factor f

Load ratio λ (% allowable)	Beam	Column LID > 10	Column LID ≤ 10
λ ≥ 75	1,0	1,0	1,2
75 > λ ≥ 50	1,1	1,1	1,3
λ ≤ 50	1,3	1,3	1,5

Source: Lie, T.T. (1977) *Canadian Journal of Civil Engineering*, 4, 161–169. With permission.

Example 10.4: Fire performance of beam in example 10.1
                        using Lie's formula

Determination of f:

$k = 0,156$ as above (Example 10.3), so from Table 10.2, $f = 1,3$.
From Equation (10.27):

$$t_{fi,d} = 0,1fB\left(4 - \frac{B}{D}\right) = 0,1 \times 1,3 \times 75\left(4 - \frac{75}{250}\right) = 36\,\text{min}$$

Lie's equation was derived for a charring rate of 0,6 mm/min. Thus a time of 36 min will be an overestimate.

10.2.3 ASCE empirical approach for beams

In *Standard Calculation Methods for Structural Fire Protection* (ASCE, 2007), the standard fire resistance $t_{fi,d}$ for beams with a lesser nominal dimension are given by the following:
Beams heated on four sides:

$$t_{fi,d} = \gamma z B\left(4 - \frac{2B}{D}\right) \tag{10.28}$$

Beams heated on three sides:

$$t_{fi,d} = \gamma z B\left(4 - \frac{B}{D}\right) \tag{10.29}$$

where γ takes a value of 0,1 mm/min and the load factor z is given by

$$z = 0,7 + \frac{30}{r} \leq 1,3 \tag{10.30}$$

where r, the ratio of applied load to allowable load, is expressed as a percentage. It should be noted that the method is a modified version of that proposed by Lie.

Example 10.5: Checking beam of example 10.1 using ASCE approach

Determine r on the basis of the service loading (ignoring the self-weight which is negligible).

In the fire limit state, $q_{fi} = 0,3 \times (0,6 \times 2,5) + 0,2 = 0,65$ kN/m.
In the serviceability limit state (corresponding to working load design), $q = 0,6 \times 2,5 + 0,2 = 1.70$ kN/m.

$$r = 100\frac{q_{fi}}{q} = 100\frac{0,65}{1,70} = 38,2\%$$

From Equation (10.30),

$$z = 0,7 + \frac{30}{r} = 0,7 + \frac{30}{38,2} = 1,49$$

As this is greater than the limiting value of 1,3, $z = 1,3$. From Equation (10.29):

$$t_{fi,d} = \gamma z B\left(4 - \frac{B}{D}\right) = 0,1 \times 1,3 \times 75\left(4 - \frac{75}{250}\right) = 36 \text{ min}$$

The required fire period is 30 min. Thus the beam is satisfactory. Note that in this case the result is identical to that determined from Lie's approach.

10.2.4 Empirical determination of fire endurance for columns

For short timber columns whose strength is determined by the pure compressive strength of the timber (i.e., no buckling), the following equation can be derived for the relationship between the pre- and post-fire dimensions

$$\left(\frac{k}{\alpha}\right)\left(\frac{\frac{B}{D}}{\frac{d}{D} - \left(1 - \frac{B}{D}\right)}\right) = \left(\frac{d}{D}\right) \tag{10.31}$$

For extremely slender columns whose strength is dependent entirely on buckling, the load capacity can then be derived from the basic Euler equation for buckling strength. The equation relating section dimensions, assuming the effective length of column is the same in both the ambient and fire limit states and that the temperature dependence of the modulus of elasticity of the timber can be taken as that for loss in strength, is then given by

$$\left(\frac{k}{\alpha}\right)\left(\frac{\frac{B}{D}}{\frac{d}{D} - \left(1 - \frac{B}{D}\right)}\right) = \left(\frac{d}{D}\right)^3 \tag{10.32}$$

The assumption of similar behaviour with respect to temperature for both the modulus of elasticity and compressive strength is reasonable when deriving an empirical equation (Gerhards, 1982).

However, most columns are neither short nor extremely slender and have slenderness ratios such that failure results from a combination of squashing and buckling. Lie (1977) thus suggested that for general cases, the relationship between the column dimensions in the determination of the fire resistance of an axially loaded column when subjected to fire on all four sides could be taken as

$$\left(\frac{k}{\alpha}\right)\left(\frac{\dfrac{B}{D}}{\dfrac{d}{D}-\left(1-\dfrac{B}{D}\right)}\right)=\left(\frac{d}{D}\right)^{n}$$

(10.33)

where k is the ratio between the load applied during a fire to the column strength at ambient allowing for the enhancement of allowable stresses, α is the reduction in compressive strength of the residual section, and n is a parameter with limiting values of 1 for short columns and 3 for long columns. Lie suggested that for most practical columns a value of $n = 2$ could be taken. The fire resistance $t_{fi,d}$ based on standard classification is then calculated from

$$t_{fi,d} = \frac{D}{2\beta}\left(1-\frac{d}{D}\right)$$

(10.34)

For square columns B/D equals unity, and Equation (10.33) reduces to

$$\frac{d}{D} = \left(\frac{k}{\alpha}\right)^{\frac{1}{n+1}}$$

(10.35)

and the fire performance $t_{fi,d}$ is then given directly by

$$t_{fi,d} = \frac{D}{2\beta}\left(1-\left(\frac{k}{\alpha}\right)^{\frac{1}{n+1}}\right)$$

(10.36)

A simplified empirical relationship, also due to Lie, gives the fire resistance directly for axially loaded columns as

$$t_{fi,d} = 0{,}10 f D\left(3-\frac{D}{B}\right)$$

(10.37)

where f is a factor dependent upon the level of loading and the slenderness ratio and is given in Table 10.2. The dimensions are in millimetres with the charring rate taken as 0,6 mm/min.

For a square column, Equation (10.37) reduces to

$$t_{fi,d} = 0,20fD \tag{10.38}$$

Example 10.6: Check column of example 10.2 using both Lie approaches

Determination of k: The ultimate load-carrying capacity assuming medium-term duration load is 150 kN. The load in the fire limit state is 45 kN and $k = 45/150 = 0,30$. As the column is of medium slenderness ($\lambda \approx 100$, see Example 10.2), take $n = 2$. With $\alpha = 0,87$, Equation (10.36) becomes

$$t_{fi,d} = \frac{D}{2\beta}\left(1-\left(\frac{k}{\alpha}\right)^{\frac{1}{n+1}}\right) = \frac{150}{2\times 0,8}\left(1-\left(\frac{0,3}{0,87}\right)^{\frac{1}{1+2}}\right) = 28\,\text{min}$$

The column does not achieve 30 min. This is in line with the result from the reduced cross-section method.

Lie's approximate method [Equation (10.38)]: As the load level is below 50%, $k = 1,3$ (Table 10.2). $t_{fi,d} = 0,20fD = 0,2\times 1,3\times 150 = 39\,\text{min}$.

This result will be conservative as the charring rate is assumed as 0,6 mm/min, rather than 0,8 in the calculations.

10.2.5 ASCE empirical approach for columns

From Standard Calculation Methods for Structural Fire Protection (ASCE, 2007), the fire endurance period $t_{fi,d}$ is given by the following:

Column exposed on four sides:

$$t_{fi,d} = \gamma zD\left(3-\frac{D}{B}\right) \tag{10.39}$$

Column exposed on three sides (unexposed face is smaller dimension):

$$t_{fi,d} = \gamma zD\left(3-\frac{D}{2B}\right) \tag{10.40}$$

where γ takes a value of 0,1 mm/min, and the load factor z is given by the following:
For $\overline{L_e}/D \leq 11$:

$$z = 0,9+\frac{30}{r}\leq 1,5 \tag{10.41}$$

For $\overline{L_e}/D > 11$:

$$z = 0{,}7 + \frac{30}{r} \leq 1{,}3 \tag{10.42}$$

where $\overline{L_e}$ is the effective (or system) length and r is the ratio of applied load to allowable load expressed as a percentage.

> **Example 10.7: Checking column in example 10.2 using ASCE approach**
>
> The effective length $\overline{L_e}$ is 3 m and thus $\overline{L_e}/D = 3000/150 = 20$. Equation (10.42) should be used. As for Example 10.6, assume the column is fully loaded in the fire limit state, giving a value of r as 100%.
>
> $$z = 0{,}7 + \frac{30}{r} = 0{,}7 + \frac{30}{100} = 1{,}0$$
>
> Use Equation (10.39) to determine the fire resistance:
>
> $$t_{fi,d} = \gamma z D\left(3 - \frac{D}{B}\right) = 0{,}1 \times 1{,}0 \times 150\left(3 - \frac{150}{150}\right) = 30\,\text{min}$$
>
> This just satisfies the design requirement of 30 min.

10.2.6 Approach of Stiller

The full analytical work behind the equations developed by Stiller (1983) will not be covered here except to note that equations derived are based on a large body of research that attempted to fit calculated results for the performance of timber beams and axially loaded columns to the results from a large number of tests. The equations to determine the loss of section that allow the rounding of arrises and any strength losses in the core to be ignored were determined using regression analyses on various hypotheses until acceptable fits were found.

10.2.6.1 Beams

Stiller gives the depth of charring $d_{char,s}$ in millimeters after an exposure of $t_{fi,d}$ min to the standard furnace curve on the side faces of the beam as

$$d_{char,s} = 0{,}753 t_{fi,d} + 8{,}92 \tag{10.43}$$

and the depth of charring on the upper or lower face $d_{char,b}$ as

$$d_{char,b} = 1{,}472 d_{char,s} - 0{,}12 \tag{10.44}$$

The elastic section modulus of the section is calculated on the original section reduced by the charring depths calculated from Equations (10.43) and (10.44) and the bending strength determined on the reduced section.

10.2.6.2 Columns

For axially loaded columns exposed on all four sides, the charring depth $d_{char,col}$ is given by

$$d_{char,col} = 0,59t_{fi,d} + 6,4 \qquad (10.45)$$

The load-carrying capacity is then determined using the residual section with the axial compressive strength $\sigma_{w,t}$ determined using a strut buckling interaction equation which is given by

$$\frac{\sigma_{w,t}}{\sigma_d} = \frac{1}{2} + \frac{1+\eta}{2}\frac{\sigma_E}{\sigma_d} - \sqrt{\left(\frac{1}{2} + \frac{\sigma_E}{\sigma_d}\frac{1+\eta}{2}\right)^2 - \frac{\sigma_E}{\sigma_d}} \qquad (10.46)$$

where σ_d is the design compressive strength of the timber, E is the Young's modulus, σ_E is the Euler buckling stress calculated using the slenderness ratio λ determined on the reduced section, and the imperfection factor η is given by

$$\eta = 0,1 + \frac{\lambda}{200} \qquad (10.47)$$

Example 10.8: Determination of fire performance of beam in example 10.1 using Stiller approach

A timber beam (Grade C22) is 250 mm deep × 75 mm wide and is simply supported over a span of 4,5 m. The beam system is at 600 mm spacing and carries a permanent action on the beam of 0,2 kN/m and a total variable load of 2,5 kPa. The required fire resistance is 30 min.

Load in the fire limit state is $0,3 \times (0,6 \times 2,5) + 0,2 = 0,65$ kN/m. Maximum bending moment $M_{Ed,fi} = 0,65 \times 4,5^2/8 = 1,65$ kNm.

Calculate the loss of section for the side faces from Equation (10.43):

$$d_{char,s} = 0,753 \times 30 + 8,92 = 31,5 \text{ mm}$$

Calculate the loss of section from the bottom face from Equation (10.44):

$$d_{char,b} = 1,472 \times 31,5 - 0,12 = 46,2 \text{ mm}$$

Reduced width $b = B - 2d_{char,s} = 75 - 2 \times 31,5 = 12$ mm
Reduced depth $d = D - d_{char,b} = 250 - 46,2 = 203,8$ mm
  (i) Flexure:
      Elastic section modulus $Wel = bd2/6 = 0,083 \times 103$ mm3.
      Use 20% fractile strengths and stiffnesses determined in
        Example 10.1.
      $M_{Rd,fi} = 0,083 \times 27,5 = 2,28$ kNm which exceeds $M_{Ed,fi}$..
  (ii) Shear:
      $V_{Rd,fi} = 1,5f_{v,20}bd = 1,5 \times 3,0 \times 12 \times 203,8 = 11,0$ kN ($V_{Ed,fi} =$
      1,46 kN).
  (iii) Deflection:
      $I = 8,46 \times 10^6$ mm⁴.
      $\delta_b = 5ML^2/48EI = 5 \times 1,65 \times 4,5^2/(48 \times 8,5 \times 10^{-6} \times 8,46 \times 10^6)$
      $= 0,048$ m.
Span-deflection ratio $= 4,5/0,048 = 94$.

Although this is low, it would be acceptable in an accidental limit state.

Example 10.9: Column design using Stiller approach

Check the fire performance of a 150 mm square 3 m high column to
last 30 min carrying an axial load in the fire limit state of 45 kN. Use
Equation (10.45) to determine charring depth:

$d_{char,col} = 0,59 \times 30 + 6,4 = 24,1$ mm.
$d = D - 2d_{char,col} = 150 - 2 \times 24,1 = 101,8$ mm.
$i = d/2\sqrt{3} = 29,4$ mm.
$\lambda = L_e/i = 3000/29,4 = 102$.

$$\sigma_E = \frac{\pi^2 E_{20}}{\lambda^2} = \frac{8,5 \times 10^3 \pi^2}{102^2} = 8,06 \text{ MPa}$$

Take the design compressive stress σ_d as 20 MPa. From Equation (10.47):

$$\eta = 0,1 + \frac{\lambda}{200} = 0,1 + \frac{102}{200} = 0,61$$

From Equation (10.46):

$$\frac{\sigma_{w,t}}{\sigma_d} = \frac{1}{2} + \frac{1+\eta}{2}\frac{\sigma_E}{\sigma_d} - \sqrt{\left(\frac{1}{2} + \frac{\sigma_E}{\sigma_d}\frac{1+\eta}{2}\right)^2 - \frac{\sigma_E}{\sigma_d}}$$

$$= \frac{1}{2} + \frac{1,61}{2}\frac{8,06}{20} - \sqrt{\left(\frac{1}{2} + \frac{8,06}{20}\frac{1,61}{2}\right)^2 - \frac{8,06}{20}} = 0,298$$

or $\sigma_{w,t} = 0,298 \times 20 = 5,96$ MPa. Multiply this by 1,25 to give the 20% fractile:

$$N_{Rd,fi} = 1,25 \times \sigma_{w,t}d^2 = 1,25 \times 5,96 \times 101,8^2 = 77,2 \text{ kN}$$

This exceeds the design fire load.

10.3 TIMBER FLOORS AND PROTECTED TIMBER SYSTEMS

10.3.1 Timber floors

Where the joists of timber floors are directly exposed to fire, the methods outlined in the previous sections of this chapter may be used. Where there is plasterboard or similar protection to the underside of a floor or where additional insulation is between the joists, the above methods are not applicable. Either the method mentioned in the following section should be used or reference made to a publication about fire protection of timber floors by the Association of Specialist Fire Protection Contractors and Manufacturers (1993). This latter reference is also useful where existing floors must be upgraded or checked for increased fire resistance periods.

10.3.2 Protected timber systems

The earliest available data on the performance of stud walls were from Meyer-Ottens (1967) and included details of German fire resistance tests. The results developed the concept that the contributions of various parts of a system produced additive contributions to fire performance. For timber members in unfilled voided construction, the start of charring is delayed by a time t_{ch} given by

$$t_{ch} = t_f \tag{10.48}$$

where t_f is given:

For fire protective claddings and wood-based panels, thickness h_p:

$$t_f = \frac{h_p}{\beta_0} - 4 \tag{10.49}$$

For type A and H gypsum plasterboard for walls or floors with joist spacing less than 400 mm:

$$t_f = 2,8h_p - 11 \tag{10.50}$$

For type A and H gypsum plasterboard for walls or floors with joist spacing between 400 and 600 mm:

$$t_f = 2,8h_p - 12 \tag{10.51}$$

For the separating function of wall and floor assemblies to be satisfied, $t_{ins} > t_{req}$:

$$t_{ins} = \sum_i t_{ins,0,i} k_{pos} k_j \tag{10.52}$$

where $t_{ins,0,i}$ is the basic insulation value of layer i, k_{pos} is a position coefficient, and k_j is a joint coefficient. The values of $t_{ins,0,i}$ are given for a number of materials or finishes:

Plywood ($\rho \geq 450$ kg/m^3):

$$t_{ins,0,i} = 0,95h_p \tag{10.53}$$

Particleboard or fibreboard ($\rho \geq 600$ kg/m^3):

$$t_{ins,0,i} = 1,1h_p \tag{10.54}$$

Wood panelling ($\rho \geq 400$ kg/m^3):

$$t_{ins,0,i} = 0,5h_p \tag{10.55}$$

Gypsum plasterboard Types A, F, R, and H:

$$t_{ins,0,i} = 1,4h_p \tag{10.56}$$

If cavities are completely or partially filled to an insulation thickness of h_{ins}, then
Rock fibre:

$$t_{ins,0,i} = 0,2h_{ins}k_{dens} \tag{10.57}$$

Table 10.3 Values of k_{dens}

	Density (kg/m³)			
Material	15	20	26	50
Glass fibre	0,9	1,0	1,2	
Rock fibre			1,0	1,1

Source: Table E.2 of EN 1995-1-2. © British
Standards Institution.

Glass fibre:

$$t_{ins,0,i} = 0,1 h_{ins} k_{dens} \tag{10.58}$$

where the values of k_{dens} are related to the material type and density and are given in Table 10.3. For a void cavity with depths from 45 to 200 mm, $t_{ins,0}$ is taken as 5,0 min. The position coefficient k_{pos} is taken as appropriate from Tables E.3 through E.5 of EN 1995-1-2.

Chapter 11

Masonry, aluminium, plastics, and glass

This chapter covers the design of structural elements and structures fabricated from or involving the use of masonry, aluminium, or plastics. The plastics term includes the use of plastic-based composites. Notes about the behaviour of glass are also included.

11.1 MASONRY

There has been little development in the calculation of the performance of masonry in a fire, partly because one main use of masonry is for nonload-bearing fire-separating walls to form isolated compartments within large open plan structures. Often where masonry is used as external cladding, the wall carries no applied vertical load and it is only to resist the effect of horizontal wind loading.

In the UK, at least, load-bearing masonry designed to carry imposed vertical loads is used in housing (where the fire resistance requirements are low) or in single-storey structures such as sports halls (where the walls are often of diaphragm- or fin-type construction with far greater stability than conventional cavity wall systems and therefore tend not to be a problem). The other reason for so little calculation data on the effect of fire on masonry is the paucity of test data on the compressive and tensile strengths of masonry at elevated temperatures. Masonry generally acts as a separating element to ensure that the temperature on an unexposed face does not exceed the limit prescribed in the standard fire test of 140°C.

EN 1996-1-2 recommends three methods for assessing the fire resistance of masonry walls: testing, the use tabulated data, and calculation. The UK National Annex, however, does not support the use of calculation methods. Testing is required where no data are available for a particular combination of masonry unit and mortar. When using the tabulated data given in EN 1996-1-2, it is necessary to know the details of the masonry units used (density etc.), the grouping into which the units fall (see Table 3.1 in EN 1996-1-1), the type of mortar used, and applied finish.

If an element is load bearing, it is necessary to verify its load-bearing capacity in accordance with EN 1996-1-1 and carry out an additional check on the slenderness ratio of the wall. Having verified the load-bearing adequacy, the minimum thickness criterion to achieve the stated period of fire resistance for the material used can be obtained from the tables in Annex B to EN 1996-1-2. To use these tables for load-bearing walls, the level of loading determined by capacity calculation and masonry materials used must be known. The design load is compared with the design resistance of the wall to ascertain the percentage loading. If the percentage exceeds 60%, the tabulated values for 100% loading are used; otherwise the 60% loading tabulated values should be used (ISE, 2008).

11.1.1 Masonry construction insulation requirements

Tabulated requirements for the limit state of insulation for masonry units are given by specifying the wall thickness needed for a given type of brick and finish, e.g., plaster. Data on these requirements may be found in EN 1996-1-2 and ISE (2008).

Theoretically, it is also possible to perform a calculation procedure for determining the fire resistance of masonry walls. Heat transfer theory indicates that when a wall made of a given material is exposed to a heat source that maintains a constant temperature at the surface of the exposed side and the unexposed side is protected against heat loss, the unexposed side will attain a given temperature rise inversely proportional to the square of the wall's thickness. In the standard fire test, however, the fire temperature increases as the test proceeds. Therefore the time required to attain a given temperature rise on the unexposed side will be different from the time when the temperature on the exposed side remains constant. Based on data obtained from many fire tests, the following equation has been proposed to relate the fire resistance of a wall to its thickness (BIA, 2008),

$$R = (cV)^{1,7} \tag{11.1}$$

where R in hours is the fire resistance, c is a coefficient depending on the material and design of the wall, and V in cubic millimetres per square millimetre or mm^3/mm^2 is the volume of solid material per unit area of wall surface. The exponent coefficient 1,7 in Equation (11.1), instead of the theoretical value of 2, reflects the influence of rising temperature on the exposed surface, leading to shorter fire resistance.

For a wall composed of leaves or layers of multiple materials, the fire resistance of the wall can be calculated as follows (Brick Industry Association, 2008),

$$R = \left(R_1^{1/1,7} + R_2^{1/1,7} + \cdots + R_n^{1/1,7} \right)^{1,7} \tag{11.2}$$

where R_1, R_2, and R_n are the fire resistances of individual leaves.

For a wall with multiple leaves of brick, concrete masonry, or concrete, the fire resistance of the wall can be calculated as follows (ASCE/SFPE, 2003)

$$R = \left(R_1^{1/1,7} + R_2^{1/1,7} + \cdots + R_n^{1/1,7} + A_1 + A_2 + \ldots + A_m \right)^{1,7} \tag{11.3}$$

where $A_1 = A_2 = \ldots = A_m = 0{,}30$ are the air factors for each continuous air space having a distance of 12,7 to 89 mm between leaves.

The fire resistance of a wall made of a single material defined in Equation (11.1) is normally presented in terms of the equivalent thickness of the wall defined as follows:

$$t_{eff} = \frac{V_n}{LH} = \frac{WLH(1 - P_v)}{LH} = (1 - P_v)W \tag{11.4}$$

where t_{eff} is the equivalent thickness of the wall, V_n is the net volume of the wall, L, H, and W are the specified length, height, and width of the wall, respectively, and P_v is the percent void of the wall unit. Figure 11.1 to Figure 11.4 show the recommended fire resistance periods of masonry walls

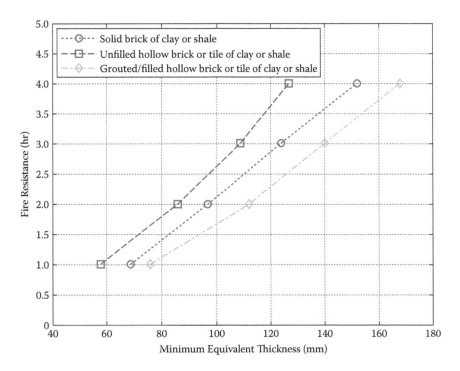

Figure 11.1 Fire resistance periods of clay masonry walls. (Data from Brick Industry Association, USA, 2008. With permission.)

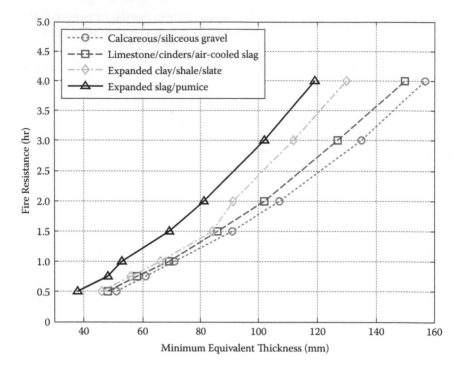

Figure 11.2 Fire resistance periods of concrete masonry walls. (Data from Brick Industry Association, USA, 2008. With permission.)

made of various materials as a function of the equivalent wall thickness by BIA (2008). Alternatively, a simplified equation for calculating the fire resistance of a concrete masonry wall with density less than 1100 kg/m³ has been proposed by de Vekey (2004) as follows:

$$R = (0,026 + 0,01P)t_{eff} \qquad (11.5)$$

where P is a coating factor taken as 0 for bare walls and 1 for plastered or rendered walls with a minimum layer thickness of 12 mm. The units of t_{eff} and R in Equation (11.5) are millimeters and hours, respectively.

Example 11.1: Fire resistance of cavity wall

Determine the fire resistance of a cavity wall with air space consisting of a leaf of 92-mm nominal solid brick units and cored at 25%, a 51-mm air space, and a leaf of 203-mm nominal concrete masonry unit

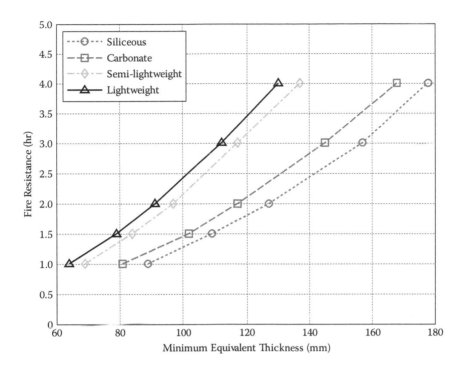

Figure 11.3 Fire resistance periods of normal weight concrete panel walls. (Data from Brick Industry Association, USA, 2008. With permission.)

made of calcareous gravel. The concrete masonry unit has dimensions of $194 \times 194 \times 397$ mm and is 53% solid.

From Equation (11.4), the equivalent thickness of the solid brick is

$$t_{eff} = (1 - P_v)W = (1 - 0,25) \times 92 = 69 \text{ mm}$$

From Figure 11.1, the fire resistance of the clay unit is

$$R_1 = 1,0 \text{ hr}$$

For a 51-mm air space:

$$A_1 = 0,30$$

From Equation (11.4), the equivalent thickness of the concrete masonry unit is

$$t_{eff} = (1 - P_v)W = 0,53 \times 194 = 103 \text{ mm}$$

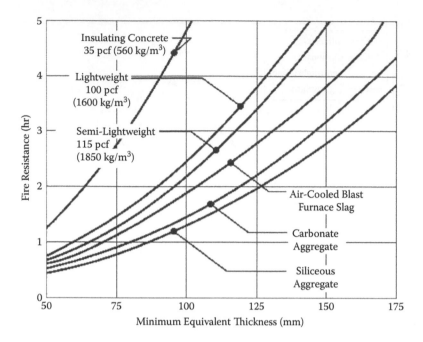

Figure 11.4 Fire resistance periods of other concrete panel walls (*From:* Brick Industry Association, USA, (2008. With permission.)

From Figure 11.2, the fire resistance of the concrete masonry unit is

$R_2 = 1,8$ hr

From Equation (11.3), the fire resistance of the entire wall assembly is

$$R = \left(R_1^{1/1,7} + R_2^{1/1,7} + A_1 \right)^{1,7} = \left(1,0^{1/1,7} + 1,8^{1/1,7} + 0,3 \right)^{1,7} = 5,5 \text{ hr.}$$

Example 11.2: Fire resistance of multilayered composite wall

Determine the fire resistance of a multi-leaf composite wall consisting of 92-mm nominal hollow brick with a mortared collar joint and 102-mm siliceous aggregate concrete wall. The gross volume of the hollow brick includes a 36% void.

From Equation (11.4), the equivalent thickness of the hollow brick is

$$t_{eff} = (1 - P_v)W = (1 - 0,36) \times 92 = 58,8 \text{ mm}$$

From Figure 11.1, the fire resistance of the hollow brick is

$R_1 = 1,0$ hr

From Figure 11.4, the fire resistance of the concrete wall with siliceous aggregate is

$$R_2 = 1,3 \text{ hr}$$

From Equation (11.3), the fire resistance of the entire wall assembly is

$$R = \left(R_1^{1/1,7} + R_2^{1/1,7} \right)^{1,7} = (1,0^{1/1,7} + 1,3^{1/1,7})^{1,7} = 3,7 \text{ hr.}$$

11.1.2 Thermal bowing

When a wall is subjected to temperature gradient across its thickness, the wall may be considered as a one-dimensional member (i.e., no lateral boundary affects unrestrained bowing). The calculation of bowing is given by Cooke (1987a and b) and Cooke and Morgan (1988):

For a cantilever:

$$\delta_{bow,c} = \frac{\alpha_m h_{wall}^2 \Delta\theta}{2d_{wall}} \tag{11.6}$$

where $\delta_{bow,c}$ is the deflection of the cantilever, $\Delta\theta$ is the temperature gradient across the leaf, d_{wall} is the thickness, α_m is the coefficient of thermal expansion, and h_{wall} is the height of the wall.

For a simply supported beam:

$$\delta_{bow,b} = \frac{\alpha_m h_{wall}^2 \Delta\theta}{8d_{wall}} \tag{11.7}$$

where $\delta_{bow,b}$ is the central defection of the beam.

For a cantilever element, the deflection is four times that for a simply supported beam element with the same temperature gradient, thickness, and coefficient of thermal expansion. Ideally, separating walls should be held at the top and bottom by the supporting structure. However, the deflections calculated from Equation (11.6) or Equation (11.7) are unlikely to be attained in practice. The aspect ratio of a tall fire-separating wall is likely to be such that the vertical restraint along the sides of wall (by the supporting structure) will cause the wall to span horizontally and reduce the deflection substantially.

A wall with no vertical or top restraint will show some bowing, although this will be reduced in the initial stages by the effects of any vertical loading applied to the wall. The vertical load is likely to cause problems only after a substantial period when the wall may tend to become laterally unstable with the vertical load-inducing high tensile stresses from moments induced by lateral deformations. The design guide issued in 1992 by the Canadian Concrete and Masonry Codes Council provides information on the forces to be resisted by horizontal members to counter sagging effects caused by the bowing of walls during a fire.

11.1.3 Load-bearing cavity walls

The most work on the testing and analysis of loaded cavity walling appears to have been carried out in Australia, although recent work was done at Ulster University (Laverty et al., 2000/1; Nadjai et al., 2001) where it was observed that the effect of increasing load (i.e., increasing compressive stress levels) may offset the potential deleterious effect of increasing slenderness.

A comprehensive series of tests on 230 mm and 270 mm thick cavity wall with various load levels were carried out in Australia (Gnanakrishnan et al., 1988). The results are given in Table 11.1. It was observed that loading on the leaf adjacent to the fire appeared far more critical than load levels on unheated or external leaves. The application of loading on a hot leaf reduces lateral deflection or thermal bowing. The application of loading on an external or cold leaf appears to stabilize a wall and also reduce deflections. Measurements during the tests indicated that the temperature rises in the external leaf were relatively low with a substantial gradient across the 50-mm cavity. Some attempt was also made to model the behaviour of

Table 11.1 Experimental results of fire tests on cavity walls

Wall type*	Loading[†]		Maximum Deflection (mm)	Time to Failure (min)
	Internal	External		
230/90	125	0	74	68
230/90	75	0	87	47[‡]
230/90	25	0	87	39[‡]
230/90	160	0	58	22
230/90	125	125	60	183[§]/240[‡]
230/90	0	160	18	240
270/110	0	240	–	300
270/110	80	0	75	50
270/110	100	0	70	34

*First figure for wall type is the overall thickness (mm) and the second is the thickness (mm) of each leaf built from clay bricks.

†Loading is given as a percentage of working (or service) loading.

‡Indication that failure occurred by excessive deflection.

§Indicates initial failure of internal leaf before whole wall failed.

Source: Adapted from Gnanakrishnan, N. et al. (1988) In *Proceedings of Eighth International Brick and Block Masonry Conference*, pp. 981–990. With permission.

the cavity walls, but due to the assumptions made and the paucity of high strength data on masonry, the results show only correct trends.

11.2 ALUMINIUM

The calculation procedure to determine the fire performance of aluminium is very similar to that for steelwork. However, compared to steel, aluminium has high thermal conductivity, low thermal capacity, and low melting temperature (between 580 and 650°C). These factors make it less favourable in fire conditions. In addition, the limiting temperature for aluminium is around 200°C (Bayley, 1992). Above this temperature, the strength loss is such that any factor of safety in the ambient design is completely eroded. The limiting temperature is taken as a function of the exact aluminium alloy in use as the temperature related to strength loss is very dependent on the amounts and types of alloying constituents.

11.2.1 Temperature calculations

Based on the fact that aluminium heats faster than steel, one can use the assumption of uniform temperature distribution to calculate the temperature of an aluminium member exposed to a fire. EN 1999-1-2 provides two equations to calculate the temperature for unprotected [Equation (11.8)] and protected [Equation (11.9)] aluminium members:

$$\Delta\theta_{al} = \frac{k_{sh}}{c_{al}\rho_{al}} \frac{A_m}{V} \times (\dot{h}_{net}\Delta t) \tag{11.8}$$

$$\Delta\theta_{al} = \max\left\{ \frac{\lambda_p/d_p}{c_{al}\rho_{al}} \frac{A_p}{V} \left[\frac{1}{1+\phi/3} \right] (\theta_g - \theta_{al})\Delta t - (e^{\phi/10} - 1)\Delta\theta_g, \quad 0 \right\} \tag{11.9}$$

in which

$$\phi = \frac{c_p\rho_p}{c_{al}\rho_{al}} \frac{d_p A_p}{V} \tag{11.10}$$

The notation used here is the same as that used in Chapter 8 for steel except for the *al* subscript that represents aluminium. The determination of section factors A_m/V and A_p/V is also the same as that used for structural steel. Section factors of typical structural aluminium sections can be found in EN 1999-1-2. It should be noted that the member temperature calculated using Equation (11.8) or (11.9) is uniform within a cross section

even when the member is not exposed on all sides. In reality, the temperature at the unexposed side will be lower than that at the exposed side.
Also, the thermal properties of insulation materials may be temperature-
dependent. In this case, the approach of using aluminium temperature to
calculate the thermal properties of insulation materials must be validated
by tests.

Example 11.3: Bridge metal temperatures

A bridge is required to link an accommodation platform to a process
platform. The bridge span is of the order of 100 m and consists of a
triangular truss with apex along the top chord. The truss uses either
circular steel tubular members with outside diameters of 457 mm and
thickness of 20 mm or circular aluminium tubular members with outside diameters of 432 mm and thickness of 35 mm to produce an equal
moment-carrying capacity at mid-span. Calculate the temperatures in
steel and aluminium for the bridge when there is a sea pool fire under
the bridge.

Assume that the fire temperature θ_g is described by Equation (3.5).
The temperature in the structural member θ_k thus can be calculated
using Equation (11.8) as follows:

$$
\Delta\theta_k = \frac{k_{sh}}{c_k\rho_k}\frac{A_m}{V}\times(\dot{h}_{net}\Delta t) = \frac{k_{sh}}{c_k\rho_k}\frac{A_m}{V}
$$
$$
\times\left\{\alpha_c(\theta_g-\theta_k)+\varepsilon_k\sigma[(\theta_g+273)^4-(\theta_k+273)^4]\right\}\Delta t
$$

(11.11)

Figure 11.5 plots the evolution of temperatures in steel and aluminium
members obtained by solving Equation (11.11) numerically with the
parametric values defined in Table 11.2. The figure indicates that the
temperatures of the steel and aluminium of clean surfaces are very
close, but both are lower than the temperature of the aluminium of
sooted surfaces. The lower temperature of aluminium occurs because
its low thermal capacity is compensated by its low emissivity, while
the sooted aluminium has an identical emissivity as the steel. Based on
the figure, the aluminium bridge will reach the limiting temperature of
200°C after 9 min if the members are sooted and after 13 min if the
members are clean. This indicates that fire protection is needed for the
bridge if it uses aluminum members.

Example 11.4: Aluminium bridge temperature history

Calculate the temperature history of the aluminium bridge described
in Example 11.3 if all members are insulated by glass wool with
a thickness d_p = 10 mm. The properties of the glass wool are:
$\lambda_p = 2,7\times10^{-7}\theta_{al}^2 +1,0\times10^{-4}\theta_{al} +0,031$ W/m°C, ρ_p = 60 kg/m³, and
c_p = 1030 J/kg°C.

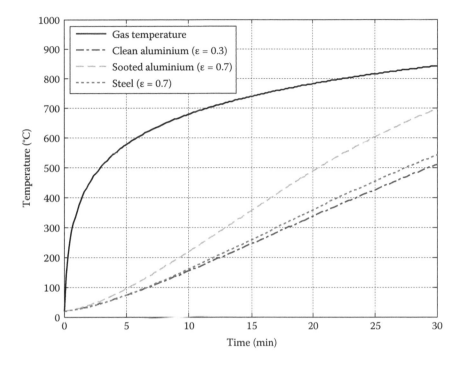

Figure 11.5 Temperature-time histories of aluminium and steel bridges in a fire.

Use Equations (11.9) and (11.10) by assuming $A_p/V = A_m/V = 1/t = 1/0,035$. The numerical results of bridge temperatures are plotted in Figure 11.6. For the purpose of comparison, the temperature of the bridge without insulation is also superimposed on the figure. The insulated aluminium bridge will be at only about 140°C after an hour of the fire, whereas it would have melted without insulation.

Table 11.2 Parametric values used in examples 11.3 and 11.4

Parameter	Aluminium	Steel
k_{sh}	1	1
c_k (J/kg°C)	Equation (5.28)	Equations (5.4)–(5.7)
ρ_k (kg/m³)	2700	7850
A_m/V (1/m)	1/0,035	1/0,02
α_c (W/m²°C)	25	25
ε_k	0,3/0,7*	0,7
σ (W/m²°C)	5,67 × 10⁻⁸	5,67 × 10⁻⁸
θ_g (°C)	Equation (3.5)	Equation (3.5)

*ε_k = 0,3 for clean surfaces and 0,7 for sooted surfaces.

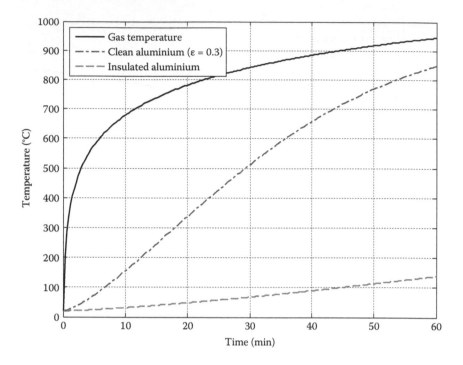

Figure 11.6 Temperature–time histories of noninsulated and insulated aluminium bridges in a fire.

11.2.2 Structural design of aluminium members

Structural design of aluminium members can be carried out using advanced calculation models such as finite element analysis models or simple calculation models such as those recommended in various design standards. Normally, the simple calculation models can be applied only when a structure is divided into a number of separated members and each member is analysed individually (EN 1999-1-1, EN 1999-1-2).

A member in tension should be checked to determine whether the design resistance of the gross cross section, taking into account the reduction of material properties at elevated temperatures, is greater than the design load. EN 1999-1-2 gives rules both for uniformly and nonuniformly distributed temperatures.

In the case of nonuniform temperature distribution, the contribution of each part of the cross section with a specific temperature is taken into account by using the 0,2% proof strength at that temperature. In the case of welding, the strength of the heat-affected zone should be determined by multiplying the 0,2% proof strength of the parent material at elevated temperature with the reduction coefficient for the weld at room temperature.

For a member in bending, the design moment resistance of a cross section with a uniform temperature distribution can be calculated based on the ambient design moment resistance by multiplying the 0,2% proof strength reduction factor at that temperature. If the temperature is not uniformly distributed within the cross section, the calculation of the design moment resistance depends on the class of the cross section.

For class 1 or 2 cross sections, the design moment resistance needs to be calculated by considering the contribution of each part of the cross section with a specific temperature and using the 0,2% proof strength at that temperature. For class 3 or 4 cross sections, the design moment resistance can be calculated still based on the ambient design moment resistance by multiplying the 0,2% proof strength reduction factor at the temperature equal to the maximum temperature of the cross-section reached at time t.

For a member in compression or bending, flexural, torsional and lateral-torsional buckling checks should also be carried out. For aluminium, the ratio between strength and Young's modulus decreases with increased temperature. Therefore, the relative slenderness used in the fire design according to EN 1999-1-2 is determined with the material properties at ambient temperature. The buckling resistance with this relative slenderness is also determined by using the buckling curve obtained at ambient temperature. This approach applies to both members in compression (flexural or torsional buckling) and in bending (lateral-torsional buckling).

For considering the effect of creep on lateral deflection during buckling, the buckling resistance of an aluminium column at elevated temperature is further reduced by a factor of 1,2, i.e., it equals the buckling resistance obtained at ambient temperature by multiplying the 0,2% proof strength reduction factor at that temperature and dividing by 1,2.

11.3 PLASTICS AND PLASTIC-BASED COMPOSITES

Plastics and especially plastic-based composites are structurally very efficient due to their weight-to-strength ratios but perform poorly when exposed to the effects of fire. This means that such materials need extensive levels of protection to retain load-carrying capacity at elevated temperatures. This implies that the insulation thicknesses should ensure that the temperatures within the plastic elements must be kept close to ambient. There is an additional problem in that some plastics decompose with temperature and emit highly inflammatory or toxic gases.

With increasing use of plastic-based composites, knowledge of their fire performance becomes an important issue. This is particularly so for composites applied in the aircraft, marine, and petroleum industries. The heat from a fire may weaken polymers and cause eventual creep and structural failure as the temperature exceeds the glass transition temperature.

In addition, a polymer may ignite and spread flames, releasing further heat and potentially toxic smoke.

However, composites are by their nature inherently fire resistant. The inert fibre reinforcement displaces polymer resin during a fire and thus removes fuel. When the outermost layers of a composite laminate lose their resin, they act as an insulating layer, slowing heat penetration. With careful design and the right choice of resins, additives, and fillers, plastic-based composites can be used to build structures with better fire resistance. Typical examples are the phenolic compounds used in firewalls.

Bishop and Sheard (1992) report tests carried out under exposure to hydrocarbon rather than the cellulosic standard curve for fire protection of pultruded phenolic resin systems. They note that even though the levels of protection needed are higher than those for steelwork, the resultant strength-to-weight ratio is still more favourable for resins than for steel. Unfortunately, the temperature levels for which the insulation was designed when applied to the extruded sections is not stated.

Wong (2003) reports test results on glass fibre-reinforced polymer (GFRP) stringer wall systems and material properties. The compressive strength drops from around 275 MPa at ambient to around 85 MPa at 90°C and 20 MPa at 250°C. The values of Young's modulus at the same temperatures are 22,25, 15,6 and 6,6 GPa, respectively. Results comparing the elevated temperature tensile strengths of carbon fibre-reinforced plastic (CFRP) and glass fibre-reinforced plastic are also presented. Both materials show strength losses of about 84% at 500°C, although in absolute terms CFRP is stronger. No size effects were noted on GFRP specimens between 9,5 and 12,7 mm in diameter.

11.4 GLASS

Glass is commonly used in modern buildings. Conventional glass softens around 700 to 800°C although shattering can occur below these temperatures where load is applied through expansion of the system used to support glazing. Heated glass usually fails by the strains induced by heating too quickly on one side. It is generally considered that any glazing will fail in the very early stages of fire, and thus any openings normally glazed can be considered effective ventilation sources. Roof lights will allow a fire to vent.

Where glass is required to resist the effects of fire or acts as a fire-rated barrier, for example, in fire doors, fire-resistant glazed systems should be used. Examples include wired glass (wire mesh embedded within the glass), modified toughened soda lime glass, glass ceramic, toughened borosilicate, and fire-resistant laminated interlayer (clear or wired) glass. In all cases, specific performance details should be obtained directly from the manufacturer, since performance varies from one glazed system to another.

Fire-resistant glazed systems play a major role in modern fire protection strategies for buildings because of the importance of glass in modern transparent architecture. This role includes compartmentation and fire separation, protection of escape and access corridors, and use in fire doors and facades designed to limit external fire spread from floor to floor or building to building. Applications can be vertical, horizontal, or inclined and can cover the following situations (Wood, 2004):

- Internal and external fire doors
- Interior partitions and compartment walls
- Roofs and floors or ceilings
- Façade glazing
- Escape and access corridors
- Stairways, lobbies, and enclosures of protected shafts

The application of fire-resistant glazed systems in England and Wales is covered by the guidance provided in Part B of Schedule 1 to the building regulations (or its equivalent in Scotland and Northern Ireland). The regulations require fire resistance performance to be determined by reference to national tests such as the BS 476: Part 10 (2009) or European tests such as BS EN-1364 (1999).

Manzello et al. (2007) conducted a fire performance test of a nonloadbearing wall assembly of four glass sections, two of which were fitted with tempered double-pane glass and the other two with tempered single-pane glass. The wall assembly was exposed to an intensive real-scale compartment fire. The fallout temperature and corresponding heat flux at the exposed surfaces of the single-pane glass panels were found to be about 400°C and 50 kW/m^2, respectively—similar to values reported in the literature (Shields et al., 2001a and 2001b). The fallout temperature of the unexposed surface of the double-pane glass panel was around 80 to 120°C, and the measured total heat flux at the unexposed surface of the double-pane gel filled glass sections was only around 1 to 2 kW/m^2.

Chapter 12

Frames

Whereas the previous chapters dealt with the design aspects of isolated structural elements when exposed to the effects of fire, this chapter is concerned with how those elements behave when connected to form part of a structure and how the continuity effects generated when elements are connected can be taken advantage of in a design. Rather than repeat recent material from a number of papers, tests on the various frames at Cardington are covered here to place the emphasis on the implications for design. Additionally, the performance of connections and portal frames will be highlighted.

12.1 TESTS ON ISOLATED FRAMES

The earliest large-scale test reported by Cooke and Latham (1987) was on a frame heated by a natural compartment fire. The frame was unprotected steelwork, and the stanchion webs were filled in by nonload-bearing blockwork that had already been shown to increase the fire resistance of bare steelwork (Building Research Establishment, 1986; Robinson and Latham, 1986). The stanchions had pinned feet and were 203 × 203 × 52 UC (3,53 m long). The rafter, which was laterally unrestrained and connected to the columns by flexible end plates, was a 406 × 178 × 54 UB (4,55 m long). Both members were of Grade 43A (S275) steel and carried full service loading comprising 39,6 kN vertical loads applied at fifth points on the rafter and axial loads of 552 kN applied at the top of each stanchion.

The test had to be terminated due to the inability of the load to be applied because excessive deflections occurred after 22 min in a natural fire deemed equivalent to 32,5 min of a standard furnace test. The column alone had a fire resistance of 36 min, and the beam that failed from the occurrence of plastic hinges had a fire resistance of 22 min. This test surely started to demonstrate the beneficial effects of continuity and led to the concept of testing on a full-size multi-storey, multi-bay structure at Cardington.

327

12.2 TESTS ON LARGE FRAME STRUCTURES AT CARDINGTON

12.2.1 Timber frame structures

The timber frame structure at Cardington was subjected to two fire tests. The first one was effectively designed to evaluate the fire spread (and compartmentation) and the second was to determine the performance of stairs subject to arson.

In the first test, a timber crib fire was ignited in a corner flat on the level 3 storey. Peak temperatures reached around 1000°C in the compartment. The fire lasted about 64 min including a pre-flashover of 24 min, and was terminated after the fixity of the ceiling boards was lost and the joists were subjected to about 8 min of direct fire exposure. The temperatures outside the fire compartment did not exceed 50°C except for some areas in the cavities that reached 100°C. The results from this test allowed a relaxation of the England and Wales building regulations on timber frame construction and brought the Scotland and Northern Ireland regulations into line.

Preliminary tests in a purpose-built test rig demonstrated that the stair system needed fire-retardant treatment (i.e., raw timber burnt out too quickly). The test demonstrated no fire spread from the ignition sources. The fire lasted 31 min, and the stairs were perfectly usable (Enjily, 2003).

12.2.2 Concrete frame structures

A single concrete frame structure was constructed. It was slightly unusual in that it was a flat slab with diagonal steel flats acting as bracing. Also, the columns were kept at the same size throughout the structure, and the concrete strength was increased in the columns in the lower storeys. The structure was designed for testing the optimization of the execution process rather than investigating fire performance although fire tests were envisaged from the initiation of the project (Chana and Price, 2003).

The structure had seven storeys with five sets of columns in one direction and four in the other, all at 7,5 m spacing. Each storey was 3750 mm high from soffit to soffit. The slab was 250 mm thick (Grade C37 concrete with flint-type aggregate). The internal columns were 400 mm square, and the external columns 400 mm × 250 mm. The columns in the lower storeys were constructed using Grade 85 high strength concrete with limestone aggregate and microsilica. They also contained 2,7 kg/m^3 polypropylene fibres.

The test was carried out in September 2001 shortly before the Cardington Test Facility was closed. The technical data are taken from Bailey (2002). Since the authors were present at the test and able to inspect the structure shortly after the test, the interpretations of the results are those of the authors.

The compartment chosen for the fire test involved four bays of the lowest storey of the frame surrounding one of the square internal columns. The design fire was parametric fire with an opening factor of 0,08 m^{0,5} and an insulation factor of 1104 J/m^{2}s^{0,5}K that yields a maximum temperature around 1050°C at about 35 min. The fire load was 40 kg/m^2 timber cribs, equivalent to 720 MJ/m^2.

At 18 min, the instrumentation failed when the ceramic blanket that was shot-fired to the soffit of the slab at the top of the blockwork and plaster-board detached due to the severe spalling of the slab. Part of the reason for the slab spalling was the aggregate type (flint), but the spalling was exacer-bated by the high strength of the concrete (61 MPa cube strength or around 50 MPa cylinder strength at 28 days), the high moisture content of 3,8% by weight, and a permeability of 6,75 × 10^{-17} m^2. The structure was around 3,5 years old at the time of the test. Using strength data gain for normal cements from EN 1992-1-2, the cylinder strength at the time of test may have been around 1,35 × 50 = 67 MPa. This would put the slab concrete into the high strength category.

Figure 9 in Tenchev and Purnell (2005) indicates spalling in the slab would be expected at around 8 to 10 min at a depth of 10 mm if the fire exposure met the standard furnace curve. In the actual test, spalling com-menced at 6 to 7 min after ignition. The depth of spalling was around 20 to 25 mm, and much of the bottom reinforcement was exposed. The test also indicated the possible need to consider both maximum and minimum strength requirements for concrete.

The central column (Grade C85) with a cube strength of 103 MPa (cyl-inder strength 80 MPa) at 28 days did not spall. The moisture content was 4,4%, and the permeability was 1,92 × 10^{-19} m^2 with an estimated cylinder strength at the time of test slightly over 100 MPa. After the test, cracking was observed at the corners of the column. When the cracking occurred is not known. It should be noted that the moisture contents were high because the building was unable to dry out as a normal structure in open air would.

The structural variable fire load was 3,25 kPa above the fire compart-ment. This gave loads of 925 kN in the central column and 463 kN in the edge column.

At the time of failure of the instrumentation, the slab deflections were between 10 and 60 mm, with the higher values generally in the centres of the slabs or where severe spalling occurred. The residual deflections in the slab on average were around 60 mm.

More worrying was the lateral deflection of the edge column that caused buckling of the steel flats forming the bracing. The maximum deflection during the test exceeded 100 mm. The residual deflection of this column at first floor level was 67 mm. Cracking was also observed at the column-slab interface at first floor level.

Although the Cardington test was useful because it showed that the type of frame tested could satisfactorily resist a compartment fire, a number of issues remain unresolved:

1. A single fire test is inadequate to allow full calibration of any computer simulation. This is in part due to the unfortunate loss of temperature and deformation data in the test due in part to the need to consider other severity and position fire scenarios.
2. The reasonable flexibility of the structure in the horizontal direction may have alleviated the effects of spalling in the slab. A much stiffer structure with lift shafts and stairwells may have induced higher compressive stresses into the slab and hence increased spalling levels.

12.2.3 Composite steel frames

The eight-storey structure high plan with for five bays (each 9 m long) by three bays (each 6 m wide) surrounding a central bay of 9 m with an approximate floor area of 945 m². The soffit-to-soffit height was 4 m. The primary beams were $356 \times 171 \times 51$ UB Grade S355 on the 6-m spans and $610 \times 229 \times 101$ UB Grade S275 on the 9-m spans. The secondary beams for all nine spans were $305 \times 165 \times 40$ UB Grade S755. The perimeter beams were also $356 \times 171 \times 51$ UB Grade S355. The upper-storey columns were $254 \times 254 \times 89$ UC Grade S355 and $305 \times 305 \times 137$ UC Grade S355. The composite deck was lightweight concrete (cube strength 47 MPa, cylinder strength 38 MPa).

Six main tests involved applied loading in each test of 2,663 kPa (Moore and Lennon, 1997) with a subsequent test on an edge compartment with a static fire load of 5,0 kPa. The summary below is in part taken from Moore and Lennon, with additional information from Bailey et al. (1999); Izzuddin and Moore (2002); O'Connor et al. (2003); Newman et al. (2006), and Wang et al. (2013).

Test 1: Restrained beam — The test was carried out on one of the secondary composite beams with a gas-fired furnace surrounding the central 8 m of the 9-m span of a beam at level 7 with a floor area of 24 m². The heating rate was 3 to 10°C/min. At a maximum temperature of 875°C in the lower flange, a maximum vertical displacement of 232 mm was recorded. The residual deflection was 133 mm. Local buckling of the bottom flange at the ends of the beams and failure of the end plate connections at both ends were observed. It appeared that the failure of the end-plates was gradual and due to the high tensile forces induced during cooling.

Test 2: Plane frame — This test was conducted on a complete plane frame across the structure at level 4 with a floor area of 53 m² using a purpose-built furnace measuring 21 m long by 3 m high. The floor beams and connections were unprotected. The columns were protected to within 200 mm below the connections. At a steel temperature of 800°C, a maximum vertical displacement of 445 mm was recorded. This test also produced distortional buckling of the column heads with the columns shortening by around 200 mm. This led to fire protection of the columns and connections in later tests. The secondary beams were heated over a length of around 1 m. An inspection after the test indicated shearing of the bolts in the fin plate connections. Again, this was due to the high tensile forces induced during cooling.

Test 3: Corner compartment No. 1 — This was conducted under a natural fire regime with timber cribs providing a fire load of 45 kg/m². Ventilation was controlled by a moveable shutter on one face. The opening factor was initially 0,031 $m^{0,5}$ and increased slightly to 0,034 $m^{0,5}$. It was estimated that the heating regime had a time equivalent around 85 min at level 2 with a floor area of 76 m². The internal compartment wall was placed slightly eccentric to the 9-m internal beam and under the 6-m internal beam. In each case, a 15-mm deflection allowance was inbuilt. Little damage to the block-work compartment walls was observed. The perimeter beams and columns were fire protected but the internal beams were left bare. Temperatures exceeding 1000°C were recorded with an associated maximum displacement of 425 mm. The maximum slab deflection was 269 mm in the centre of the compartment. After cooling, this deflection recovered to 160 mm.

Test 4: Corner compartment No. 2 — This was carried out on a corner compartment 9 m × 6 m at level 3 with a fire load of 40 kg/m². The compartment walls were constructed of fire-resistant boarding running between the boundary columns. The columns were fire protected up to and including the connections. One external face (length 9 m) was formed using double-glazed aluminium screen. The fire initially did not flashover due to lack of oxygen and two panes of glass had to be removed to achieve flashover. Finally, the maximum compartment temperature was around 1050°C; the temperature recorded on the lower flange of the unprotected beam was 903°C. The maximum slab deflection was 270 mm. The stud walls suffered severe damage owing to the beam deflections. Additionally, local buckling of the beams at the connections and end plate fractures in the connections during cooling were observed. A substantial amount of cracking in upper surface of the composite slab around the column was also observed.

Test 5: Large compartment — The compartment was designed to be representative of an open plan office with dimensions of 18 m × 21 m at level 3. The fire load was 40 kg/m² timber cribs. The compartment was bounded by erecting a fire-resistant wall across the whole structure. The lift shaft was also provided with additional protection. Double glazing was installed on two sides of the building with the middle third of the open area on each side left open for ventilation. All the beams were left unprotected, whilst internal and external columns were protected up to and including the connections. The fire started sluggishly. However, because of a cross draught, the maximum temperatures were reduced. The recorded maximum temperature was 760°C, with the steelwork some 60°C lower. The maximum slab deflection was 557 mm which recovered to 481 mm after cooling. Extensive local buckling was found at the beam-to-beam connections with a number of failures of end plates. One case of a complete fracture between the beam web and end plate was noted. The shear extended to the composite floor above the connection causing large cracks in the floor but no collapse.

Test 6: Demonstration compartment — The compartment had an area of 180 m² at level 2 and was filled with office furniture (and timber cribs) to give a fire load of 45 kg/m² timber equivalent. Ventilation was provided by a combination of blank openings and windows. All the beams were left unprotected. Internal and external columns were protected up to and including the connections. At around 10 min, the temperature at the rear of the compartment exceeded 900°C and reached an eventual maximum of 1213°C. The recorded maximum steel temperatures were in excess of 1100°C in the floor beams and 1012 to 1055° C in the primary beams. The maximum deflections were around 640 mm on one of the secondary beams. The residual deflection after cooling was 540 mm. Cracking of the floor around one of the column heads was also observed due, in part, to inadequate lapping of the mesh in the top of the slab.

Test 7: Edge compartment — The compartment on the fourth storey was 11 m × 7 m (floor area of 77 m²) and was symmetrically arranged around two edge columns and two internal columns. The fire load was 40 kg/m² timber equivalent. The maximum gas temperature of 1108°C was reached after 55 min, and the maximum steel temperature reached was 1088°C. The maximum deflection exceeded 1 m with a corresponding residual deflection of 925 mm.

Observations — The following are the observations from the tests.

1. Clearly the overall behaviour of members in a frame is far superior to that predicted by isolated furnace tests. This is due to continuity and also to alternative load paths or load-carrying mechanisms.
2. Temperature levels were far higher than those that would be permitted by design codes, even allowing for the load levels under applied variable loads of around one third of the ambient variable imposed load (O'Connor et al., 2003).
3. These loads were high but did not cause failure. The values attained only confirm that the deflection limits imposed in the standard furnace test are not applicable to structural behaviour.
4. The values of residual deflections are high; i.e., the deflections in the test were effectively plastic, and hence irrecoverable. Some of the residual deflections of the slabs were due to the beams. Photographic evidence from the tests indicates large residual deflections of the beam system.
5. Although the connections performed well during the heating phase, it was clear that the cooling phase may have caused problems due to high tensile forces induced into the beam systems. These tensile forces may cause failures of end plates, welds and/or bolt shear.

The performance of a frame under test conditions clearly has implications for design, but these implications can only be realized if designers understand how a frame actually behaves. This can be achieved only through computer simulation. Much effort has been expended on this endeavor and the results from the Cardington Tests can be reproduced (Bailey et al., 1996; Plank et al., 1997; Bailey, 1998; Huang et al., 1999; Sanad et al., 2000).

One immediate conclusion was that secondary beams could go unprotected subject to limits on position and fire endurance (Bailey and Newman, 1998; Newman et al., 2006). This work also led to the identification of additional load-carrying mechanisms, e.g., beside acting in flexure, a slab also generates membrane action (Bailey and Moore, 2000a and b; Huang et al., 2003a and b). The membrane action is generated by a compression ring around the edges of the slab and a central tension zone. Bailey (2001) produced a design guide presenting a simplified approach and later developed a fully designed approach (2003) that may be outlined as follows.

The total load capacity is based on unprotected beams within the area considered plus loading on the slab determined by yield-line response. However, the yield-line load capacity may be enhanced due to membrane action. Thus, the total load $q_{p\theta}$ that may be carried is given by

$$q_{p\theta} = e\left(\frac{WD_{slab,int}}{WD_{floor,ext}}\right) + \frac{WD_{beam,int}}{WD_{floor,ext}} \qquad (12.1)$$

where e is an enhancement factor due to membrane action, $WD_{slab,int}$ is the internal work done by the slab, $WD_{beam,int}$ is the internal work done by the beam, and $WD_{floor,ext}$ is the work done by the loading. Equation (12.1) can be simplified to

$$q_{p\theta} = eq_{p\theta,slab} + q_{p\theta,udl} \qquad (12.2)$$

where $q_{p\theta,slab}$ is the load carried by the slab and $q_{p\theta,udl}$ is the uniformly distributed load (per unit area) supported by the beam. From conventional yield-line analysis of a rectangular slab under sagging moments, $q_{p\theta,slab}$ is given by

$$q_{p\theta,slab} = \frac{24m_{p\theta}}{(\alpha L)^2 \left[\sqrt{3+\alpha^2} - \alpha\right]^2} \qquad (12.3)$$

where L is the longer side of the slab, α is the aspect ratio of the slab (<1,0), and $m_{p\theta}$ is the temperature-reduced value of the sagging moment that may be based solely on the strength loss of the reinforcement as the concrete temperature will be relatively low (since the neutral axis depth is small), and any contribution from the profile sheet steel decking is ignored.

The value of $q_{p\theta,beam}$ over the area supported by an internal beam is determined using a temperature-modified flexural strength M_θ based on the temperature of the lower flange and that the beam may be treated as simply supported, i.e.,

$$q_{p\theta,beam} = \frac{8M_\theta}{L^2} \tag{12.4}$$

The enhancement factor e is a function of the vertical displacement to effective depth ratio (w/d), the inverse of the aspect ratio ($1/\alpha$), and a parameter g_0 related to the depth of the compression block d_c as follows:

$$g_0 = 1 - \frac{2d_c}{d} \tag{12.5}$$

where d is the effective depth of the slab. The maximum allowable vertical deflection w is given by

$$w = \frac{\alpha_c(\theta_2 - \theta_1)l^2}{19,2h} + \sqrt{\frac{3L^2}{8} \frac{0,5f_{y,20}}{E_{s,20}}} \leq \frac{\alpha_c(\theta_2 - \theta_1)l^2}{19,2h} + \frac{l}{30} \tag{12.6}$$

where α_c is the coefficient of thermal expansion of concrete ($18 \times 10^{-6}/°C$ for normal weight concrete and $8 \times 10^{-6}/°C$ for lightweight concrete), $\theta_2 - \theta_1$ is the temperature difference between top and bottom of the slab (for design fire resistance, 770°C for up to 90 min fire resistance and 900°C for 120 min), L and l are the longer and shorter spans of the slab, and h is the thickness that may be taken from the mid-height of trough decking and the overall height for dovetail decking.

Example 12.1: Composite slab with membrane action

A composite slab 110 mm thick (concrete Grade 25/30) with Richard Lees Holorib decking has H10 bars at 150-mm centres in each direction. The slab element is 4 m × 5 m and is bounded by protected beams on the perimeter and supported by an unprotected single beam at midpoint of the 5-m side (Figure 12.1).

Determination of the load-carrying capacity for a standard fire exposure of 60 min:

Beam capacity — A 203 × 133 × 30 beam has a flange thickness of 9,6 mm. Use data for a 10-mm flange from Table 10 of BS 5950 Part 8 to determine a lower flange temperature of 938°C. From Table 5.6, $k_{y,\theta} = 0,0524$. $M_\theta = k_{y,\theta}M_{pl,Rd} = 0,0524 \times 200 = 10,48$ kNm. From Equation (12.4):

$$q_{p\theta,beam} = \frac{8M_\theta}{L^2} = \frac{8 \times 10,48}{4^2} = 5,24 \text{ kN/m}$$

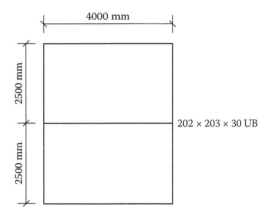

Figure 12.1 Slab and beam layout for Example 12.1.

This carries a udl over a width of 2,5 m. Thus $q_{p\theta,udl} = q_{p\theta,beam}/2,5 = 5,24/2,5 = 2,10$ kPa.

To determine the temperature-reduced moment capacity $m_{p\theta}$ of the slab, it will be conservative to ignore the contribution of the decking. The sagging reinforcement can be designed as if for a slab, but with the steel strength taken as $k_y(\theta)f_y$. Place the required reinforcement level with the top of the dovetail; i.e., the effective depth is h_1 or 59 mm.

To determine θ_s using Equation (9.7), the values of A/L_r, u_3, z, h_1, l_3, and α are taken from Example 9.1. Thus

$$\theta_s = c_0 + c_1 \frac{u_3}{h_1} + c_2 z + c_3 \frac{A}{L_r} + c_4 \alpha + c_5 \frac{1}{l_3}$$

$$= 1191 - 250\frac{51}{51} - 240 \times 2,56 - 5,01 \times 26,2 + 1,04 \times 104 - \frac{925}{38} = 279°C$$

From Table 5.6, $k_y(\theta) = 1,0$, $A_s = 524$ mm²/m. It can be noted that for reinforced concrete design in the fire limit state, the values of γ_c, γ_s and α_{cc} are all 1,0. From Equations (6.11) and (6.12) of Martin and Purkiss (2006),

$$\frac{A_s k_y(\theta)f_y}{bdf_{ck}} = \frac{524 \times 1,0 \times 500}{1000 \times 59 \times 25} = 0,178$$

$$\frac{x}{d} = 1,25\frac{A_s k_y(\theta)f_y}{bdf_{ck}} = 1,25 \times 0,178 = 0,223$$

$$\frac{M_{Rd}}{bd^2 f_{ck}} = \frac{A_s k_y(\theta)f_y}{bdf_{ck}}\left(1 - \frac{0,8}{2}\frac{x}{d}\right) = 0,178(1 - 0,4 \times 0,223) = 0,162$$

or

$$M = 0,162bd^2f_{ck} = 0,162 \times 1 \times 0,059^2 \times 25 \times 10^6 = 14,1 \, \text{kNm/m}$$

To determine $q_{p\theta,slab}$ from Equation (12.3), with $\alpha = 0,8$ and $L = 5$ m:

$$q_{p\theta,slab} = \frac{24m_{p\theta}}{(\alpha L)^2 \left[\sqrt{3+\alpha^2} - \alpha \right]^2} = \frac{24 \times 14,1}{(0,8 \times 5)^2 \left[\sqrt{3+0,8^2} - 0,8 \right]^2} = 17,2 \text{kPa}$$

Depth of the compression block is $0,8x = 0,8 \times 0,223 \times 59 = 10,5$ mm. From Equation (12.5):

$$g_0 = 1 - \frac{2d_c}{d} = 1 - \frac{2 \times 10,5}{59} = 0,64$$

To determine deflection w from Equation (12.6):

$$w = \frac{\alpha_c (\theta_2 - \theta_1) l^2}{19,2h} + \sqrt{\frac{3L^2}{8} \frac{0,5f_{y,20}}{E_{s,20}}} = \frac{18 \times 10^{-6} \times 770 \times 4^2}{19,2 \times 0,110} + \sqrt{\frac{3 \times 5^2}{8} \frac{0,5 \times 500}{210 \times 10^3}}$$

$$= 0,211\text{m}$$

Limiting value —

$$w = \frac{\alpha_c (\theta_2 - \theta_1) l^2}{19,2h} + \frac{l}{30} = \frac{18 \times 10^{-6} \times 770 \times 4^2}{19,2 \times 0,11} + \frac{4}{30} = 0,238\text{m}$$

The calculated deflection is less than the limiting value, hence, $w/d = 0,211/0,059 = 3,6$.

The aspect ratio $1/\alpha = 1/0,8 = 1,25$. From Bailey (2001), $e = 1,77$. From Equation (12.2),

$$q_{p\theta} = eq_{p\theta,slab} + q_{p\theta,udl} = 1,77 \times 17,2 + 5,1 = 35,5\text{kPa}$$

12.3 FIRE BEHAVIOURS OF CONNECTIONS

12.3.1 Tests on beam and column assemblies

The earliest tests reported by Lawson (1990a, 1990b) were carried out on stub beams (305×165×40 UB Grade S275) bolted with varying types of connections to stub columns (203×203×52 UC Grade S275). The beams were loaded to give shear forces and hogging moments typical of those in conventional construction.

The tests utilized varying types of structural fire protection. Where full protection was envisaged, a spray protection was designed to give 60-min fire resistance at a critical steel temperature of 550°C. In certain cases, the column web was infilled with nonload-bearing concrete blockwork, and in other columns no protection was applied.

The results show that the connections were capable of adsorbing a high degree of moment before undue deformations occurred and thus a significant transfer of moment could occur from the beam into the column during a fire. It should be noted that the tests were on assemblies that modelled internal columns where the net moment on the column would sensibly be zero owing to the balanced nature of the loading. This will not be the case for an external column where, if moment transfer from the beam is allowed during a fire, either substantially higher amounts of fire protection or a larger column for a given amount of fire protection will be needed as the load ratio on the column is increased.

The tests also provided data on the temperatures reached in the bolted connections. These data enable capacity checks to be carried out on connections. Owing to the local buckling observed close to the ends of beams in the Cardington tests and the tensile forces induced during the cooling phase, the results from Lawson are now considered unsafe for use in the design of frames.

Unfortunately, these tests were carried out only at a single load level and could not therefore be used to determine moment rotation characteristics for connections subject to variable load levels as would be needed for computer analysis of fire-affected steel frameworks.

12.3.2 Moment–curvature relationships

Experimental work (Leston-Jones et al., 1997) suggested that the moment–curvature relationships for flush end plate connections may be expressed using an equation similar to the Ramberg-Osgood (1943) equation for the ductile stress-strain curve, i.e.,

$$\phi = \frac{M}{A} + 0{,}01\left(\frac{M}{B}\right)^{n} \tag{12.7}$$

where ϕ is the rotation under a moment M, A is related to the stiffness of the connection, B to its strength, and n defines the nonlinear shape of the curve. These tests were carried out in the absence of axial loads. The results from the Cardington tests demonstrated that axial loads can significantly affect the behaviour of connections in a fire.

12.3.3 Whole connection behaviours

Much of this section is based on Wang et al. (2010, 2013). Two aspects of whole connection behaviour were revealed by the Cardington tests. First, due to moment transfer away from the sagging zones from loss of moment capacity of the beam and slab system, the connections became subject to moments that may not have been considered in the ambient design. Second, during the cooling phase, the beam went into catenary action and induced tensile forces into the beam and, more importantly, into the connection.

It is convenient to discuss connection behaviour by type of connection. For the first four types, the tests were conducted on isolated connections tested at ambient, 450°C, 550°C, and 650°C. For the last temperature, the tests were carried out on beam–column subassemblies heated to failure.

12.3.3.1 Fin plates

Failure of Grade 8.8 M20 bolts was generally due to shear in the bolts at both ambient and elevated temperatures. When stronger bolts (Grade 10.9 or M24) were used, failure occurred by web fracture at ambient and by bolt shear at elevated temperatures.

12.3.3.2 Web cleats

These generally had high rotation capacities. Failure at ambient temperature occurred when bolt heads punched through the angles connected to the column flanges. At 450°C and 500°C, failure was due to fracture of the angle cleat close to the heel of the joint with lower deformation than at ambient. At 650°C, ductility appeared improved, but with eventual failure due to bolt shear.

12.3.3.3 Partial end plates

These exhibited low rotation capacity and generally failed by fracture.

12.3.3.4 Full end plates

In early tests, these failed by stripping of the threads on the bolts. In later tests, two nuts were used to assess the rotational behaviour of the connection. At ambient and 450°C, failure was by shear in the end plate, whereas at 550°C and 650°C, the bolts failed in tension.

12.3.3.5 Extended end plates

These were tested as part of a project using beam and column subassemblies. Failure of a $254 \times 254 \times 73$ UC member was due to excessive buckling in the lower flange of the beam, whereas failure of a $152 \times 152 \times 23$ UC member resulted from large deformations in the column flanges with resultant plastic hinges. Both examples failed around 750°C.

To summarize, fin plate connections (designed as pinned) should not be used for primary members. This probably also applies to partial end plate connections. Web cleat connections probably offer the best connection behaviour in fire due to their relatively large rotational capacities.

12.4 PITCHED ROOF PORTALS

EN 1993-1-2 imposes no specific requirement to check whether a portal frame erected close to a boundary will collapse inward rather than outward and cause damage to adjacent property or injuries to firefighters and other persons, although such evaluation is required by UK regulatory authorities.

The rafters of a single-storey portal frame are generally unprotected and lose their strength rapidly in a fire. This means that they are unable to carry any of the remaining dead load from the roof and/or supply any propping restraint for the stanchions or external cladding. Since the eaves connection at ambient conditions is rigid, any attempt by the rafters to deflect will cause rotation and lateral movement to the tops of the stanchions. This lateral movement must not be sufficient to cause the stanchion to topple outward (Fig. 12.2).

The principle is to ensure that a stanchion will rotate about its foot into the space occupied by the structure when movement occurs. This means that although the base of the stanchion was designed to be pinned at ambient conditions, there must be some degree of fixity during a fire to ensure a degree of stability. The full background of the calculations is given in Simms and Newman (2002).

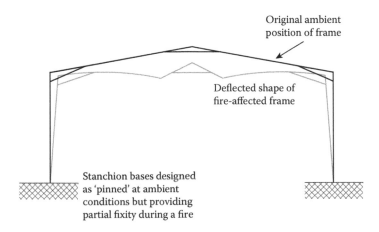

Original ambient
position of frame

Deflected shape of
fire-affected frame

Stanchion bases designed
as 'pinned' at ambient
conditions but providing
partial fixity during a fire

Figure 12.2 Collapse behaviour of pitched roof portal frame in a fire.

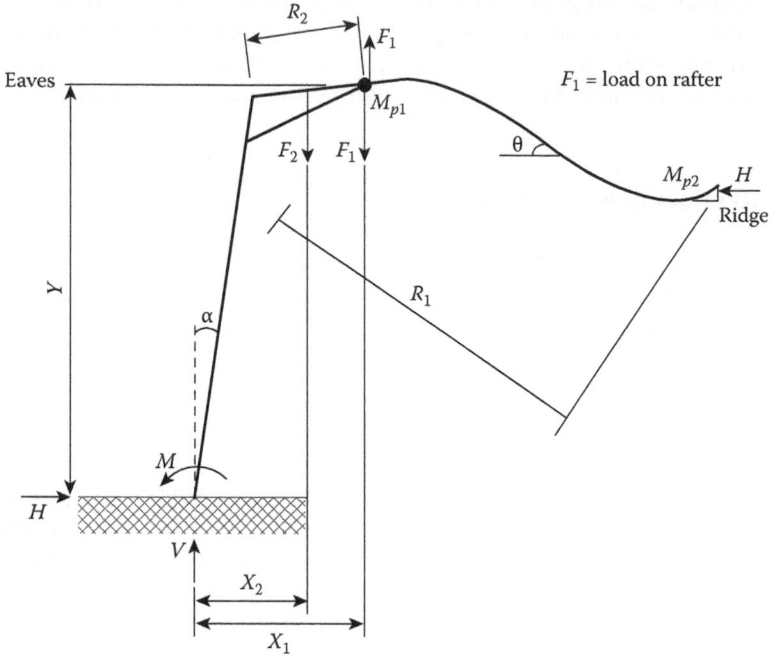

Figure 12.3 Geometry of portal frame and actions on frame in a fire.

At collapse, hinges tend to occur at either side of the apex connection and in the rafter at the end of the haunch. With these assumed positions of plastic hinges, it is then possible to determine the moment and other forces required to provide stability at the foot of the stanchion (Figure 12.3). To carry out the calculations, assumptions are also needed for the resultant frame geometry in the fire. The vertical reaction V_{Sd} is given by

$$V_{Sd} = F_1 + F_2 \tag{12.8}$$

where F_1 is the force on the main rafter of length R_1 between haunches and F_2 is the force on the length R_2 equal to the haunch length. The horizontal reaction H_{Sd} is given by

$$H_{Sd} = \frac{F_1 R_1 \cos\theta - 2(M_{pl,1} + M_{pl,2})}{2R_1 \sin\theta} = \frac{F_1}{2\tan\theta} - \frac{M_{pl,1} + M_{pl,2}}{R_1 \sin\theta} \tag{12.9}$$

From consideration of the loading,

$$F_1 = 0{,}4q_f SG \tag{12.10}$$

where q_f is the load on the rafter, S is the frame spacing, G is the distance between haunches, $M_{pl,1}$ and $M_{pl,2}$ are the temperature-reduced plastic moments of the rafter at the end of the haunch and the rafter, respectively, and θ is the slope of the rafter at collapse.

R_1 can be determined in terms of G, initial rafter pitch θ_0, and elongation, and is given by

$$R_1 = 1,02 \frac{G}{2\cos\theta_0} \qquad (12.11)$$

where an elongation of 2% has been assumed. If $M_{pl,1}$ and $M_{pl,2}$ are both taken as $0,065 M_{pl,rafter}$, H_{Sd} becomes

$$H_{Sd} = \frac{q_f SG}{4\tan\theta} - \frac{0,255 M_{pl,rafter}\cos\theta_0}{G\sin\theta} \qquad (12.12)$$

and the moment at the base of the stanchion M_{Sd} is

$$M_{Sd} = H_{Sd}Y + F_1 X_1 + F_2 X_2 + M_{pl,1} \qquad (12.13)$$

where Y is the height to the end of the haunch. For a column rotation angle around 1 degree,

$$X_1 \approx \frac{Y}{60} + \frac{L-G}{2} \qquad (12.14)$$

and

$$X_2 \approx \frac{Y}{60} + \frac{L-G}{4} \qquad (12.15)$$

Thus

$$F_1 X_1 + F_2 X_2 = q_f SGY\left(\frac{L}{120G} + \frac{L^2 - G^2}{8GY}\right) \qquad (12.16)$$

and

$$HY = \frac{q_f SGY}{4\tan\theta} - M_{pl,rafter}\left(\frac{0,255Y}{G}\frac{\cos\theta_0}{\sin\theta}\right) \qquad (12.17)$$

If the haunch length is around 10% of the span, as a further approximation

$$\frac{L}{120G} = \frac{1}{96} \tag{12.18}$$

The rafter sag angle θ is given for single bay portal frames by

$$\theta = \arccos\left(\frac{L - 2X_1}{2R_1}\right) \tag{12.19}$$

or

$$\theta = \arccos\left(\frac{\left(G - \dfrac{Y}{30}\right)\cos\theta_0}{1,02G}\right) \tag{12.20}$$

A parametric study indicated that Equation (12.20) can be simplified to:

$$\theta = \arccos(0,97\cos\theta_0) \tag{12.21}$$

The value of M_{Sd} may be further simplified to:

$$M_{Sd} = q_f SGY\left(A + \frac{B}{Y}\right) - M_{pl,rafter}\left(\frac{CY}{G} - 0,065\right) \tag{12.22}$$

where

$$A = \frac{1}{4\tan\theta} + \frac{1}{96} \tag{12.23}$$

$$B = \frac{L^2 - G^2}{8G} \tag{12.24}$$

$$C = 0,255\frac{\cos\theta_0}{\sin\theta} \tag{12.25}$$

The horizontal reaction H_{Sd} is then given by

$$H_{Sd} = q_f SGA - \frac{CM_{pl,rafter}}{G} \tag{12.26}$$

The applied loading q_f should be taken as the dead load of the roofing material and purlins less any combustible material. Further guidance can be found in Table 2.1 of Simms and Newman (2002). Some further considerations need to be covered before undertaking an example set of calculations. Provisions must be made for longitudinal stability, and any one of the following conditions may be satisfied:

1. The provision of four equal-sized holding-down bolts in each base plate, symmetrically distributed about the section in the longitudinal direction at a minimum spacing of 70% of the width of the stanchion flange,
2. The provision of a masonry wall properly tied to the stanchion that restrains the column in the plane of the wall at a height not less than 75% of the height to eaves,
3. The provision of suitably designed horizontal restraint members that must be fire protected,
4. The provision of any protected area of wall horizontal steel members having a combined tensile strength of $0{,}25VSd\Sigma$ (height of unprotected area)/(height to eaves) where the summation is taken over all the frames.

Example 12.2: Portal frame stability check

The portal frame illustrated in Figure 12.4 is to be checked for stability requirements because it is adjacent to a boundary. Assume that all cladding is non-combustible. The five load per unit area is taken as 0,42 kPa, which includes the weight of the frame.

Fire load per unit run $q_f = 9 \times 0{,}42 = 3{,}78$ kN/m.
Length between haunches $G = 25 - 2 \times 2{,}73 = 19{,}54$ m.
Height to end of haunch $Y = 5 + 2{,}73\tan 10 = 5{,}48$ m.

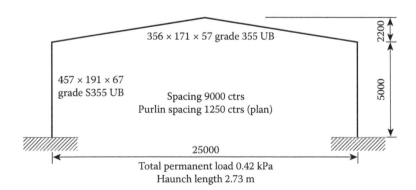

Figure 12.4 Design data for determination of fire performance of portal frame (Example 12.2).

From Equation (12.20),

$$\theta = \arccos\left(\frac{\left(G - \frac{Y}{30}\right)\cos\theta_0}{1,02G}\right) = \arccos\left(\frac{\left(19,54 - \frac{5,48}{30}\right)\cos 10}{1,02 \times 19,54}\right) = 17,0°$$

The approximate Equation (12.21) gives

$$\theta = \arccos(0,97\cos\theta_0) = \arccos(0,97\cos 10) = 17,2°$$

When the angles are almost identical, use the value of $\theta = 17°$. To determine parameters A, B and C from Equations (12.23) to (12.25),

$$A = \frac{1}{4\tan\theta} + \frac{1}{96} = \frac{1}{4\tan 17} + \frac{1}{96} = 0,828$$

$$B = \frac{L^2 - G^2}{8G} = \frac{25^2 - 19,54^2}{8 \times 19,54} = 1,56$$

$$C = 0,255\frac{\cos\theta_0}{\sin\theta} = 0,255\frac{\cos 10}{\sin 17} = 0,859$$

Note that

$$M_{pl,rafter} = W_{pl}f_y = 1010 \times 355 \times 10^{-3} = 359 \text{kNm}$$

From Equation (12.22),

$$M_{Sd} = q_f SGY\left(A + \frac{B}{Y}\right) - M_{pl,rafter}\left(\frac{CY}{G} - 0,065\right)$$

$$= 0,42 \times 9 \times 19,58 \times 5,48\left(0,828 + \frac{1,56}{5,48}\right) - 359\left(\frac{0,859 \times 5,48}{19,58} - 0,065\right)$$

$$= 388,3 \text{kNm}$$

The minimum value of the overturning moment $M_{Sd,min}$ should be $M_{pl,column}/10$, so

$$M_{Sd,min} = (1471 \times 355 \times 10^{-3})/10 = 52,2 \text{kNm}$$

To determine the horizontal reaction from Equation (12.26):

$$H_{Sd} = q_f SGA - \frac{CM_{pl,rafter}}{G} = 0,42 \times 9 \times 19,54 \times 0,828 - \frac{0,859 \times 359}{19,54} = 45,4 \text{kN}$$

The minimum value of $H_{Sd,min}$ is $M_{pl,column}/10Y$, so

$$H_{Sd,min} = (1471 \times 355 \times 10^{-3})/(10 \times 5) = 10,4 \text{kN}$$

Again this condition is satisfied. The vertical reaction is simply half the load on the frame.

The preceding chapters considered the design of elements or frames under the effects of fire. The evaluations of structures after damage from fire remain to be considered.

12.5 FINITE ELEMENT ANALYSIS OF FRAMES

Traditionally, structural design for fire has been based on structural element behaviour observed in furnace tests. Early research focused on the fire behaviours of individual structural elements such as beams, columns, slabs, and connections. However, it is evident from fire tests on frame structures that the overall behaviours of members in a frame are very different from those predicted by isolated furnace tests. Although a number of fire tests on frame structures have been reported in the literature, the available results are still limited because of the large number of parameters involved in the tests. Due to the high cost of large compartment fire tests, efforts have been made to develop alternative methods. In recent years, computer simulation has become popular for predicting the fire behaviours of structural elements in frame structures.

EN 1992-1-2 states that the advanced calculation methods for structural fire analysis should provide a realistic analysis of structures subjected to fire, based on fundamental physical behaviour leading to reliable approximations of the expected behaviours of relevant structural components exposed to fire. These methods can be applied to a whole structure, part of a structure, or to a single element. Advanced calculation methods generally involve two steps. One is thermal analysis to determine the temperature evolution in structural elements, and the other is mechanical stress analysis taking into account the effects of temperature, as described in Chapter 6.

12.5.1 Steel frames

Early work on fire analysis of steel frames using finite element methods was reported by Wang (2000b), who presented the results of an analysis of the global structural behaviour of an eight-storey steel framed building at Cardington during two BRE large-scale fire tests. The results of these tests were analysed using a specialist finite element computer program called FIREFRAME.

Due to difference in fire intensities, the two BRE tests revealed markedly different structural behaviours of the steel frame. The results demonstrated the ability of FIREFRAME in simulating flexural bending behaviour. However, in order to simulate the Cardington frame behaviour during the corner test, a more advanced numerical procedure was needed to deal with slab tensile membrane behaviour at large deflections. Both the test and computer simulation results suggest that columns may attain large moments as a result of pushing from adjacent heated beams, but as the test column temperatures were low, it was not possible to assess the column failure behaviour. Furthermore, the simulation results indicate that large sagging moments may develop in heated beams during cooling.

Tan et al. (2002) analyzed two-dimensional steel frames at both ambient and elevated temperatures. The analysis model considered the geometric and material nonlinearities along with creep. The model can be applied to steel frames subjected to increasing external loads at ambient temperature or constant external loads at elevated temperatures. It can predict the collapse load or critical temperature.

The model was validated using a number of experimental and analytical benchmark tests. Following these benchmark tests, Tan et al. further investigated creep effects on the buckling of heated columns. It was found that creep was dominant for temperatures over 400°C, indicating that the simplified method for calculating the critical temperature in EN 1993-1-2 may be unsafe for columns of intermediate slenderness under both uniform and nonuniform heating.

Liew and Chen (2004) presented a numerical approach for inelastic transient analysis of steel frame structures subjected to explosion loading followed by fire. The approach adopts the use of beam–column elements and fiber elements to enable a realistic modeling of the overall framework subjected to localized explosion and fire. Elastoplastic material models at elevated temperatures and high-strain rates are employed. Verification examples are provided. The influence of blast loads on the fire resistance of a multistory steel frame was also examined.

Using a generalized plastic hinge concept, Junior and Creus (2007) proposed a simplified procedure for analysing three-dimensional frames under elevated temperatures, where normal force and bending moments considered in the plastification process are extended to include temperature effects. Additional refinements to allow for the gradual spread of yielding on the member are included by means of stiffness reduction factors. The model was compared with experimental results and more refined models using examples that covered different situations. The results demonstrated that the simplified procedure can produce decent predictions of internal forces for two- and three-dimensional frames on fires at low computational cost.

Mossa et al. (2009) described a study of the fire behaviours of steel portal frame buildings at elevated temperatures using the finite element code SAFIR. The finite element analysis is three dimensional and covers several support conditions at the column bases including axial restraints provided by end walls, different fire severities in a building, various levels of out-of-plane restraints of columns, and the effects of concrete encasements of columns. A number of analyses indicated that the bases of steel portal frames at foundations must be designed and constructed with some level of fixity to ensure that a structure will deform acceptably during a fire without outward collapse of the walls. The analyses also showed that to avoid side sway (outward collapse), it is not necessary for steel portal frame columns to be fire protected unless the designer wishes to ensure that the columns and wall panels remain standing during and after a fire.

12.5.2 Reinforced concrete frames

Huang et al. (2006) analyzed the behaviour of a reinforced concrete building and floor slabs exposed to fire. The results showed that slab behaviour during fire exposure is highly influenced by both tensile and compressive membrane actions. The presence of adjacent cold slab areas had a significant influence on the behaviour of the structure within the fire compartment, because the cold part of the structure provided an increase of the fire endurance of the whole system through structural continuity.

The analyses performed were halted due to buckling of the fire-exposed columns, indicating the importance of designing reinforced concrete columns for high fire resistance periods to prevent column failures.

Mossa et al. (2008) presented a numerical modelling of the fire behaviour of two-way reinforced concrete slabs in a multi-storey multi-bay building. The building was square, with three bays in each direction. The concrete slab was supported by a perimeter frame, four internal columns, and no internal beams. It is assumed that all nine bays of the concrete slab at one level were subjected to fire defined by two fire curves. One was the standard ISO 834 fire for a 4-hr duration and the other was a parametric fire based on ISO 834 fire for 1 hr with temperatures decaying to ambient in another 2 hr. The effects of the fires are described in relation to the redistribution of bending moments and the development of tension field action in the slab. The results showed that the tensile membrane forces of the slab are limited by the strength loss in the reinforcing steel bars as they heat up and by increasing vertical deflections.

Kodur et al. (2009) proposed a macroscopic finite element model for tracing the fire responses of reinforced concrete (RC) structural members. The model accounts for critical factors to be considered for performance-based fire resistance assessment of RC structural members. Fire-induced spalling, various strain components, high temperature material properties, restraint effects, various fire scenarios, and failure criteria were all incorporated in the model. The numerical model was validated using full-scale fire resistance test data. Case study examples were provided to demonstrate the use of the computer program of the model to trace the responses of RC members under standard and design fire exposures. The case studies demonstrated that a fire scenario has a significant effect on the fire resistance of RC columns and beams. Further, macroscopic finite element models are capable of predicting the fire responses of RC structural members with sufficient accuracy for practical applications.

Han et al. (2012) presented a numerical investigation of the fire performance of concrete-filled steel tubular (CFST) columns and reinforced concrete beam frames. The planar composite frame consisted of two CFST columns and a RC beam with slab. A finite element analysis model was developed and verified against the test results of the same composite frames. The model was then used to investigate the behaviours of similar framed structures in fire. Stresses, strains, internal forces, and displacements were analysed to identify the failure mechanisms of the frames.

The numerical models described above did not consider the effects of thermally induced pore pressure on the development of mechanical stresses and the occurrence of material failure. Strong evidence indicates that the pore pressure build-up during temperature rise can cause concrete spalling. Phan (2008) conducted an experimental study that measured thermally induced pore pressure and corresponding concrete temperatures in both HSC and NSC to quantify the effects of factors influencing pore pressure build-up and potential for explosive spalling. The specimens were $100 \times 200 \times 200$ mm concrete blocks heated to a maximum temperature of 600°C at 5°C/min and 25°C/min. It was found that pore pressure developments are directly related to the moisture transport process and exert significant influence on the occurrence of explosive spalling. The factors affecting pore pressure build-up and explosive spalling include water-to-cement ratio, curing conditions, heating rates, and polypropylene fibres.

Fire analysis is much more complicated in concrete structures than in steel structures. Apart from the pore pressure, another difficulty in dealing with concrete is the material softening caused by concrete crush. Markovic et al. (2012) investigated the effects of material softening on the finite element analyses of RC structures at high temperatures.

Analyses of a two-storey, two-bay RC frame subjected to both mechanical and fire loads revealed that the strains were spatially oscillating around the localized region if continuous finite elements that account for strain softening of material wherever it occurs were used. This sometimes yields a totally false result or loss of convergence of the global Newton method. The width of the localized band tended to zero with increasing mesh density, indicating that the results were highly mesh sensitive. However, if the finite elements utilized constant-strain crack band elements for the localized regions and continuous elements for the remaining regions, computationally stable solutions were achieved which predicted accurate post-instability responses for a wide range of crack band element widths. The important point is, however, obtaining a reliable estimate of the width of the crack band element at high temperatures. This requires extensive experimental testing with hot concrete in compression.

Chapter 13

Assessment and repair of fire-damaged structures

Often the initial response when looking over a fire-damaged structure is one of despair and horror at the extent of damage. This situation is exacerbated by the amount of nonstructural debris lying around and the acrid smells of many combustion products. In most cases the damage is not as severe as initially thought, even though immediate decisions must be taken on whether the short-term safety of the structure temporary propping is necessary or, indeed, whether demolition work is required. Such decisions may need to be taken very quickly after a fire and will generally be based on a visual survey and expert judgement.

The assessment of fire-damaged structures is very much a 'black art' in that it relies heavily on experience. Furthermore, the insurance company of the owner or occupier will become involved. Even if a structure is capable of being saved, economics will determine whether it should be repaired or demolished and completely rebuilt. This question can often be answered after a thorough visual inspection has been carried out.

13.1 VISUAL INSPECTION

The aim of a visual inspection is to determine: (1) the short-term stability of a structure and (2) the extent and severity of the fire.

13.1.1 Stability

If possible, the original drawings for the structure should be consulted. A review of drawings will allow assessment of how the structure transmits applied loading and identification of the principal load-carrying members and members providing structural stability. The inspection needs to check excessive deformation, deflection, or cracking of main load-carrying members and integrity of the connections of main members. It is also necessary to consider structure stability if excessive bowing has occurred in masonry cladding or internal compartment walls.

In the case of concrete construction, attention should be given to damage due to spalling on beams and columns as this may reduce the load-carrying capacity of members due to excessive temperature rises in any reinforcement. Where the fire has only affected only part of a structure, it is essential that the inspection extends to any part of the structure not damaged directly by the fire. It is possible that a substantial redistribution of forces can occur in unaffected parts of a structure. This redistribution of forces has been noted in theoretical work on concrete frames by Kordina and Krampf (1984). They found that moments in the frame remote from the fire-affected compartment exceeded the design moments. In the Broadgate fire (SCI, 1991), the structure behaved during the fire in a totally different manner from the way it was designed in that forces were redistributed away from the fire by columns acting in tension to transmit forces to the relatively cool upper floors of the structure.

13.1.2 Estimation of fire severity

The first method of obtaining a rough estimate of fire severity is a review of the fire brigade records to determine types and numbers of appliances called out, time needed to fight the fire, the length of time between reporting the fire and the arrival of the brigade, the operation of any automatic fire detection or fire fighting equipment, and the degree of effort required to fight the fire.

The second approach is to estimate the temperature reached in the fire by studying the debris left by the fire. It is thus important that no debris is removed until such a study is carried out to prevent loss of vital evidence. The identification of debris materials may indicate the fire temperature reached since most materials have known specific melting or softening temperatures. Some typical data are given in Table 13.1 (Parker and Nurse, 1956). Care should be exercised, however, when using these data because temperatures vary over the height of a fire compartment. For this reason, the original positions of artefacts are important. Also, this method only gives an indication that particular temperatures were reached but does not show the durations of exposure to temperatures.

A third method that is available to give an estimate in terms of either standard furnace test duration or a known fire, is to measure charring depth on any substantial piece of timber known to have been exposed to the fire from its initiation. The charring depth can be related back to the standard furnace exposure since timber of known or established density can be assumed to char at a constant rate between 30 and 90 min of standard exposure (Section 5.2.4). The position of a timber specimen in a compartment should also be noted. An estimate of actual fire exposure can be obtained using Equations (10.5) to (10.10).

A fourth method is to calculate fire severity from estimates of compartment size, fire load density and areas of openings (ventilation factor) using the equations presented in Section 4.4. It should be remembered that these equations assume the whole fire load ignites instantaneously and that the

Table 13.1 Melting point data

Material behaviour	Approximate Temperature (°C)
Softening or collapse of polystyrene	120
Shrivelling of polythene	120
Melting of polythene	150
Melting of polystyrene	250
Darkening of cellulose	200 to 300
Soldered plumbing adrift	250
Melting or softening of lead plumbing	300 to 350
Softening of aluminium	400
Melting of aluminium	650
Softening of glass	700 to 800
Melting point of brass	800 to 1000
Melting point of silver	950
Melting point of copper	1100
Melting point of cast iron	1100 to 1200

Source: Adapted from Parker, T.W. and Nurse, R.W. (1956) *Investigations into Building Fires Part 1: Estimation of Maximum Temperatures Attained in Building Fires from Examination of Debris,* Technical Paper 4. London: Her Majesty's Stationery Office, pp. 1–5.

whole ventilation is available from the start of the fire. These assumptions may not be totally accurate for a large compartment fire. In practice, none of the above methods is completely reliable, and a combination of methods must be used to attain a reasonable, answer.

A visual inspection will have identified areas that must be immediately demolished because the damage is too extensive to repair and areas that may be capable of repair if sufficient strength can be ensured. An inspection will also identify areas of no damage or superficial damage. This last category merits no further discussion. The problems arising when demolition is necessary are considered in Section 13.5. If repair is considered feasible, a much more thorough investigation is required to ascertain the extent and severity of any damage and the residual strength of the structure. First it is necessary to clear all debris from the structure and clean as much smoke damage as possible to allow unimpeded examinations of all surfaces.

13.2 DAMAGE ASSESSMENT

This assessment is performed in a series of stages. The first stage involves a complete and fully detailed survey of the structure. The second stage ascertains the residual strength of both the individual members and of the complete structure.

13.2.1 Structural survey

For any structure, the first stage is to perform a full line and level survey to assess the residual deformations and deflections in the structure. The measured deflections should be compared with those for which the structure was designed. Care should be taken to note the effects of any horizontal movements due to thermal actions during the fire. Such effects of horizontal movement are often apparent away from the seat of the fire (Malhotra, 1978; Beitel and Iwankiw, 2005). Other observations required for a survey depend on the main structural material: steel, concrete, or masonry.

In concrete structures, it is necessary to note the presence of spalling and exposed reinforcement. Large amounts of spalling do not necessarily imply that the structure or its reinforcement is substantially weakened since spalling may occur late in a fire due to the action of cold water from fire fighters' hoses. It is likely that spalling occurred early in a fire where the exposed surfaces are smoke blackened.

It is useful to note the colour of the exposed concrete face because colour can give an indication of the temperature to which the element was exposed. Care is needed because spalling may nullify an accurate observation and some aggregates do not exhibit colour changes. Note should also be made of crack formations. Cracking is unlikely to be deleterious in the tension zones of reinforced concrete beams, but will indicate severe problems if found in the compression zones of beams or slabs or in columns. The fire test on the concrete frame at Cardington demonstrated the ability of the frame to remain intact in spite of large degrees of spalling, but the frame also suffered large horizontal deformations at the top of the fire test compartment (Bailey, 2002).

Since most structural steels regain most strength on cooling (see Section 13.3.1), steel structures will suffer little loss in strength. However, the resultant deformations are likely to indicate the state of a structure. It is important to assess the integrity of the connections; it is possible that bolts failed within the connections or became unduly deformed. Where the floors are composed of profile sheet steel decking and *in situ* concrete, examination should be made to discover any separations between the decking and beams. This separation can still occur even if thorough deck stud welding was used. Another potential point of failure is the shear bond between the decking and *in situ* concrete. Even with substantial damage of these types, a structure may still be intact as demonstrated after the fire tests on steel frame structures at Cardington (Bailey, 2004).

Masonry is used in low-rise load-bearing structures and as cladding to framed structures. The major cause of distress to masonry walls is expansion or movement in the structure caused by thermal actions on the frame or flooring. This is less likely in low-rise construction where substantial amounts of timber are likely to be used. Note should therefore be made of any areas that show signs of punching failure or excess deflections on the outer leaf of cavity walls. If the damage is restricted to the inner leaf, it may be possible to retain the external leaf and rebuild only the inner leaf if the wall ties are still reusable.

During a visual survey, attention should be given to the need to conduct tests on structural materials to ascertain their residual strengths. The testing methods used may be nondestructive or require the taking of samples from damaged portions of the structure and control specimens from undamaged areas.

13.2.2 Materials Testing

13.2.2.1 Concrete

The only common destructive test is to remove concrete cores, usually 40 mm in diameter, from the fire-damaged zone and test them in compression according to the relevant standard, e.g., BS 1881 Part 120 (1983) and relate the measured strength to an equivalent cube strength using appropriate empirical formulae. Great care is needed with the use of cores to assess residual strengths as it is necessary to attempt to extract cores free from reinforcement, although the presence of reinforcement can be allowed for in assessing equivalent strengths.

A further problem in heavily damaged structures is the ability to obtain cores of sufficient integrity to be tested. It is also necessary to obtain cores from an undamaged part of the structure where concrete of a similar specified grade was used. To aid the assessment of loss of strength, it is useful, if possible, to obtain the original cube or cylinder control-test records during construction. It is also useful to note any colour changes in the concrete along the length of the core because this can help assess the residual strengths of parts of the structure where it may not be possible to extract cores. A series of nondestructive test methods (described below) are available although they all present disadvantages.

13.2.2.1.1 Ultrasonic pulse velocity (UPV) measurement

Although the apparatus for this test is conveniently portable, the results obtained are not very sensitive and have the disadvantage of being comparative in that a reference is needed to establish base values of strength and pulse velocity. The test may be performed by measuring the time taken to transmit a signal through a member or by measuring the time taken for a reflected signal to travel from transmitter to receiver (Figure 13.1).

In the former case, it is necessary to be able to gain access to both sides of a member. Another limitation is that the thickness cannot exceed about

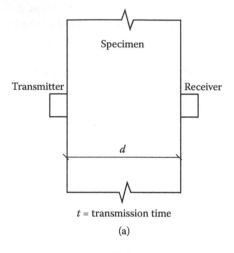

Specimen

Transmitter   Receiver

d

t = transmission time

(a)

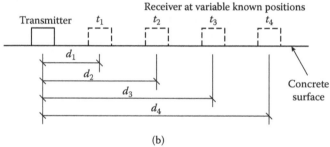

Receiver at variable known positions

Transmitter t_1 t_2 t_3 t_4

d_1

d_2

d_3 Concrete
                                    surface
d_4

(b)

Figure 13.1 Methods of UPV measurement. (a) Direct method ($V = d/t$). (b) Indirect method (velocity given as slope of distance–time graph).

200 mm. In the latter case, the surface must be good enough to allow a series of readings to be taken. A similar series of readings should be taken for the reference value. Provided reference values of both the pulse velocity and strength are known, it is possible to estimate a loss in strength if the loss in UPV is known. Test results demonstrated that the loss in strength $(1 - \sigma_{c,\theta}/\sigma_{c,20})$ is related linearly to the loss in pulse velocity $(1 - U_\theta/U_{20})$ by an equation of the form

$$\left(1 - \frac{\sigma_{c,\theta}}{\sigma_{c,20}}\right) = k_1\left(1 - \frac{U_\theta}{U_{20}}\right) + k_2 \tag{13.1}$$

where $\sigma_{c,\theta}$ and U_θ are the compressive strength and UPV at a temperature θ and $\sigma_{c,20}$ and U_{20} are the reference strength and UPV, respectively, and k_1 and k_2 are constants depending on the concrete age and composition

(Purkiss, 1984; 1985). Benedetti (1998) proposed a rather complex method capable of determining the loss in elastic modulus within a fire-damaged zone using the reflection method. Benedetti suggests a linear degradation model of elastic modulus with temperature is adequate. The method does rely on a relatively undamaged surface.

13.2.2.1.2 Schmidt hammer

This test measures only the properties of the concrete in a surface layer and requires a clean smooth surface to give reliable results. It also needs calibrating for a given concrete and is not suitable where knowledge of the concrete properties within an element is required.

13.2.2.1.3 Windsor probe and pull-out test

These tests are grouped because they both require reasonable surfaces for examination. Nene and Kavle (1992) report the use of the Windsor probe to assess the *in situ* strength of fire-damaged concrete but give few details on the results obtained.

13.2.2.1.4 Thermoluminescence test

This test requires only very small samples of mortar obtained using very small diameter cores to be subjected to testing. By studying changes of the silica within samples, it is possible to determine the temperature reached by the concrete (Placido, 1980; Smith and Placido, 1983). This testing also requires very specialized equipment that may not be readily available.

13.2.2.1.5 Differential thermal and thermogravimetric analyses

Both differential thermal analysis (DTA) and thermogravimetric analysis (TGA) can be used to evaluate changes in the structure of a concrete when exposed to heating as the pattern of the responses changes with temperature and causes physicochemical changes in the cement (Handoo et al. 1991). Short et al. (2000) indicate that DTA and TGA are satisfactory only on unblended cements. Cements with pulverized fuel ash (PFA) or ground-granulated blast furnace slag (GGBFS) show few or no changes in response when heated.

13.2.2.1.6 Petrographic analysis

In this technique, thin slices from cores are examined under a microscope and the isotropy, density, and type of cracking are observed. Riley (1991) suggested that when the temperature exceeds 500°C, the cement paste appears anisotropic under polarized light. The crack patterns also change:

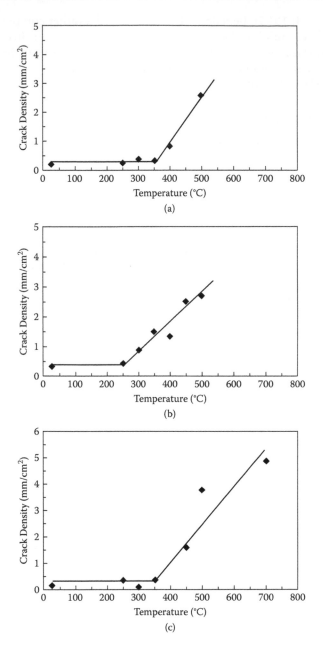

Figure 13.2 Development of crack density with temperature for concrete made with siliceous aggregate and (a) ordinary Portland cement (OPC), (b) OPC/PFA, (c) OPC/GGBS, (d) limestone, and (e) granite. (*Source:* Short, N.R., Purkiss, J.A., and Guise, S.E. (2002) *Structural Concrete*, 3, 137–143. With permission.)

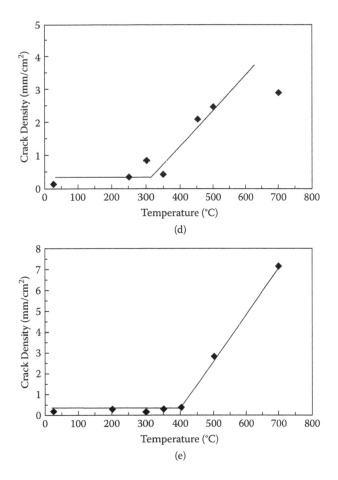

Figure 13.2 (continued) Development of crack density with temperature for concrete made with siliceous aggregate and (a) ordinary Portland cement (OPC), (b) OPC/PFA, (c) OPC/GGBFS, (d) limestone, and (e) granite. (*Source:* Short, N.R., Purkiss, J.A., and Guise, S.E. (2002) *Structural Concrete*, 3, 137–143. With permission.)

below 300°C, cracks form between the boundaries of the aggregate and the mortar matrix. Above 500°C, the cracks tend to pass through the matrix. More recently, Short et al. (2002) and Short and Purkiss (2004) demonstrated that it is possible to quantify the relationship between crack density and temperature reached due to heating (Figure 13.2).

With the unexplained exception of siliceous aggregate concrete containing PFA, the correlation between temperature θ_{cd} at which the crack density increases above base value and the temperature θ_{cs} at which compressive strength loss starts to occur is good (Table 13.2). The ability of change in

Table 13.2 Initial crack density and strength transition temperatures

Concrete type	Initial crack density, C_0 (mm/cm²)	θ_{cd} (°C)	θ_{cs} (°C)
OPC/siliceous	0,29	350	325
OPC/PFA/siliceous	0,36	250	325
OPC/GGBFS/siliceous	0,26	350	350
OPC/limestone	0,31	300	325
OPC/granite	0,24	400	400

Source: Short, N.R., Purkiss, J.A., and Guise, S.E. (2002) Structural Concrete, 3, 137–143. With permission

crack density to predict the position of a 325°C isotherm (at which compressive strength loss starts to occur) is illustrated in Figure 13.3. More recently, Felicetti (2004a) demonstrated that the process of determining colour change may be expedited using digital camera techniques.

13.2.2.1.7 Stiffness damage test

This is a type of compression test carried out on cylindrical specimens 175 mm long and 75 mm in diameter under a limited stress range of 0 to around 4,5 MPa under cyclic loading with the strains measured over the central 67 mm (Nassif et al., 1995; Nassif et al., 1999; Nassif, 2000). Measurements are then taken of variously defined elastic moduli and of hysteresis between the loading cycles. The test results from concrete uniformly heated to temperatures of 470°C confirm data on residual Young's modulus and may provide an alternative method of performance assessment at moderate temperatures. It is not clear what effect a temperature gradient along the specimen would have on the results.

13.2.2.1.8 Surface permeability test

Montgomery (1997) proposed the use of air permeability and water sorption tests on heated concrete to ascertain damage measured by surface pull-off tensile strength using an epoxy-bonded 50-mm diameter steel disc. The 150-mm cube specimens were heated by an imposed flame for 2 hr at a prescribed surface temperature. The tensile strengths will be those of the surface (as will the air permeability and water sorption), but the cube strengths will to some extent be functions of the temperature distributions in specimens. Despite the limited data reported, the method appears to give reasonable correlation of air permeability, API, and pull-off tensile strength, $f_{t,\theta}$,

$$\left(1 - \frac{f_{t,\theta}}{f_{t,20}}\right) = -0,00814\left(1 - \frac{API_\theta}{API_{20}}\right)(R^2 = 0,826) \tag{13.2}$$

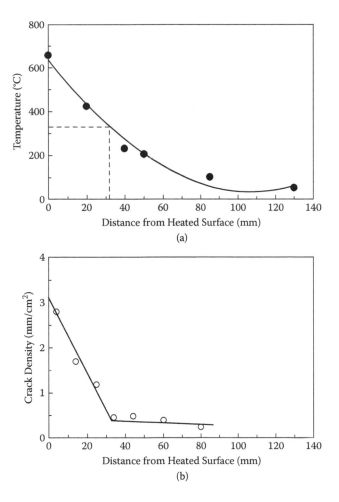

Figure 13.3 (a) Temperature distribution and (b) crack density for OPC–siliceous aggregate concrete cylinder heated from one end.

Similar equations can be determined for the other parameters but the values of R^2 in each case are around 0,5 which indicates a poorer fit.

13.2.2.1.9 Fire behaviour (FB) test

Dos Santos et al. (2002) proposed a test in which a core taken from a fire-damaged structure is sliced into 15-mm thick discs and the water adsorption is measured. They reported that the water adsorption was high near the fire-damaged face and dropped to a sensibly constant value away from the face. The percentage increase in water adsorption (WA) over the mean value with respect to temperature θ may be given by:

$$WA = 0,5 + \frac{4}{300}(\theta - 200)(R^2 = 0,831)$$ (13.3)

A similar pattern emerged for measurements of split cylinder strengths on the discs. The decrease in splitting tensile strength against temperature is given by:

$$f_{t,20} - f_{t,\theta} = 1,35 + 3,5 \times 10^{-3}(\theta - 200)$$ (13.4)

or using their reported ambient value of $f_{t,20}$ of 2,4 MPa, Equation (13.4) becomes

$$f_{t,\theta} = 1,75 - 3,5 \times 10^{-3}(\theta - 200)$$ (13.5)

Equation (13.5) implies zero tensile strength at 500°C, which may be slightly low.

13.2.2.1.10 Drilling resistance test

Using a portable hammer drill, Felicetti (2004b) observed drilling resistance (a measure of the work done in drilling) and drilling time (defined as seconds per millimeter of hole depth) to determine a correlation with residual strength loss. However, the measurements appeared to be more sensitive in normal weight concretes than in the softer lightweight concretes. Furthermore, measurements of drilling resistance indicated no strength loss below 500 to 600°C whereas the drilling time showed slightly lower values around 400°C.

13.2.2.1.11 Hammer and chisel assessment

Although not a scientific method in the generally accepted sense, this method is probably the best way to obtain a very quick, albeit crude, assessment of concrete quality and strength.

An overview of traditional non-destructive testing on fire-damaged concrete is given by Muenow and Abrams (1987).

For reinforcement, similar techniques used for structural steel are available. However, where specimens are taken from tensile steel in beams or compressive steel in columns, the elements or structure must be propped because removal of a specimen will reduce the strength of the member. It may be possible to remove samples from shear links at the mid-point of a beam or a column without propping.

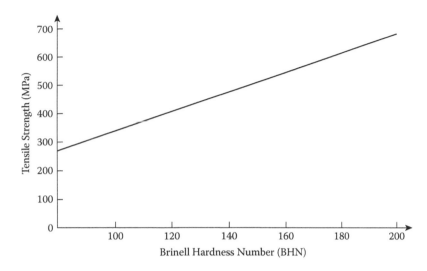

Figure 13.4 Relationship of steel strength and Brinell hardness number (BHN). (Source: Kirby, B.R., Lapwood, D.G., and Thompson, G. (1986) *The Reinstatement of Fire Damaged Steel and Iron Framed Structures.* Rotherham: Swinden Technology Centre. With permission.)

13.2.2.2 Steel

Two essential approaches may be used to assess residual steel strengths. The first is to remove test coupons or samples and subject them to a standard tensile test. Care should be taken in removing test specimens to not further weaken the damaged structure; necessary propping should be used.

The second approach is using nondestructive tests. The most suitable is the hardness indentation test that usually measures the Brinell hardness. There is a direct, sensibly linear relationship between the Brinell hardness number (BHN) and tensile strength (Figure 13.4; Kirby et al., 1986). It is important to use care with this test because a number of results are needed before the strength estimates are statistically reliable.

13.3 ASSESSMENT OF STRENGTH OF STRUCTURE

Strength assessment can be performed using material strength data derived from testing régimes described in the foregoing section or by assessment of the temperatures within the structural element and knowledge of the residual (post-heating and cooling) properties of materials. Often a combination of both approaches will be needed.

Effectively any strength assessment of an element of a structure can be undertaken using the same basic approaches outlined in previous chapters to assess structural performance at elevated temperatures. Use can also be made of experimental results from residual strength tests on fire-affected members. Such results should be used with care because data on the residual strengths of fire-affected columns were obtained from tests of columns heated with no applied load (unloaded) which is not the case in a fire (Lie et al., 1986; Lin et al., 1989).

13.3.1 Residual properties

Besides the residual properties of concrete and steel, it is also necessary to consider materials of a more historical nature such as wrought or cast iron. Fire damage is no respecter of history!

13.3.1.1 Concrete

The only essential property of concrete required for the assessment of fire damage is its residual compressive strength. Typical strength data for normal strength concrete from Malhotra (1954), Abrams (1968), and Purkiss (1984, 1985) are plotted in Figure 13.5. The data indicate that older historical concretes appear to show worse performance than more modern concretes. This may be due to the changes in cements and the resultant cement–mortar matrices that occur over time (Somerville, 1996).

Chan et al. (1999) and Poon et al. (2001) both support the data by Purkiss indicating that normal strength concrete loses about 25% of its strength at 400°C, 60% at 600°C, and 85% at 800°C. The residual strength of concrete is lower than the strength measured at elevated temperatures due to further degradation during cooling caused by differing thermal properties of the aggregate and cement matrix. This difference appears to be affected also by the type of post-firing cooling in that quenched values are lower than those obtained by air cooling (Nassif et al., 1999).

However, it is not usual to take account of any pre-load applied to specimens as it is conservative not to do so. Equally, although various researchers reported strength gains where the temperatures to which a concrete was heated were relatively low (around 200°C), it is prudent to ignore such rises in analysis of concrete elements.

Annex C of EN 1994-1-2 contains equations for calculating residual strength loss. However, the equations are inconsistent with each other and with test data. Test data indicate that below 325°C no compressive strength loss occurs although the equations in EN 1994-1-2 indicate a loss of 0,77 for siliceous aggregate and 0,82 for limestone. The equations in EN 1994-1-2 should not be used.

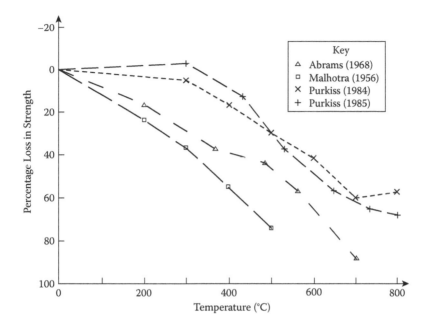

Figure 13.5 Variation of residual strengths of concrete with temperature. (*Sources:* Adapted from Malhotra, H.L. (1956) *Magazine of Concrete Research*, 8, 85-94; Abrams, M.S. (1968) In *Temperature and Concrete*, Special Publication SP-25. Detroit: American Concrete Institute, pp. 33–58; Purkiss, J.A. (1985) In Marshall, I.H., Ed., *Composites 3*. London: Elsevier Applied Science, pp. 230–241; and Purkiss, J.A. (1984) *Journal of Cement Composites and Lightweight Concrete*, 6, 179–184. With permission.)

High performance concrete — Phan et al. (2001) present data on residual compressive strength loss of high performance concretes with and without silica fume. The concretes with silica fume performed better than those without. The mixes with silica fume retained between 75 and 100% of their ambient strengths between 20 and 300°C. The concretes with lower water-to-cement ratios behaved appreciably better.

The two concretes without silica fume dropped to a strength retention of 70% at 100°C, remaining at similar values until 300°C when a further drop to 55% occurred at 450°C. Phan et al. (2001) also demonstrated no real difference in the degradation of Young's modulus in normal and high-strength concretes in that both lose around 70 to 75% of their elastic moduli at 450° in a sensibly linear fashion. Poon et al. (2001) reported a 10% gain in residual strength at 200°, around 5% loss at 400°C, 45% at 600°C, and 75% at 800°C.

Gowripalan (1998) reported tests on residual strength properties of high-strength concrete subject to varying heating and cooling régimes (3- and 24-hr exposure, tested dry and wet). The effect on heating régime for specimens tested dry was negligible at 250 or 1000°C and significant at 500°C where heating for 24 hr produced a strength loss of 70% compared with a loss of 55% when heated for 3 hr. The effect of water immersion on specimens heated for 24 hr was negligible compared to specimens tested dry.

Gowripalan also reported that residual splitting tensile strength dropped to around 25% of ambient at 750°C. For concretes of strength around 50 MPa, the effect on the compressive residual strength is that the addition of silica fume increases residual strength at a given temperature, but the pattern does not appear consistent (Saad et al., 1996). Overall, the results are similar in that strength loss does not seem to occur until 300 to 350°C. If their splitting tensile strength results are normalized, little difference appears between the concretes with almost linear degradations to 25% of ambient at 600°C. The effect of varying PFA dosage on residual compressive strength is not marked when results are normalized, although a strength gain at 200°C is noted. At around 350°C, the strength is the same as ambient before it drops around 20% at 800°C (Xu et al., 2001).

Self-compacting concrete — Persson (2003) provides much of the available data on residual properties of self-compacting concrete of strengths of between 15 and 60 MPa. The results may be summarized by the following formulae:

Residual compressive strength:

$$\frac{\sigma_{\theta,0}}{\sigma_{20,0}} = -0,0000005\theta_c^2 - 0,000729\theta_c + 1,01 \tag{13.6}$$

Residual elastic modulus:

$$\frac{E_{\theta,0}}{E_{20,0}} = -0,0000008\theta_c^2 - 0,00196\theta_c + 1,04 \tag{13.7}$$

Residual strain at peak stress:

$$\frac{\varepsilon_{\theta,0}}{\varepsilon_{20,0}} = -0,0000035\theta_c^2 - 0,000301\theta_c + 1,0 \tag{13.8}$$

Persson also provides a formula relating residual static modulus to concrete strength:

$$E_{stat,res} = f_{c,20}(90,507 - 0,000263\theta - 0,00152f_{c,20}) \tag{13.9}$$

subject to the limit $5 \leq f_{c,20} \leq 60$ MPa.

Fibre concretes — Chen and Liu (2004) report residual compressive strength behaviour of high performance concretes reinforced with various combinations of steel, carbon, and polypropylene fibres. All the concretes retained their normalized compressive strengths at 200°C. Most of them showed almost linear losses of 60 to 70% at 800°C. The two exceptions were plain high performance concrete that dropped to 55% loss at 400°C and 90% loss at 800°C and high performance concrete with polypropylene fibres that dropped to around 50% at 400°C and about 35% at 800°C (similar to other fibre concretes). Purkiss (1984) and El-Refal et al. (1992) provide data on the residual strength performance of steel fibre concrete.

Purkiss et al. (2001) report that slurry-infiltrated fibre concrete (SIFCON) retains around one third of its flexural tensile strength after cooling from 600°C compared to an almost total loss of strength for the matrix. Residual toughness indices show an approximately linear drop from ambient to 400°C, after which they are sensibly constant. The drop is around 70%. The residual dynamic modulus drops approximately linearly to around 10% of its ambient value at 800°C, compared to the matrix that shows a slower and less dramatic decline to around 50% at 800°C. The residual compressive strength of the particular mix used shows an increase at 200°C before dropping to around one third of the ambient strength at 800°C. The matrix again shows a less serious decline to about two thirds of ambient at 800°C, with a slight increase at 200°C. It is thought that the peak is due to the presence of PFA in the mix. The effect was also noted by Short et al. (2002),

13.3.1.2 Structural steel

All the results quoted here are from Kirby et al. (1986). For Grade 43A (S275) steel, there is no residual strength loss based on the 0,2% proof stress when the steel is heated to temperatures up to 600°C, but a 30% reduction occurs at 1000°C. The variation in residual strength between these temperatures is sensibly linear. The pattern for Grade 50D (S355 J2) steel is similar except that the strength loss at 1000°C is only about 15%. Kirby et al. also quote results on a U.S. ASTM A572 Grade 50 steel (equivalent to S355) which is similar to a UK Grade 50B (S355 JR) steel. Again, no reduction was noted below 600°C, but a 30% reduction occurred at 800°C. At 1000°C, the strength reduction was only around 12%.

It should be noted that in all the tests except on the American steel at 800°C, the measured tensile strengths exceeded the minimum guaranteed yield strengths. COR-TEN B structural steel behaved similarly to the U.S. Grade 50 steel.

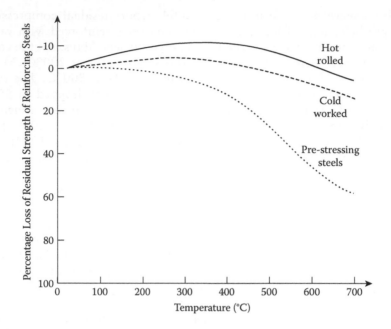

Figure 13.6 Variation of residual strengths of reinforcing and pre-stressing steels with temperature. (Source: Holmes, M., Anchor, R.D., Cooke, G.M.E. et al. (1982) *Structural Engineer*, 60, 7–13. With permission.)

13.3.1.3 Reinforcing and pre-stressing steels

Data on such steels are presented in Figure 13.6 (Holmes et al., 1982). The yield strength for reinforcing steel shows an increase above ambient strength at temperatures below about 550°C and a decrease above 550°C. Pre-stressing steels show no change in strength below 300°C, but a substantial drop after this point; at 800°C, only around 50% of strength remains.

13.3.1.4 Cast and wrought iron

Wrought iron appears to show a marginal strength increase at temperatures up to 900°C and thus appears able to perform well in a fire if excessive deformations do not occur (Kirby et al., 1986).

Cast iron will also perform reasonably well unless unduly large bending moments are applied to a member during a fire. The good fire performance in real structures is in part due to the very low stresses to which cast iron members were subjected in design. One potential problem is that brittle failure is possible if red-hot cast iron is quenched by cold water from fire fighters' hoses or if additional loads are induced during a fire (Barnfield and Porter, 1984).

13.3.1.5 Masonry

Few data focus on the residual strength of masonry, but those that are available indicate that clay bricks lose virtually no strength at 1000°C whereas concrete and calcium silicate bricks lose around 75% of their strength and mortar has no residual strength at 1000°C (Lawrence and Gnanakrishnan, 1988a and b). Türker et al. (2001) report test data on residual strengths of mortar that indicate strength losses for limestone or quartzite mortars around 20% at 500°C, 65% at 700°C, and 80% at 850°C. For lightweight aggregate (pumice) mortars, the figures are no loss, 40%, and 50%.

13.3.2 Determination of temperatures within elements

The methods used here are exactly the same as those used to assess structure performance during fires. If standard solutions based on exposure to the standard furnace test are used, the fire equivalent time will need calculating to enable such methods to yield realistic answers. Bessey (1956) and Ahmed et al. (1992) report visible colour changes in heated concrete. However such changes can be difficult to observe by eye alone because of the type of aggregate. The application of colour image analysis techniques can overcome this problem.

When concrete is heated, there is an obvious change in the frequency of occurrence of red (Short et al., 2001). The primary yellow and green colours appear to have little impact (Figure 13.7a). When a temperature range of 20 to 500°C is examined, an increase in hue values occurs around 250 to 300°C (Figure 13.7b). Where there is a thermal gradient situation, the onset of changes in hue values corresponds with a temperature around 300°C. The original work by Short et al. was carried out using an Olympus polarizing microscope and a Sight Systems Ltd. workstation. Felicetti (2004a) successfully demonstrated that a similar technique can be adapted to digital photography to make the analysis much easier.

The absence of colour change should not be taken as an indication of exposure only to comparatively low temperatures. Colour change is dependent on certain impurities in the aggregates that may not be present in specific cases. Some sources of siliceous aggregates may not produce any colour changes. The results from the work of Bessey (1956) on siliceous aggregates and Ahmed et al. (1992) on limestone aggregates are given in Table 13.3.

Following a structural assessment determining what a residual structure is strong enough to carry the imposed loads or that only minor strengthening is required, attention must turn to methods of repair. However, before methods are chosen, the economics of the situation must be considered. Estimates should be prepared of the costs and durations of repair projects

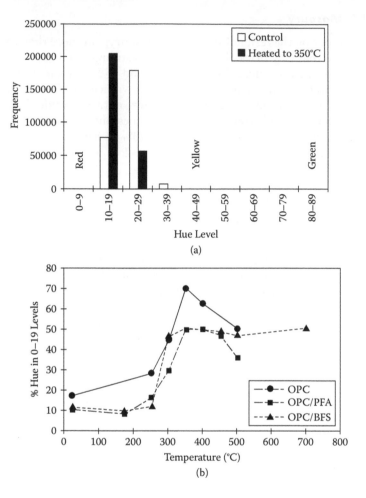

Figure 13.7 Colour changes of heated concrete. (a) Frequency of occurrence for levels of hue from 0 to 89 (control and fired samples). (b) Hue measurements for samples heated to equilibrium temperature (siliceous aggregate). (*Source:* Short, N.R., Purkiss, J.A., and Guise, S.E. (2001) *Construction and Building Materials,* 15, 9–15. With permission.)

or demolition and rebuilding. These processes should start immediately after a fire in a complex or important structure.

It is likely that repairing a structure will be more economical if only minor repairs are required, perhaps with some minor demolition and replacement. A CIB report (Schneider and Nägele, 1989) contains further information about assessing repair potential along with additional references.

Table 13.3 Colour changes in heated siliceous and limestone concrete

Colouration	Temperature (°C)	Condition
Siliceous		
Normal	0 to 300	Normal strength
Pink	300 to 600	Loss in strength
Whitish to grey	600 to 950	Weak and friable
Buff	Above 950	Weak and friable
Limestone		
Grey	0 to 200	Normal strength
Light pink	200 to 400	Loss in strength
Dull grey	400 to 600	Poor

Note: Not all siliceous or limestone aggregate concretes will show these changes because changes may be caused by impurities in sands and aggregates. Absence of or differences in color should be treated with care.

Sources: Bessey, G.E. (1956) *Investigations into Building Fires Part 2: Visible Changes in Concrete and Mortar Exposed to High Temperature.* Technical Paper 4. London: BRE, pp. 6–18; Ahmed, A.E., Al-Shaikh, A.H., and Arafat, T.I. (1992) *Magazine of Concrete Research,* 44, 117–125. With permission.

13.4 METHODS OF REPAIR

As far as steelwork is concerned, any repair will be in the form of partial replacement if the original structure has deformed beyond the point at which it can be reused. Where the steelwork is still intact, it is almost certain that the fire protection system used will need partial or total replacement. Any intumescent paint systems will certainly need renewing. For further information on the reinstatement of steel structures, see Smith et al. (1981).

In the case of superficial damage to masonry, it may well be sufficient to apply cosmetic repairs with plaster-based products, although the integrity of any cavity insulation should be checked before considering this technique. In other cases, replacement of one or more leaves is likely to be necessary.

Timber structures will generally need total replacement. This certainly applies to modern roofing systems that have small member thicknesses and no residual sections. In the case of older and more historic structures, timber roofing and flooring systems will be far more substantial and thus repairs may be feasible (Dixon and Taylor, 1993).

Concrete structures generally provide the greatest scope for repair and strengthening and engineers have a large number of choices in such cases. This section gives only an overview of the situation. More detailed guidance is given in a Concrete Society Report (1990). Any repair must satisfy all the original design criteria of a structure including strength, deflection, durability, and fire resistance. If a structure needs strengthening, it is essential that the new sections are capable of carrying the forces within the new sections and also capable of transmitting the forces from the existing sections of the structure. For example, it will be necessary to ensure sufficient lap length between existing and new reinforcement. Repairs can be effected by concrete spraying (gunite), resin restoration, or overcladding.

For gunite repairs, it is essential that all exposed concrete faces are thoroughly cleaned to ensure that the gunite bonds fully to the existing concrete. It will often be necessary to place very light mesh within the depth of a repair to aid integrity unless the area is very small (Figure 13.8).

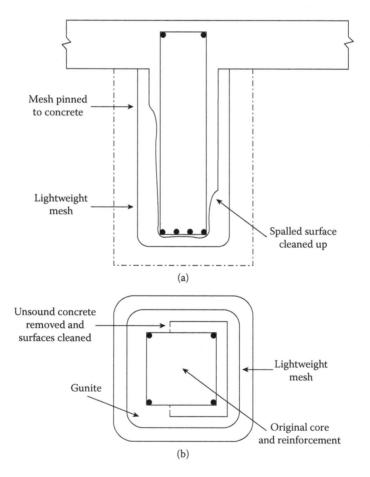

Figure 13.8 Gunite repairs to fire-damaged concrete: (a) typical beam repair and (b) typical column repair.

Resin repairs are usually applicable only to lightly damaged areas where spalling is shallow. However, most resins soften around 80°C, and thus their integrity in a fire may be suspect (CIRIA, 1987). Plecnik and Fogerty (1986) confirm from tests that epoxy-repaired beams lose strength and stiffness rapidly at uniform temperatures above 120°C. It is thus necessary to be very careful when specifying resin repairs and consider the provision of additional fire protection using plaster finishes.

Overcladding can take two forms. For slightly damaged walls or slab soffits, the damage can be covered by plasterboard and battens. For columns, glass-reinforced cement panels can be used. The gaps between the panels and the original columns should be filled with mortar or concrete.

Case studies on the assessment, repair and reinstatement of fire-damaged concrete building structures are provided in Morales (1992) and Nene and Kavle (1992). For repair of a concrete bridge structure (part of which needed demolition), see Boam and Cropper (1994).

13.5 DEMOLITION

Clearly, the same safety hazards that exist for structures being demolished for reasons other than fire damage exist for those so damaged, except that problems of stability are exacerbated for fire-damaged structures as the structure itself is inherently weaker, often to such an extent that little physical effort may be made for demolition. Additional issues also warrant consideration (Purkiss, 1990b):

1. Some products of combustion may be toxic and thus a structure may need clearing and full venting before demolition can take place.
2. Asbestos will be present in some structures as lagging to hot water pipes and tanks or in older steel structures as fire protection to beams. In both cases, the asbestos will need specialized handling before demolition can start.
3. A basement is likely to be partially or fully filled with water following fire fighting and will need pumping out. It should be noted that toxic combustion products may be present and care must be taken in deciding where such effluents can be discharged.

References

Abdel-Rahman, A.K. and Ahmed, G.N. (1996) Computational heat and mass transport in concrete walls exposed to fire. *Numerical Heat Transfer (Part A)*, 29, 373–395.

Abrams, M.S. (1968) Compressive strength of concrete up to 1600°F (871°C). In *Temperature and Concrete*, Special Publication SP-25. Detroit: American Concrete Institute, pp. 33–58.

Ahmed, G.N. and Hurst, J.P. (1997) An analytical approach for investigating the cases of spalling of high-strength concrete at elevated temperatures. In *International Workshop on Fire Performance of High-Strength Concrete*. Gaithersburg, MD: NIST, pp. 95–108.

Ahmed, A.E., Al-Shaikh, A.H. and Arafat, T.I. (1992) Residual compressive and bond strengths of limestone aggregate concrete subjected to elevated temperatures. *Magazine of Concrete Research*, 44, 117–125.

Aldea, C.M., Franssen, J.M. and Dortreppe, J.C. (1997) Fire test on normal and high strength reinforced concrete columns. In *International Workshop on Fire Performance of High-Strength Concrete*. Gaithersburg, MD: NIST, pp. 109–124.

Ali, F.A., O'Connor, D. and Abu-Tair, A. (2001) Explosive spalling of high-strength concrete columns in fire. *Magazine of Concrete Research*, 53, 197–204.

Ali, H.M., Senseny, P.E. and Alpert, R.L. (2004) Lateral displacement and collapse of single-story steel frames in uncontrolled fires. *Engineering Structures*, 26, 593-607.

Allen, B. (2006) Intumescent fire protection of tall buildings. *New Steel Construction*, 14, 24–28.

American Concrete Institute Committee 363. (1984) State-of-the-art report on high-strength concrete. *ACI Journal*, 81, 364–411.

American Society of Civil Engineers. (2007) *Standard Calculation Methods for Structural Fire Protection*. Report ASCE/SEI/SFPE 29-05.

Anderberg, Y. (2004) The effects of the constitutive models on the prediction of concrete mechanical behaviour and on the design of concrete structures exposed to fire. In: *Fire Design of Concrete Structures: What Now? What Next?* Proceedings of Workshop, Politechnico di Milano, pp. 37–48.

Anderberg, Y. (1988) Modelling steel behaviour. *Fire Safety Journal*, 13, 17–26.

Anderberg, Y. (1983) *Properties of Materials at High Temperatures-Steel*. RILEM Report, University of Lund, Sweden.

Anderberg, Y. (1978a) *Armeringsståls Mekaniska Egenskaper vid Höga Temperaturer.* Bulletin 61, University of Lund, Sweden.

Anderberg, Y. (1978b) Analytical fire engineering design of reinforced concrete structures based on real fire characteristics. In *Proceedings of Eighth Congress of Fédération Internationale de la Précontrainte.* London: Concrete Society, pp. 112–123.

Anderberg, Y. and Thelandersson, S. (1976) *Stress and Deformation Characteristics of Concrete 2: Experimental Investigation and Material Behaviour Model.* Bulletin 54, University of Lund, Sweden.

Anon (1993) Goodnight Vienna. *Fire Prevention,* 257, 40–41.

Anon (1990) Birmingham fire may force rebuild. *New Civil Engineer,* Sept. 6, 11.

Anon (1988) Fenell report finds fundamental errors made by senior underground management. *Fire Prevention,* 215, 12.

Anon (1987) King's Cross tragedy. *Fire Prevention,* 205, 5–6.

Anon (1986) News scene. *Fire Prevention,* 188, 5.

Anon (1985) Fire tragedy at football stadium. *Fire Prevention,* 81, 5–6

Anon (1983) An FPA review of the official report on the Stardust disco fire. *Fire Prevention,* 158, 12–20.

Anon (1980) Report of the Woolworth's fire, Manchester. *Fire Prevention,* 138, 13–24.

Anon (1973) The Summerland fire. *Fire Prevention,* 1–16.

Arup. (2005) Madrid Windsor fire: the Arup view. http://www.arup.com/fire/feature. cfm?pageid=6150

Ashton, L.A. (1966) Effect of restraint of longitudinal deformation or rotation. In *Feuerwiderstandsfähigheit von Spannbeton.* Braunschweig: Bauerlag GmbH, pp. 15–19.

Ashton, L.A. and Bate, S.C.C. (1960) The fire-resistance of pre-stressed concrete beams. *Proceedings of the Institution of Civil Engineers,* 17, 15–38.

ASFPCM. (1993) *Fire Protection of Timber Floors.* Aldershot: Association of Specialist Fire Protection Contractors and Manufacturers.

Babrauskas, V. and Williamson, R.B. (1978a) The historical basis of fire resistance testing I. *Fire Technology,* 14, 184–189 and 205.

Babrauskas, V. and Williamson, R.B. (1978b) The historical basis of fire resistance testing II. *Fire Technology,* 14, 304–316.

Baddoo, N.R. and Burgan, B.A. (2001) *Structural Design of Stainless Steel.* Publication P291, Steel Construction Institute.

Bahrends, J.F.B. (1966) Brandversuche an durchlaufenden Spannbetonbauteilen. In *Feuerwiderstandsfähigheit von Spannbeton.* Braunschweig: Bauerlag GmbH, pp. 20–23.

Bailey, C.G. (2004) Structural fire design: core or specialist subject. *The Structural Engineer,* 82, 32–38.

Bailey, C.G. (2003) *New Fire Design Method for Steel Frames with Composite Floor Slabs.* FBE Report 5. Garston: BRE.

Bailey, C.G. (2002) Holistic behaviour of concrete buildings in fire. *Proceedings of the Institution of Civil Engineers, Buildings and Structures,* 152, 199–212.

Bailey, C.G. (2001) *Steel Structures Supporting Composite Floor Slabs: Design for Fire.* Digest 462. Garston: BRE.

Bailey, C.G. (2000) Effective lengths of concrete-filled steel square hollow sections in fire. *Proceedings of the Institution of Civil Engineers, Structures and Buildings,* 140, 169–178.

Bailey, C.G. (1998) Development of computer software to simulate the structural behaviour of steel-framed building in fire. *Computers and Structures*, 67, 421–438.

Bailey, C.G., Burgess, I.W., and Plank, R.J. (1996) Computer simulation of a full-scale structural fire test. *The Structural Engineer*, 74, 93–100.

Bailey, C.G. and Ellobody, E. (2009) Comparison of unbonded and bonded post-tensioned concrete slabs under fire conditions. *The Structural Engineer*, 87, 23–31.

Bailey, C.G. and Lennon, T. (2008) Full-scale tests on hollow core floors. *The Structural Engineer*, 86, 33–39.

Bailey, C.G., Lennon, T., and Moore, D.B. (1999) The behaviour of full-scale steel-framed buildings subjected to compartment fires, *The Structural Engineer*, 77, 15–21.

Bailey, C.G. and Moore, D.B. (2000a) The structural behaviour of steel frames with composite floor slabs subject to fire. Part 1: Theory. *The Structural Engineer*, 78, 19–27.

Bailey, C.G. and Moore, D.B. (2000b) The structural behaviour of steel frames with composite floor slabs subject to fire. Part 2: Design. *The Structural Engineer*, 78, 28–33.

Bailey, C.G. and Newman, G.M. (1998) The design of steel framed buildings without applied fire protection. *The Structural Engineer*, 76, 77–81.

Baker, G. and de Borst, R. (2005) An anisotropic thermomechanical damage model for concrete at transient elevated temperatures. *Philosophical Transactions of Royal Society A*, 363, 2603–2628.

Baldwin, R. (1975) Structural fire protection: the economic option. In *Proceedings of Jubilee Conference of Midlands Branch of Institution of Structural Engineers*. Birmingham: Institution of Structural Engineers, pp. 115–137.

Baldwin, R. and North, M.A. (1973) A stress–strain relationship for concrete at high temperature. *Magazine of Concrete Research*, 24, 99–101.

Baldwin, R. and Thomas, P.H. (1973) *Passive and Active Fire Protection: Optimum Combination*. Fire Research Note 963. Garston: BRE.

Bali, A. (1984) *The Transient Behaviour of Plain Concrete at Elevated Temperatures*. PhD Thesis, University of Aston, Birmingham.

Barnfield, J.R. (1986) *The UK Procedure for the Appraisal of Fire Protection to Structural Steelwork*. EGOLF Seminar, Brussels, Nov.

Barnfield, J.R. and Porter, A.M. (1984) Historic buildings and fire, fire performance of cast-iron structural elements. *The Structural Engineer*, 62A, 373–380.

Bayley, M.J. (1992) The fire protection of aluminium in offshore structures. In *Proceedings of IMechE Conference on Materials and Design against Fire*, Paper C438/019/92. London: Mechanical Engineering Publications, pp. 113–120.

Bažant, Z.P. (1997) Analysis of pore pressure, thermal stress and fracture in rapidly heated concrete. In *Proceedings of International Workshop on Fire Performance of High-Strength Concrete*. Gaithersburg, MD: NIST, pp. 155–164.

Bažant, Z.P., Ed. (1988) *Mathematical Modelling of Creep and Shrinkage of Concrete*. Chichester: John Wiley & Sons.

Bažant, Z.P. (1983) Mathematical model for creep and thermal shrinkage of concrete at high temperatures. *Nuclear Engineering and Design*, 76, 183–191.

Bažant, Z.P., Chern, J.C. and Thonguthai, W. (1981) Finite element program for moisture and heat transfer in heated concrete. *Nuclear Engineering and Design*, 68, 61–70.

Bažant, Z.P. and Kaplan, M.F. (1996) *Concrete at High Temperature*. London: Longman.

Bažant, Z.P. and Thonguthai, W. (1978) Pore pressure and drying of concrete at high temperature. *Journal of Engineering Mechanics Division*, 104, 1059–1079.

Bažant, Z.P. and Thonguthai, W. (1979) Pore pressure in heated concrete walls: theoretical prediction. *Magazine of Concrete Research*, 31, 67–76.

Becker, J., Bizri, H. and Bresler, B. (1974) *Fires-T2: A Computer Programme for the Fire Response of Structures–Thermal*. Report UCB-FRG 74-1. Department of Civil Engineering, University of California, Berkeley.

Becker, J. and Bresler, B. (1972) *Fires-RC: A Computer Program for the Fire Response of Structures–Reinforced Concrete Frames*. Report UCB-FRG 74-3, Department of Civil Engineering, University of California, Berkeley.

Beitel, J.J. and Iwankiw, N.R. (2005) Historical survey of multi-storey building collapses in fire. *International Fire Protection*, 24, 43–46.

Belytschko, T., Liu, W.K. and Moran, B. (2000) *Nonlinear Finite Element Methods for Continua and Structures*. Chichester: John Wiley & Sons.

Benedetti, A. (1998) On the ultrasonic pulse propagation into fire damaged concrete. *ACI Structural Journal*, 95, 259–271.

Bessey, G.E. (1956) *Investigations into Building Fires Part 2: Visible Changes in Concrete and Mortar Exposed to High Temperature*. Technical Paper 4. London: BRE, pp. 6–18.

Bishop, G.R. and Sheard, P.A. (1992) Fire-resistant composites for structural sections. *Composite Structures*, 21, 85–89.

Bishop, P. (1991) Blaze adds to insurer unease. *New Civil Engineer*, 4, Aug. 15.

Boam, K. and Cropper, D. (1994) Midlands link motorway viaducts: rehabilitation of a fire damaged structure. *Proceedings of Institution of Civil Engineers: Structures and Buildings*, 104, 111–123.

Bobrowski, J. and Bardhan-Roy, B.K. (1969) A method of calculating the ultimate strength of reinforced and pre-stressed concrete beams in combined flexure and shear. The *Structural Engineer*, 47, 3–15.

Boström, L. (2004) *Innovative Self-Compacting Concrete-Development of Test Methodology for Determination of Fire Spalling*. Swedish National Testing and Research Institute, Report 06.

Boström, L. (2002) *The Performance of Some Self-Compacting Concretes When Exposed to Fire*. Swedish National Testing and Research Institute Report 23.

Brandes, A. (1993) *Smithells Metals Reference Book*. Amsterdam: Butterworth.

BRE. (1987) *Smoke Control in Buildings: Design Principles*, Digest 260. Garston: Building Research Establishment.

BRE. (1986) *Fire Resistant Steel Structures: Free Standing Blockwork Filled Columns and Stanchions*, Digest 317. Garston: Building Research Establishment.

Brick Industry Association. (2008) Fire resistance of brick masonry. *Technical Notes on Brick Construction*, 16, 1–16.

BS 476: Part 3: 1975 Fire tests on building materials and structures; Part 3 External fire exposure roof test. British Standards Institution.

BS 476: Part 6: 1989 Fire tests on building materials and structures; Part 6 Method of test for fire propagation for products. British Standards Institution.

BS 476: Part 7: 1987 Fire tests on building materials and structures; Part 7 Method for classification of the surface spread of flame of products. British Standards Institution.

BS 476: Part 8: 1972 Fire tests on building materials and structures; Part 8 Test methods and criteria for the fire resistance of elements of building construction. British Standards Institution.

BS 476: Part 10: 2009 Fire tests on building materials and structures. Guide to the principles, selection, role and application of fire testing and their outputs. British Standards Institution.

BS 476: Part 20: 1987 Fire tests on building materials and structures; Part 20 Method for the determination of the fire resistance of elements of construction (general principles). British Standards Institution.

BS 476: Part 21: 1987 Fire tests on building materials and structures; Part 21 Methods for the determination of the fire resistance of loadbearing elements of construction. British Standards Institution.

BS 476: Part 22: 1987 Fire tests on building materials and structures; Part 22 Methods for the determination of the fire resistance of non-loadbearing elements of construction. British Standards Institution.

BS 1881: Part 120: 1983 Methods of testing concrete; Part 120 Method for the determination of the compressive strength of concrete cores. British Standards Institution.

BS 5950: Structural use of steelwork in buildings, Part 8 Code of Practice for fire resistant design. British Standards Institution.

BS 9999: Code of Practice for Fire Safety in the Design, Management and Use of Buildings. British Standards Institution.

Buchanan, A.H. (2001) *Structural Design for Fire Safety*. Chichester: John Wiley & Sons.

Building Employers Confederation and Loss Prevention Council. (1992) *Fire Prevention on Construction Sites*. Coventry: National Contractors Group.

Burgess, I.W., El-Rimawi, J.A. and Plank, R.J. (1990) Analysis of beams with non-uniform temperature profile due to fire exposure. *Journal of Constructional Steel Research*, 16, 169–192.

Byrd, T. (1992a) Sprinklers dry as Expo burned. *New Civil Engineer*, Feb. 27, 5.

Byrd, T. (1992b) Burning question. *New Civil Engineer*, Feb. 27, 12–13.

Canter, D. (1985) *Studies of Human Behaviour in Fire: Empirical Results and their Implication for Education and Design*. Garston: BRE.

Castillo, C. and Durani, A.J. (1990) Effect of transient high temperature on high-strength concrete. *ACI Materials Journal*, 87, 47–53.

Castle, G.K. (1974) The nature of various fire environments and the application of modern material approaches for fire protection of exterior structural steel. *Journal of Fire and Flammability*, 5, 203–222.

CCMCC. (1992) *Firewalls, A Design Guide*. Ottawa: Canadian Concrete and Masonry Codes Council.

Chan, Y.N., Peng, G.F. and Anson, M. (1999) Residual strength and pore structure of high-strength concrete and normal strength concrete after exposure to high temperatures. *Cement and Concrete Composites*, 21, 23–27.

Chana, P. and Price, W. (2003) The Cardington fire test. *Concrete*, 37, 28, 30–33.

Chen, B. and Liu, J. (2004) Residual strength of hybrid-fiber-reinforced high strength concrete after exposure to high temperatures. *Cement and Concrete Research*, 34, 1065–1069.

Chen, J. and Young, B. (2007) Cold-formed steel lipped channel columns at elevated temperatures. *Engineering Structures*, 29, 2445–2456.

Chen, W.F. (1982) *Plasticity in Reinforced Concrete*. New York: McGraw Hill.

Chitty, R., Cox, G., Fardell, P.J. et al. (1992) *Mathematical Fire Modelling and Its Application to Fire Safety Design*, Report 223. Garston: BRE.

CIB W14 Report (1983) A conceptual approach towards a probability based design guide on structural fire safety. *Fire Safety Journal*, 6, 1–79.

Clayton, N. and Lennon, T. (2000) High-grade concrete columns in fire. *Concrete*, 34, 51.

Cockcroft, D. (1993) Architect's perspective on the Windsor Castle fire. *Fire Prevention*, 22, 4–7.

Concrete Society. (1990) *Assessment and Repair of Fire Damaged Concrete Structures*, Technical Report 33. Wexham Springs: Concrete Centre.

Connolly, R.J. (1997) The spalling of concrete. *Fire Engineers Journal*, January, 38-40.

Connolly, R.J. (1995) The Spalling of Concrete in Fires. PhD Thesis, Aston University, Birmingham.

Conserva, M., Donzellii, G. and Trippodo, R. (1992) *Aluminium and Its Applications*. Montichiari, Italy: Alfin-Edimet.

Consolazio, G.R., McVay, M.C. and Rish, J.W. (1997) Measurement and prediction of pore pressure in cement mortar subjected to elevated temperature. *International Workshop on Fire Performance of High-Strength Concrete*, NIST Gaithersburg, MD, pp. 125–148.

Construction Industry Research and Information Association. (1987) *Fire Tests on Ribbed Concrete Slabs*, Technical Note 131. London: CIRIA.

Cooke, G.M.E. (1987a) Fire engineering of tall fire separating walls (Part 1). *Fire Surveyor*, 16, 13–29.

Cooke, G.M.E. (1987b) Fire engineering of tall fire separating walls (Part 2). *Fire Surveyor*, 16, 19–29.

Cooke, G.M.E. and Latham, D.J. (1987) The inherent fire resistance of a loaded steel framework. *Steel Construction Today*, 1, 49–58.

Cooke, G.M.E., Lawson, R.M. and Newman, G.M. (1988) Fire resistance of composite deck slabs. *The Structural Engineer*, 66, 253–261 and 267.

Cooke, G.M.E. and Morgan, P.B.E. (1988) *Thermal Bowing in Fire and How It Affects Building Design*, Information Paper IP21/88. Garston: BRE.

Copier, W.J. (1979) The spalling of normal weight and lightweight concrete on exposure to fire. *Heron*, 24, 2.

Cox, G. (1995) Compartment Fire Modelling (Chapter 6). In *Combustion Fundamentals of Fire*. London: Academic Press.

Cruz, C.R. (1962) An optical method of determining the elastic constants of concrete. *Journal of Portland Cement Association Research and Development Laboratories*, 5, 24–32.

Davis, J.R. (1993) *ASM Specialty Handbook, Aluminium and Aluminium Alloys*. Cincinnati: ASM International.

Day, T. (1994) Fire safety in major retail outlets. *Fire Safety Engineering*, 1, 18–20.

Dayan, A. and Gluekler, E.L. (1982) Heat and mass transfer within an intensely heated concrete slab. *International Journal of Heat Mass Transfer*, 25, 1461–1467.

Deiderichs, U. (1987) Modelle zur Beschreibung der Betonverformung bei instantionären Temperaturen. In *Abschlußkolloquium-Bauwerke unter Brandeinwirkung*. Braunschweig: Technische Universität, pp. 25–34.

de Sa, C. and Benboudjema, F. (2011) Modelling of concrete nonlinear mechanical behaviour at high temperatures with different damage-based approaches. *Materials and Structures*, 44, 1411–1429.

de Vekey, R. (2004). Structural Fire Engineering Design: Materials Behaviour. *Masonry Digest* 487 Part 3. Garston, Building Research Establishment.

Department of the Environment. (1992a) *Building Regulations: Approved Document B, Fire Safety*. London: Her Majesty's Stationery Office.

Department of the Environment. (1992b) *Building Regulations and Fire Safety: Procedural Guidance*. London: Her Majesty's Stationery Office.

Department of the Environment. (1991) *Standard Fire Precautions for Contractors*. London: Her Majesty's Stationery Office.

Dhatt, G., Jacquemier, M., and Kadje, C. (1986) Modelling of drying refractory concrete. In *Proceedings of Fifth International Symposium on Drying*, McGill University, pp. 94–104.

Dixon, R. and Taylor, P. (1993) Hampton Court: restoration of the fire-damaged structure. *Structural Engineer*, 71, 321–325.

Dorn, J.E. (1954) Some fundamental experiments on high temperature creep. *Journal of Mechanics and Physics of Solids*, 3, 86–116.

Dortreppe, J.C. and Franssen, J.M. (undated) Fire attack on concrete columns and design rules under fire conditions. In *Proceedings of Symposium on Computational and Experimental Methods in Mechanical and Thermal Engineering*, University of Ghent, unpaginated.

Dortreppe, J.C., Franssen, J.M. and Vanderzeypen, Y. (1995) A straightforward calculation method for the fire resistance of reinforced concrete columns. In *First European Symposium on Fire Safety Science*, Zurich (poster paper).

Dos Santos, J.R., Branco, F.A. and de Brito, J. (2002) Assessment of concrete structures subjected to fire: the FB test. *Magazine of Concrete Research*, 54, 203–208.

Dougill, J.W. (1966) The relevance of the established method of structural fire testing to reinforced concrete. *Applied Materials Research*, 5, 235–240.

Dounas, S. and Golrang, B. (1982) *Ståhls Mekaniska Egenskaper vid Höger Temperaturer*. University of Lund, Sweden.

Dowling, J. (2005) Steel in fire: the latest news. *New Steel Construction*, 13, 30–32.

Drysdale, D. (1998) *An Introduction to Fire Dynamics*, 2nd ed. Chichester: John Wiley & Sons.

ECCS. (1985) *Design Manual on the European Recommendations for the Fire Safety of Steel Structures*. Brussels: ECCS Technical Committee 3.

ECCS. (1983) *Calculation of the Fire Resistance of Load Bearing Elements and Structural Assemblies Exposed to the Standard Fire*. Amsterdam: Elsevier.

Ehm, C. and Schneider, U. (1985) The high temperature behaviour of concrete under biaxial conditions. *Cement and Concrete Research*, 15, 27–34.

Ehm, H. (1967) Ein Betrag zur reichnerischung Bemessung von brandbeanspructen balkenartigen Stahlbetonbauteilen. Braunschweig: Technische Hochschule.

Ehm, H. and von Postel, R. (1966) Versuche an Stahlbetonkonstruktionen mit Durchlaufwirkung unter Feuerangriff. In *Feuerwiderstandsfähigheit von Spannbeton*. Braunschweig: Bauerlag GmbH, pp. 24–31.

El-Dieb, A.S. and Hooton, R.D. (1995) Water permeability measurement of high performance concrete using a high-pressure triaxial cell. *Cement and Concrete Research*, 25, 1199–1208.

El-Refal, F.E., Kamal, M.M., and Bahnasawy, H.H. (1992) Effect of high temperature on the mechanical properties of fibre reinforced concrete cast using superplasticizer. In Swamy, R.N., Ed., *Fibre Reinforced Cement and Concrete*. London: E. & F.N. Spon, pp. 749–763.

EN 13381-4: Test Methods for Determining the Contribution to the Fire Resistance of Structural Members – Part 4: Applied Passive Protection Steel Members. Comité Européen de Normalisation/British Standards Institution.

EN 1363: Fire Resistance Tests. Comité Européen de Normalisation/British Standards Institution.

EN 1363-1: Fire Resistance Tests Part 1: General Requirements. Comité Européen de Normalisation/British Standards Institution.

EN 1364: Part 1 (1999) Fire resistance tests for non-loadbearing elements. British Standards Institution.

EN 1999-1-2: Eurocode 9: Design of aluminium structures. Part 1.1 General rules for structural fire design. Comité Européen de Normalisation/British Standards Institution.

EN 1999-1-1: Eurocode 9: Design of aluminium structures. Part 1.1 General rules and rules for buildings. Comité Européen de Normalisation/British Standards Institution.

EN 1996-1-2: Eurocode 6: Design of masonry structures. Part 1.2: Supplementary rules for structural fire design. Comité Européen de Normalisation/British Standards Institution.

EN 1996-1-1: Eurocode 6: Design of masonry structures. Part 1.1: General rules for reinforced and unreinforced masonry structures. Comité Européen de Normalisation/British Standards Institution.

EN 1995-1-2: Eurocode 5: Design of timber structures. Part 1.2: Supplementary rules for structural fire design. Comité Européen de Normalisation/British Standards Institution.

EN 1995-1-1: Eurocode 5: Design of timber Structures–Part 1.1: General-Common Rules and Rules for buildings. Comité Européen de Normalisation/British Standards Institution.

EN 1994-1-2: Eurocode 4: Design of composite steel and concrete structures. Part 1.2: Supplementary rules for structural fire design. Comité Européen de Normalisation/British Standards Institution.

EN 1994-1-1: Eurocode 4: Design of composite steel and concrete structures. Part 1.1: General rules and rules for buildings. Comité Européen de Normalisation/British Standards Institution.

EN 1993-1-2: Eurocode 3: Design of steel structures. Part 1.2: Supplementary rules for structural fire design. Comité Européen de Normalisation/British Standards Institution.

EN 1993-1-1: Eurocode 3: Design of steel structures. Part 1.2: General rules and rules for buildings. Comité Européen de Normalisation/British Standards Institution.

EN 1992-1-2: Eurocode 2: Design of concrete structures. Part 1.2: Supplementary rules for structural fire design. Comité Européen de Normalisation/British Standards Institution.

EN 1992-1-1: Eurocode 2: Design of concrete structures. Part 1.1: General rules and rules for buildings. Comité Européen de Normalisation/British Standards Institution.

EN 1991-1-2: Eurocode 1: Basis of design and actions on structures. Part 1.2: Actions on structures exposed to fire. Comité Européen de Normalisation/ British Standards Institution.

EN 1990: Eurocode–Basis of Structural Design. Comité Européen de Normalisation/ British Standards Institution.

EN 13501: Fire Classification of Construction Products and Building Elements. Comité Européen de Normalisation/British Standards Institution.

Enjily, V. (2003) The fire performance of a six-storey timber frame building at BRE Cardington. Paper presented at Guidance for Structural Fire Engineering Conference, Feb. 13.

Fackler, J.P. (1959) Concernant la résistance au feu des éléments de construction. *Cahiers du Centre Scientifique et Technique du Bâtiment*, 37, 1–20.

Fairyadh, F.I. and El-Ausi, M.A. (1989) Effect of elevated temperature on splitting tensile strength of fibre concrete. *International Journal of Cement Composites and Lightweight Concrete*, 11, 175–8.

Fédération Internationale de la Précontrainte and Comité Euro-International du Béton (1978) *FIP/CEB Report on Methods of Assessment of Fire Resistance of Concrete Structural Members*. Wexham Springs : Cement and Concrete Association.

Felicetti, R. (2004a) Digital camera colorimetry for the assessment of fire-damaged concrete. In *Fire Design of Concrete Structures: What Now? What Next?* Proceedings of Workshop, Politechnico di Milano, pp. 211–220.

Felicetti, R. (2004b) The drilling resistance test for the assessment of thermal damage in concrete. In *Fire Design of Concrete Structures: What now? What next?* Proceedings of Workshop, Politechnico di Milano, pp. 241–248.

Felicetti, R. and Gambarova, P.G. (2003) Heat in concrete: special issues in materials testing. *Studies and Researches, Politechnico di Milano*, 24, 121–138.

Fellinger, J.H.H. (2004) *Shear and Anchorage Behaviour of Fire Exposed Hollow Core Slabs*. Delft: DUP Science.

Fields, B.A. and Fields, R.J. (1989) *Elevated Temperature Deformation of Structural Steel*, NISTIR 88-3899. Gaithersburg, MD: NIST.

Forsén, N.E. (1982) *A Theoretical Study on the Fire Resistance of Concrete Structures*, Report STF A82062. Trondheim: Norwegian Technical University.

Fowler, D. and Doyle, N. (1992) Stability probe leads castle salvage. *New Civil Engineer*, Nov. 26, 4–5.

Franssen, J.M. (2003) SAFIR: a thermal/structural program modeling structures under fire. In *Proceedings of American Institute for Steel Construction*, Baltimore.

Fredlund, B. (1988) A Model for Heat and Mass Transfer in Timber Structures during Fire: A Theoretical, Numerical and Experimental Study, Report LUTUDG/ CTVBB-1003. Lund Institute of Technology, Sweden.

Fu, Y.F., Wong, Y.L., Poon, C.S. et al. (2005) Stress–strain behaviour of high-strength concrete at elevated temperatures. *Magazine of Concrete Research*, 57, 535–544.

Fujita, K. (undated) *Characteristics of a Fire in a Non-Combustible Room and Prevention of Fire Damage*, Report 2(2). Tokyo: Japanese Ministry of Construction Building Research Institute.

Furamura, F. (1966) Stress-strain curve of concrete at high temperatures, Abstract 7004. *Transactions of Architectural Institute of Japan*, p. 686.

Furamura, F., Oh, C.H., Ave, T. et al. (1987) Simple formulation for stress-strain relationship of normal concrete at elevated temperature. *Report of Research Laboratory of Engineering Materials, Tokyo Institute of Technology*, 12, 155–172.

Gale, W.F. and Totemeier, T.C. (2003) *Smithells' Metals Reference Book*, 8th ed. Amsterdam: Elsevier-Butterworth-Heinemann.

Gawin, D., Pesavento, F., and Schrefler, B.A. (2004) Modelling of deformations of high strength concrete at elevated temperatures. *Materials and Structures: Concrete Science and Engineering*, 37, 218–236.

Gerhards, C.C. (1982) Effect of moisture content and temperature on the mechanical properties of wood: an analysis of immediate effects. *Wood and Fiber*, 14, 4–36.

Gernay, T. (2012) A Multi-Axial Constitutive Model for Concrete in the Fire Situation Including Transient Creep and Cooling Down Phases. PhD Thesis, University of Liége, Belgium.

Gillen, M.P. (1997) The behaviour of high performance lightweight concrete at elevated temperatures. In *Third ACI International Symposium on Advances in Concrete Technology*, 171-5. American Concrete Institute, pp. 131–155.

Gnanakrishnan, N., Lawrence, S.J. and Lawther, R. (1988) Behaviour of cavity walls exposed to fire. In *Proceedings of Eighth International Brick and Block Masonry Conference*, pp. 981–990.

Gnanakrishnan, N. and Lawther, R. (1989) Some aspects of the fire performance of single leaf masonry construction. In *Proceedings of International Symposium on Fire Engineering for Building Structures and Safety*. Melbourne: Institution of Engineers of Australia, pp. 93–99.

Gong, Z., Song, B., and Mujumdar, A. (1991) Numerical simulation of drying of refractory concrete. *Drying Technology*, 9, 479–500.

Gowripalan, N. (1998) Mechanical properties of high-strength concrete subjected to high temperatures. In *Concrete under Severe Conditions 2*, London: E. & F.N. Spon, pp. 774–784.

Guise, S.E. (1997) The Use of Colour Image Analysis for Assessment of Fire Damaged Concrete. PhD thesis, Aston University, Birmingham.

Hadvig, S. (1981) *Charring of Wood in Building Fires — Practice, Theory, Instrumentation, Measurement*. Lyngby: Technical University of Denmark.

Haksever, A. and Anderberg, Y. (1981/2) Comparison between measured and computed structural response of some reinforced concrete columns in fire. *Fire Safety Journal*, 4, 293-297.

Han, L.H., Wang, W.H., and Yu, H.X. (2012) Analytical behaviour of RC beam to CFST column frames subjected to fire. *Engineering Structures*, 36, 394–410.

Handoo, S.K., Agarwal, S., and Maiti, S.C. (1991) Application of DTA/TGA for the assessment of fire damaged concrete structures. In *Proceedings of Eighth National Symposium on Thermal Analysis*, pp. 437–443.

Hansell, G.O. and Morgan, H.P. (1994) *Design Approaches for Smoke Control in Atrium Buildings*, Report BR258. Garston: BRE.

Harmathy, T.Z. (1993) *Fire Safety Design and Concrete*. Harlow: Longman.

Harmathy, T.Z. (1969) Design of fire test furnaces. *Fire Technology*, 5, 140–150.

Harmathy, T.Z. (1967) Deflection and failure of steel supported floors and beams in fire. In *Symposium on Fire Test Methods-Restraint and Smoke*, STP422. Philadelphia: American Society for Testing and Materials, pp. 40–62.

Harmathy, T.Z. and Mehaffey, J.R. (1985) Design of buildings for prescribed levels of structural fire safety. In *Fire Safety: Science and Engineering*, STP882. Philadelphia: American Society for Testing and Materials, pp. 160–173.

Harmathy, T.Z. and Stanzak, W.W. (1970) Elevated temperature tensile and creep properties of some structural and pre-stressing steels. In *Symposium on Fire Test Performance, American Society of Testing and Materials*, STP 464. Philadelphia: American Society for Testing and Materials, pp. 186–208.

Hassen, S. and Colina, H. (2006) Transient thermal creep of concrete in accidental conditions up to 400°C. *Magazine of Concrete Research*, 58, 201–208.

He, Z.J. and Song, Y.P. (2010) Multiaxial tensile–compressive strengths and failure criterion of plain high-performance concrete before and after high temperatures. *Construction and Building Materials*, 24, 498–504.

Heinfling, G. (1998) *Contribution à la modélisation numérique du comportement du béton et des structures en béton armé sous sollicitations thermo-mécaniques à hautes températures*. PhD thesis. INSA, Lyon.

Hertz, K.D. (2006) Quenched reinforcement exposed to fire. *Magazine of Concrete Research*, 58, 43–48.

Hertz, K.D. (2005) Concrete strength for fire safety design. *Magazine of Concrete Research*, 57, 445–453.

Hertz, K.D. (2004) Reinforcement data for fire safety design. *Magazine of Concrete Research*, 56, 453–459.

Hertz, K.D. (1992) Danish investigations on silica fume concretes at elevated temperatures. *ACI Materials Journal*, 89, 345–347.

Hertz, K.D. (1985) *Analyses of Pre-stressed Concrete Structures Exposed to Fire*. Report 174. Lyngby: Technical University of Denmark.

Hertz, K.D. (1981a) *Stress Distribution Factors*, 2nd ed. Report 158 (reissued as Report 190 in 1988). Lyngby: Technical University of Denmark.

Hertz, K.D. (1981b) *Simple Temperature Calculations of Fire Exposed Concrete Constructions*. Report 159, Lyngby: Technical University of Denmark.

Heselden, A.J.M. (1984) *The Interaction of Sprinklers and Roof Venting in Industrial Buildings: The Current Knowledge*. Garston: BRE.

Heselden, A.J.M. (1968) Parameters determining the severity of fire. In *Symposium on Behaviour of Structural Steel in Fire*. London: Her Majesty's Stationery Office, pp. 19–28.

Hicks, S.J. and Newman, G.M. (2002) *Design Guide for Concrete Filled Columns*. Rotherham: Tata Steel Swinden Technology Center.

Hinkley, P.L., Hansell, G.O., Marshall, N.R. et al. (1992) Sprinklers and vents interaction. *Fire Surveyor*, 21, 18–23.

Hinkley, P.L. and Illingworth, P.M. (1990) *The Ghent Fire Tests: Observations on the Experiments*. Havant: Colt International.

Holman, J.P. (2010) *Heat Transfer*. London: McGraw Hill.

Holmes, M., Anchor, R.D., Cooke, G.M.E. et al. (1982) The effects of elevated temperatures on the strength properties of reinforcing and pre-stressing steels. *The Structural Engineer*, 60, 7–13.

Hopkinson, J.S. (1984) *Fire Spread in Buildings*, Paper IP 21/84. Garston: BRE.

Hosser, D., Dorn, T., and Richter, E. (1994) Evaluation of simplified calculation methods for structural fire design. *Fire Safety Journal*, 22, 249–304.

Huang, Z., Burgess, I.W. and Plank, R. J. (2006) Behaviour of reinforced concrete structures in fire. In *Proceedings of Fourth International Workshop on Structures in Fire*, Aveiro, pp. 561–572.

Huang, Z., Burgess, I.W. and Plank, R.J. (2003a) Modelling membrane action of concrete slabs in composite buildings in fire. I: Theoretical development. *Journal of Structural Engineering*, 129, 1093–1102.

Huang, Z., Burgess, I.W. and Plank, R.J. (2003b) Modelling membrane action of concrete slabs in composite buildings in fire. II: Validations. *Journal of Structural Engineering,* 129, 1103–1112.

Huang, Z., Burgess, I.W. and Plank, R.J. (1999) Three-dimensional modelling of two full-scale fire tests on a composite building. *Proceedings of the Institution of Civil Engineers: Structures and Buildings*, 134, 243–255.

Huebner, K.H., Thornton, E.A. and Byrom, T.G. (1995) *The Finite Element Method for Engineers*, 3rd ed. New York: John Wiley & Sons.

Hughes, B.P. and Chapman, G.P. (1966) Deformation of concrete and microconcrete in tension and compression. *Magazine of Concrete Research*, 18, 19–24.

Hughes, T.R.J. (1977) Unconditionally stable algorithm for nonlinear heat conduction. *Computer Methods in Applied Mechanical Engineering*, 10, 135–139.

Iding, R.H. and Bresler, B. (1990) Effect of restraint conditions on fire endurance of steel-framed construction. In *Proceedings of 1990 ASIC National Steel Construction Conference*.

Iding, R.H. and Bresler, B. (1987) *FASBUS II User's Manual*. Wiss, Janney, Elstner Associates.

Iding, R.H., Bresler, B. and Nizamuddin, Z. (1977a) *Fires-T3: A Computer Program for the Fire Response of Structures-Thermal*. Report UCB FRG 77-15. Berkeley: University of California.

Iding, R.H., Bresler, B. and Nizamuddin, Z. (1977b) *Fires-RC II: Structural Analysis Program for Fire Response of Reinforced Concrete Frames*. Report UCB-FRG. Berkeley: University of California, pp. 77–78.

Ingberg, S.H. (1928) Tests of the severity of building fires. *US National Fire Protection Quarterly*, 22, 43–61.

ISE (2008) *Manual for The Design of Plain Masonry in Building Structures to Eurocode 6*. London: Institution of Structural Engineers.

ISE (2002) *Safety in Tall Buildings and Other Buildings with Large Occupancy*. London: Institution of Structural Engineers.

ISE and Concrete Society (1978) *Design and Detailing of Concrete Structures for Fire Resistance*. London: Institution of Structural Engineers.

ISO 834. (1975) *Fire Resistance Tests-Elements of Building Construction*. Geneva: International Organization for Standardization.

Issen, L.A., Gustaferro, A.H. and Carlson, C.C. (1970) *Fire Tests of Concrete Members: Improved Method of Estimating Thermal Restraint Forces*, Publication SP 464. Philadelphia: American Society for Testing and Materials, pp. 153–165.

Izzuddin, B.A. and Moore, D.B. (2002) Lessons from a full-scale fire test. *Proceedings of the Institution of Civil Engineers, Structures and Buildings*, 152, 319–329.

Jansson, R. and Boström, L. (2004) Experimental investigation on concrete spalling in fire. In *Fire Design of Concrete Structures: What Now? What Next?* Proceedings of Workshop, Politechnico di Milano, pp. 109–114.

Ju, J.W. and Zhang, Y. (1998) Axisymmetric thermo-mechanical constitutive and damage modeling for airfield concrete pavement under transient high temperature. *Mechanics of Materials*, 29, 307–23.

Junior, S.V. and Creus, G.J. (2007) Simplified elastoplastic analysis of general frames on fire. *Engineering Structures*, 29, 511–518.

Kammer, C. (2002) *Aluminium Taschenbuch 1. Grundlagen und Werkstoffe*, 16th ed. Dusseldorf: Aluminium Verlag.

Kankanamge, D. and Mahendran, M. (2012) Behaviour and design of cold-formed steel beams subject to lateral-torsional buckling at elevated temperatures. *Thin-Walled Structures*, 61, 213–228.

Kaufman, J.G. (1999) *Properties of Aluminium Alloys: Tensile, Creep and Fatigue Data at High and Low Temperatures*. Metals Park: ASM International.

Kawagoe, K. (1958) *Fire Behaviour in Rooms*, Report 27. Tokyo: Building Research Institute.

Kelly, F. and Purkiss, J.A. (2008) Reinforced concrete structures in fire: a review of current rules. *The Structural Engineer*, 86, 33–39.

Kersken-Bradley, M. (1993) Eurocode 5, Part 10, Structural Fire Design. In *Proceedings of Oxford Fire Conference*. Risborough: Timber Research and Development Association.

Kersken-Bradley, M. (1986) Probabilistic concepts in fire engineering. In Anchor, R.D. et al., Eds., *Design of Structures against Fire*. London: Elsevier, pp. 21–39.

Khalafallah, B.H. (2001) Coupled Heat and Mass Transfer in Concrete Exposed to Fire. PhD Thesis, Aston University, Birmingham.

Khennane, A. and Baker, G. (1993) Uniaxial model for concrete under variable temperature and stress. *Journal of Engineering Mechanics*, 119, 1507–1525.

Khennane, A. and Baker, G. (1992) Thermo-plasticity model for concrete under transient temperature and biaxial stress. *Proceedings of Royal Society*, 439, 59–80.

Khoury, G.A. (1992) Compressive strength of concrete at high temperatures: a reassessment. *Magazine of Concrete Research*, 44, 291–309.

Khoury, G.A, Grainger, B.N. and Sullivan, P.J.E. (1986) Strain of concrete during first cooling from 600°C under load. *Magazine of Concrete Research*, 38, 3–12.

Khoury, G.A, Grainger, B.N. and Sullivan, P.J.E. (1985a) Transient thermal strain of concrete: literature review, conditions within specimen, and behaviour of individual constituents. *Magazine of Concrete Research*, 37, 131–144.

Khoury, G.A, Grainger, B.N. and Sullivan, P.J.E. (1985b) Strain of concrete during first heating to 600°C under load. *Magazine of Concrete Research*, 37, 195–215.

Kirby, B.R. (1986) Recent developments and applications in structural fire engineering design: a review. *Fire Safety Journal*, 11, 141–179.

Kirby, B.R. (undated) *Fire Engineering in Sports Stands*. Cleveland, UK: Tata Steel.

Kirby, B.R., Lapwood, D.G. and Thompson, G. (1986) *The Reinstatement of Fire Damaged Steel and Iron Framed Structures*. Rotherham: Swinden Technology Centre.

Kirby, B.R., Newman, G.M., Butterworth, N. et al. (2004) A new approach to specifying fire resistance periods. *The Structural Engineer*, 82, 34–37.

Kirby, B.R. and Preston, R.R. (1988) High temperature properties of hot-rolled structural steels for use in fire engineering design studies. *Fire Safety Journal*, 13, 27–37.

Kirby, B.R., Wainman, D.E., Tomlinson, L.H. et al. (1994) *Natural Fires in Large Scale Compartments*. Fire Research Station Collaborative Project. Rotherham: Swinden Technology Center.

Kodur, V.K.R. (1999) Performance-based fire resistance design of concrete filled steel columns. *Journal of Constructional Steel Research*, 51, 21–36.

Kodur, V.K.R., Dwaikat, M. and Raut, N. (2009) Macroscopic FE model for tracing the fire response of reinforced concrete structures. *Engineering Structures*, 31, 2368–2379.

Kodur, V.K.R. and Lie, T.T. (1995) Fire performance of concrete-filled hollow steel columns. *Journal of Fire Protection Engineering*, 7, 89–98.

Kordina, K. and Henke, V. (1987) Sicherheitstheoretische Untersuchungen zur Versagenwahrscheinlichkeit von brandbeanspruchten Bauteilen bzw, Bauwerksabschnitten, Sonderforschungsbereich 148. Braunschweig: Technische Universität.

Kordina, K. and Krampf, L. (1984) Empfehlungen für brandschutztechnisch richtiges Konstruieren von Betonbauwerkern. Deutscher Auschuss für Stahlbeton, 352, 9–33.

Krampf, L. (undated) Investigations on the Shear Behaviour of Reinforced Concrete Beams Exposed to Fire. Braunschweig: Technische Universität.

Kruppa, J. (1992) Aging effect of fire insulation of steel structures. In *Proceedings of IMechE Conference on Materials and Design against Fire*, Paper C438/007/92. London: Mechanical Engineering Publications, pp. 129–134.

Latham, D.J., Kirby, B.R. and Thompson G. (1987) The temperatures attained by unprotected structural steelwork in experimental natural fires. *Fire Safety Journal*, 12, 139–152.

Laverty, D., Nadjai, A. and O'Connor, D. (2000/1) Modelling of thermo-structural response of concrete masonry walls subjected to fire. *Journal of Applied Fire Science*, 10, 3–19.

Law, M. (1997) A review of formulae for T equivalent. In *Proceedings of Fifth International Symposium on Fire Safety Science*, Melbourne.

Law, M. (1995) The origins of the 5 MW Fire. *Fire Safety Engineering*, 4, 17–20.

Law, M. (1983) A basis for the design of fire protection of building structures. *The Structural Engineer*, 61A, 25–33.

Law, M. (1978) Fire safety of external building elements. *Engineering Journal of American Institute of Steel Construction*, 59–74.

Law, M. (1973) Prediction of fire resistance. In *Symposium on Fire Resistance Requirements for Buildings: A New Approach*. London: Her Majesty's Stationery Office, pp. 16–29.

Law, M. and O'Brien, T. (1989) *Fire Safety of Bare External Structural Steel*, Report SCI-P-009. Ascot: Steel Construction Institute.

Lawrence, S.J. and Gnanakrishnan, N. (1988a) The fire resistance of masonry walls: an overview. In *Proceedings of First National Structural Engineering Conference*. Melbourne: Australia Institution of Engineers, pp. 431–437.

Lawrence, S.J. and Gnanakrishnan, N. (1988b) *The Fire Resistance of Masonry Walls*, Technical Record 531. Chatswood, Australia: National Building Technology Centre.

Lawson, R.M. and Newman, G.M. (1996) *Structural fire design to EC3 and EC4, and comparison with BS 5950*, Technical Report 159. Ascot: Steel Construction Institute.

Lawson. R.M. (1990b) *Enhancement of Fire Resistance of Beams by Beam to Column Connections*, Report TR-086. Ascot: Steel Construction Institute.

Lawson, R.M. (1990a) Behaviour of steel beam to column connections in fire. *The Structural Engineer*, 68, 263–271.

Lawson, R.M. (1985) *Fire Resistance of Ribbed Concrete Floors*, Report 107. London: Construction Industry Research and Information Association.

Lennon, T. (2003) *Precast Hollow Core Slabs in Fire*. Information Paper IP5/03. , BRE.

Lennon, T. and Clayton, N. (1999) Fire tests on high grade concrete with polypropylene fibres. In: *Proceedings of 9th Concrete Communications Conference*, British Cement Association, pp. 255–264.

Lennon, T., Rupasinghe, R., Canasius, G. et al. (2007) *Concrete Structures in Fire - Performance, Design and Analysis*. Report BR 490. Garston: BRE.

Leston-Jones, L.C., Burgess, I.W., Lennon, T. et al. (1997) Elevated temperature moment-rotation tests on steelwork connections. *Proceedings of Institution of Civil Engineers: Structures and Buildings*, 122, 410-419.

Li, L.Y. and Purkiss, J.A. (2005) Stress-strain constitutive equations of concrete at elevated temperatures. *Fire Safety Journal*, 40, 669–686.

Lie, T.T. (1977) A method of assessing the fire resistance of laminated timber beams and columns. *Canadian Journal of Civil Engineering*, 4, 161–169.

Lie, T.T. (1974) Characteristic temperature curves for various fire severities. *Fire Technology*, 10, 315–326.

Lie, T.T. and Kodur, V.K.R. (1996) Thermal and mechanical properties of steel fibre-reinforced concrete at elevated temperatures. *Canadian Journal of Civil Engineering*, 23, 511–517.

Lie, T.T. and Kodur, V.K.R. (1995) *Effect of Temperature on Thermal and Mechanical Properties of Steel Fibre-Reinforced Concrete*, Report 695. Ottawa: National Research Council of Canada.

Lie, T.T., Rowe, T.J., and Lin, T.D. (1986) *Residual Strength of Fire-Exposed Reinforced Concrete Columns*, Special Publication SP92-9. Detroit: American Concrete Institute, pp. 153–74.

Liew, J. and Chen, H. (2004) Explosion and fire analysis of steel frames using fiber element approach. *Journal of Structural Engineering*, 130, 991–1000.

Lin, C.H., Chen, S.T. and Hwang, T.Z. (1989) Residual strength of reinforced concrete columns exposed to fire. *Journal of Chinese Institute of Engineers*, 12, 557–565.

Loss Prevention Certification Board. (2012) *Red Book*. London: Loss Prevention Certification Board.

Lua, H., Zhao, X.L. and Han, L.H. (2009) Fire behaviour of high strength self-consolidating concrete filled steel tubular stub columns. *Journal of Constructional Steel Research*, 65, 1995–2010.

Luccioni, B.M., Figueroa, M.I. and Danesi, R.F. (2003) Thermo-mechanic model for concrete exposed to elevated temperatures. *Engineering Structures*, 25, 729–742.

Luikov, A.V. (1966) *Heat and Mass Transfer in Capillary Porous Bodies*. Oxford: Pergamon Press.

Majumdar, P., Gupta, A. and Marchertas, A. (1995) Moisture propagation and resulting stress in heated concrete walls. *Nuclear Engineering and Design*, 156, 159–165.

Majumdar, P. and Marchertas, A. (1997) Heat moisture transport and induced stresses in porous materials under rapid heating. *Numerical Heat Transfer A*, 32, 111–130.

Malhotra, H.L. (1994) What is fire resistance? *Fire Safety Engineering*, 1, 19–22.

Malhotra, H.L. (1993) Fire compartmentation: needs and specification. *Fire Surveyor*, 22, 4–9.

Malhotra, H.L. (1987) *Fire Safety in Buildings*. Garston: BRE.

Malhotra, H.L. (1986) A survey of fire protection developments for buildings. In *Design of Structures against Fire*. London: Elsevier Applied Science, pp. 1–13.

Malhotra, H.L. (1984) *Spalling of Concrete in Fires*, Technical Note 118. London: Construction Industry Research and Information Association.

Malhotra, H.L. (1982a) *Design of Fire-Resisting Structures*. Glasgow: Surrey University Press.

Malhotra, H.L. (1982b) Report on the work of technical committee 44PHT: properties of materials at high temperatures. *Matériaux et Constructions*, 15, 161–170.

Malhotra, H.L. (1978) Some noteworthy fires in concrete structures. In *Proceedings of Eighth Congress of the Fédération International de la Précontrainte*. Wexham Springs: Concrete Centre, pp. 86–98.

Malhotra, H.L. (1956) Effect of temperature on the compressive strength of concrete. *Magazine of Concrete Research*, 8, 85–94.

Maljaars, J., Soetens, F. and Katgerman, L. (2008) Constitutive model for aluminium alloys exposed to fire conditions. *Metallurgical and Materials Transactions A*, 39, 778–789.

Maljaars, J., Twilt, L., Fellinger, J.H.H. et al. (2010) Aluminium structures exposed to fire conditions: an overview. *HERON*, 55, 85–122.

Maréchal, J.C. (1970) Influence des températures sur la résistance, sur le fluage, sur la déformation plastique du béton. *Annales de L'Institute Technique du Bâtiment et des Travaux Publics*, 23, 123–146.

Markovic, M., Saje, M., Planinc, I. et al. (2012) On strain softening in finite element analysis of RC planar frames subjected to fire. *Engineering Structures*, 45, 349–361.

Marshall, N.R. (1992) *Smoke Control in Large Stores: an Extended Calculation Method for Slit Extraction Design*, Occasional Paper OP51. Garston: BRE.

Marshall, N.R. and Morgan, H.P. (1992) User's guide to BRE spill plume calculations. *Fire Surveyor*, 21, 14–19.

Martin, L.H. and Purkiss, J.A. (2006) *Concrete Design to EN 1992*. Oxford: Butterworth Heinemann.

Melinek, S.J. (1989), Prediction of the fire resistance of insulated steel. *Fire Safety Journal*, 14, 127–134.

Melinek, S.J. and Thomas, P.H. (1987) Heat flow to insulated steel. *Fire Safety Journal*, 12, 1–8.

Meyer-Ottens, C. (1975) *Zur Frage der Abplatzungen an Bauteilen aus Beton bei Brandbeanspruchungen*. Deutsche Ausschuß für Stahlbeton, Heft 248.

Meyer-Ottens, C. (1967) The behaviour of load bearing and non-loadbearing internal and external walls of wood and wood based materials under fire exposure. In *Symposium 3 on Fire and Structural Use of Timber*. London: Her Majesty's Stationery Office, pp. 77–89.

Ministry of Works (1952) Fire Grading of Buildings. Part II: Fire Fighting Equipment; Part III: Personal Safety; Part IV: Chimneys and Flues. Garston: BRE (1992 reissue).

Ministry of Works. (1946) Fire Grading of Buildings Part I: General Principles and Structural Precautions. Garston: BRE (1992 reissue).

Mlakar, P.F., Dusenberry, D.O., Harris, J.R. et al. (2003) Pentagon building performance report. Civil Engineering, 73, 43–55.

Montgomery, F.R. (1997) Surface permeability measurement as a means of assessing fire damage of concrete. In The Concrete Way to Development, Vol. 2. Johannesburg: Concrete Society of South Africa, pp. 821–828.

Mooney, J. (1992) Surface radiant energy balance for structural thermal analysis. Fire and Materials, 6, 61–66.

Moore, D.B. and Lennon, T. (1997) Fire engineering design of steel structures. Progress in Structural Engineering and Materials, 1, 4–9.

Morales, E.M. (1992) Rehabilitation of a fire damaged building. In Evaluation and Rehabilitation of Concrete Structures and Innovations in Design, Special Publication SP-128. Detroit: American Concrete Institute, pp. 1457–1472.

Morgan, H.P. (1993) Combining sprinklers and vents: an interim approach. Fire Surveyor, 22, 10–14.

Morgan, H.P. and Gardner, J.P. (1991) Design Principles for Smoke Ventilation in Enclosed Shopping Centres, Report BR 186. Garston: BRE.

Morris, B. and Jackman, L.A. (2003) An examination of fire spread in multi-storey building via glazed curtain wall façades. The Structural Engineer, 6, 22–26.

Morris, W.A., Read, R.E.H. and Cooke, G.M.E. (1988) Guidelines for the Construction of Fire-Resisting Structural Elements, 2nd ed. London: Her Majesty's Stationery Office.

Mossa, P.J., Dhakal, R.P., Bong, M.W. et al. (2009) Design of steel portal frame buildings for fire safety. Journal of Constructional Steel Research, 65, 1216–1224.

Mossa, P.J., Dhakal, R.P., Wang, G. et al. (2008) The fire behaviour of multi-bay, two-way reinforced concrete slabs. Engineering Structures, 30, 3566–3573.

Muenow, R.A. and Abrams, M.S. (1987) Non-destructive testing methods for evaluating damage and repair of concrete exposed to fire. In Repair and Rehabilitation of Concrete Structures, Publication SCM-16. Detroit: American Concrete Institute, pp. 284–295.

Muirhead, J. (1993) Making it safely through construction. Fire Surveyor, 22, 5–7.

Mustapha, K.N. (1994) Modelling the Effects of Spalling on the Failure Modes of Concrete Columns in Fire. PhD Thesis, Aston University, Birmingham.

Nadjai, A., Bailey, C.G., Vassart, O. et al. (2011) Full-scale fire test on a composite floor slab incorporating long span cellular steel beams. The Structural Engineer, 89, 18–25.

Nadjai, A., Laverty, D. and O'Gara, M. (2001) Behaviour of compartment masonry walls in fire situations. In Topping, B.H.V., Ed., Civil and Structural Engineering Computing. Stirling: Saxe-Coburg Publications, pp. 407–431.

Nassif, A.Y. (2000) A new classification system for fire-damaged concrete based on the strain energy dissipated in a hysteresis loop. Magazine of Concrete Research, 52, 287–295.

Nassif, A.Y., Burley, E. and Rigden, S. (1995) A new quantitative method of assessing fire damage to concrete structures. Magazine of Concrete Research, 47, 271–278.

Nassif, A.Y., Rigden, S. and Burley, E. (1999) The effects of rapid cooling by water quenching on the stiffness properties of fire-damaged concrete. *Magazine of Concrete Research*, 51, 255–261.

Nechnech, W., Meftah, F. and Reynouard, J.M. (2002) An elasto-plastic damage model for plain concrete subjected to high temperatures. *Engineering Structures*, 24, 597–611.

Nene, R.L. and Kavle, P.S. (1992) Rehabilitation of a fire damaged structure. In *Evaluation and Rehabilitation of Concrete Structures and Innovations in Design*, Vol. 2, Special Publication SP-128. Detroit: American Concrete Society, pp. 1195–1211.

Neville, A.M. (1973) *Properties of Concrete*. New York: John Wiley & Sons.

Newman, G.M. (1993) *The Fire Resistance of Shelf Angle Floor Beams to BS 5950: Part 8*, Publication 126. Ascot: Steel Construction Institute.

Newman, G.M. (1992) *The Fire Resistance of Web-Infilled Steel Columns*, Publication 124. Ascot: Steel Construction Institute.

Newman, G.M. and Lawson, R.M. (1991) *Fire Resistance of Composite Beams*, Technical Report SCI-P-109. Ascot: Steel Construction Institute.

Newman, G.M., Robinson, J.T. and Bailey, C.G. (2006) *Fire Safe Design: A New Approach to Multi-Storey Steel-Frame Buildings*, 2nd ed., Publication P 288. Ascot: Steel Construction Institute.

Newman, L.C., Dowling, J. and Simms, W.I. (2005) *Structural Fire Design: Off-Site Applied Thin Film Intumescent Coatings*, 2nd ed., Publication P 160. Ascot, Steel Construction Institute.

Nwosu, D.I., Kodur, V.K.R., Franssen, J.M. et al. (1999) *User Manual for SAFIR: A Computer Program for Analysis of Structures at Elevated Temperature Conditions*, Interim Report 782. Ottawa: National Research Council of Canada.

O'Connor, M.A., Kirby, B.R. and Martin, D.M. (2003) Behaviour of a multi-storey composite steel frame building in fire. *The Structural Engineer*, 81, 27–36.

Ödeen, K. (1969) Fire resistance of glued laminated timber structures. In *Symposium 3 on Fire and Structural Use of Timber in Buildings*. London: Her Majesty's Stationery Office, pp. 7–15.

Parker, T.W. and Nurse, R.W. (1956) *Investigations into Building Fires Part 1: Estimation of Maximum Temperatures Attained in Building Fires from Examination of Debris*, Technical Paper 4. London: Her Majesty's Stationery Office, pp. 1–5.

PD 6688-1-2. *Background Paper to UK National Annex to BS EN 1991-1-2*. London: British Standards Institution.

PD 7974-1. *Application of Fire Safety Engineering Principles to the Design of Buildings Part 1: Initiation and Development of Fire within Enclosure of Origin*. London: British Standards Institution.

PD 7974-2. *Application of Fire Safety Engineering Principles to the Design of Buildings Part 2: Spread of Smoke and Toxic Gases within and beyond Enclosure of Origin*. London: British Standards Institution.

PD 7974-3. *Application of Fire Safety Engineering Principles to the Design of Buildings Part 3: Structural Response and Fire Spread beyond Enclosure of Origin*. London: British Standards Institution.

Persson, B. (2003) *Self-Compacting Concrete at Fire Temperatures*, Report TVMB-3110. Lund Institute of Technology.

Pettersson, O., Magnusson, S.E. and Thor, J. (1976) *Fire Engineering Design of Steel Structures*, Publication 50. Stockholm: Swedish Institute of Steel Construction.

Pettersson, O. and Witteveen, J. (1979/80) On the fire resistance of structural steel elements derived from standard fire tests or by calculation. *Fire Safety Journal*, 2, 73–87.

Phan, L.T. (2008) Pore pressure and explosive spalling in concrete. *Materials and Structures*, 41, 1623–1632.

Phan, L.T. and Carino, N.J. (1998) Review of mechanical properties of HSC at elevated temperatures. *Journal of Materials in Civil Engineering*, 10, 58–64.

Phan, L.T., Lawson, J.R. and Davis, F.L. (2001) Effects of elevated temperature exposure on heating characteristics, spalling, and residual properties of high performance concrete. *Matériaux et Constructions*, 34, 83–91.

Philleo, R. (1958) Some physical properties of concrete at high temperatures. *Proceedings of the American Concrete Institute*, 54, 857–864.

Placido, F. (1980) Thermoluminescence test for fire-damaged concrete. *Magazine of Concrete Research*, 32, 112–116.

Plank, R.J., Burgess, I.W. and Bailey, C.G. (1997) Modelling the behaviour of steel-framed building structures by computer. In *Proceedings of Second Cardington Conference on Fire, Static and Dynamic Tests of Building Structures*. London: E. & F.N. Spon.

Plecnik, J.M. and Fogerty, J.H. (1986) Behavior of epoxy repaired beams under fire. *Journal of Structural Engineering*, 112, 906–922.

Plem, E. (1975) *Theoretical and Experimental Investigations of Point Set Structures*, Document D9, Copenhagen: Danish Council for Building Research.

Poon, C.S., Azhar, S., Anson, M. et al. (2001) Comparison of the strength and durability performance of normal- and high-strength pozzolanic concretes at elevated temperatures. *Cement and Concrete Research*, 31, 1291–1300.

Pope, R. (2006) Lessons from Madrid. *New Steel Construction*, 14, 26–28.

Popovics, S. (1973) A numerical approach to the complete stress-strain curve of concrete. *Cement and Concrete Research*, 3, 583–599.

Popovics, S. (1970) A review of stress-strain relationships for concrete. *ACI Journal*, 67, 243–248.

Proulx, G. (1994) Human response to fires. *Fire Research News*, 71, 1–3.

Purkiss, J.A. (2000) High-strength concrete and fire. *Concrete*, 34. 49–50.

Purkiss, J.A. (1990b) The decommissioning of fire damaged building structures. In White, I.L., Ed., *Decommissioning and Demolition*. Thomas Telford Ltd., pp. 68–72.

Purkiss, J.A. (1990a) Computer modelling of concrete structural elements exposed to fire. In *Proceedings of Interflam '90*. London: Interscience Communications, pp. 67–75.

Purkiss, J.A. (1988) A systems approach to fire safety engineering. In *Structural Safety Evaluation Based on System Identification Approaches*. Braunschweig: F. Vieweg und Sohn Verlag, pp. 394–413.

Purkiss, J.A. (1987) Thermal expansion of steel fibre reinforced concrete up to 800°C. In Marshall, I.H., Ed., *Composites 4*. London: Elsevier Applied Science, pp. 404–415.

Purkiss, J.A. (1985) Some mechanical properties of glass fibre reinforced concrete at elevated temperatures. In Marshall, I.H., Ed., *Composites 3*. London: Elsevier Applied Science, pp. 230–241.

Purkiss, J.A. (1984) Steel fibre reinforced concrete at elevated temperature. *International Journal of Cement Composites and Lightweight Concrete*, 6, 179–184.

Purkiss, J.A. (1972) *A Study of the Behaviour of Concrete Heated to High Temperatures under Restraint or Compressive Loading*. Ph.D. Thesis. University of London.

Purkiss, J.A. and Bali, A. (1988) The transient behaviour of concrete at temperatures up to 800°C. In *Proceedings of 10th Ibausil*. Weimar: Hochschule für Architektur und Bauwesen, pp. 234–239.

Purkiss, J.A., Claridge, S.L. and Durkin, P.S. (1989) Calibration of simple methods of calculating the fire resistance of flexural reinforced concrete members. *Fire Safety Journal*, 15, 245–263.

Purkiss, J.A., Maleki-Toyserkani, M. and Short, N.R. (2001) Thermal and residual mechanical behaviour of SIFCON subjected to elevated temperatures. In *11th Concrete Communications Conference*. British Concrete Association, pp. 357–366.

Purkiss, J.A., Morris, W.A. and Connolly, R.J. (1996) Fire resistance of reinforced concrete columns: correlation of analytical methods with observed experimental behaviour. In *Proceedings of Interflam '96*. London: Interscience Communications, pp. 531–541.

Purkiss, J.A. and Mustapha, K.N. (1996) An investigation into the influence of concrete constitutive models on the behaviour of reinforced concrete columns exposed to fire. *Journal of Institution of Engineers, Malaysia*, 57, 23–32.

Purkiss, J.A. and Mustapha, K.N. (1995) A study of the effect of spalling on the failure modes of concrete columns in a fire. In *Concrete 95, Toward Better Concrete Structures*. Brisbane : Concrete Institute of Australia and Fédération Internationale de la Précontrainte, pp. 263–272.

Purkiss, J.A. and Weeks, N.J. (1987) A computer study of the behaviour of reinforced concrete columns in a fire. *The Structural Engineer*, 65B, 22–28.

Ramberg, W. and Osgood, W. (1943) *Description of Stress-Strain Curves by Three Parameters*, Technical Note TN902. Washington: U.S. National Advisory Committee on Aeronautics.

Ranawaka, T. and Mahendran, M. (2009) Experimental study of the mechanical properties of light gauge cold-formed steels at elevated temperatures. *Fire Safety Journal*, 44, 219–229.

Rasbash, D.J. (1984/5) Criteria for acceptability for use with quantitative approaches to fire safety. *Fire Safety Journal*, 8, 141–158.

Read R.E.H. (ed) (1991) *External Fire Spread: Building Separation and Boundary Distances*. Garston: BRE.

Read, R.E.H. (1985) Trade-offs between sprinklers and passive fire protection. Personal communication.

Read, R.E.H. and Morris, W.A. (1993) *Aspects of Fire Precautions in Buildings*, 3rd ed. London: Department of the Environment.

Redfern, B. (2005) Lack of fire stops blamed for speed of Madrid tower inferno. *New Civil Engineer*, Feb. 17, 5–7.

Riley, M.A. (1991) Possible new method for the assessment of fire-damaged concrete. *Magazine of Concrete Research*, 43, 87–92.

Robbins, J. (1991) Running repair. *New Civil Engineer*, Aug. 15, 17–18.

Robbins, J. (1990) Baptism of fire. *New Civil Engineer*, July 12, 17–19.

Robbins, P.J. and Austen, S.A. (1992) Flexural testing of steel fibre reinforced refractory concrete at elevated temperatures. In Swamy, R.N., Ed., *Fibre Reinforced Cement and Concrete*. London: E. & F.N. Spon, pp. 153–165.

Robertson, A.F. and Gross, D. (1970) Fire load, fire severity, and fire endurance. In *Symposium on Fire Test Performance*, STP 464. Philadelphia: American Society for Testing and Materials, pp. 3–29.

Robinson, J.T. and Latham, D.J. (1986) Fire resistant steel design: the future challenge. In Anchor, R.D. et al., Eds., *Design of Structures against Fire*. London: Elsevier, pp. 225–236.

Robinson, J.T. and Walker, H.B. (1987) Fire safe structural design. *Construction and Building Materials*, 1, 40–50.

Rogowski, B.F.W. (1969) Charring of timber in fire tests. In *Symposium 3 on Fire and Structural Use of Timber in Buildings*. London: Her Majesty's Stationery Office, pp. 52–59.

Rosato, C. (1992) London Underwriting Centre. *Fire Prevention*, 248, 33–35.

Saad, M. Abo-El-Enein, S.A. Hanna, G.B. et al. (1996) Effect of temperature on physical and mechanical properties of concrete containing silica fume. *Cement and Concrete Research*, 26, 669–675.

Sanad, A.M., Lamont, S., Usmani, A.S. et al. (2000) Structural behaviour in fire compartment under different heating regimes I: Slab thermal gradients. *Fire Safety Journal*, 35, 96–116.

Sanad, A.M., Rotter, J.M., Usmani, A.S. et al. (1999) Finite element modeling of fire tests on the Cardington composite building. In *Proceedings of Interflam '99*. London: Interscience Communications.

Sano, C. (1961) Effect of temperature on the mechanical properties of wood I: Compression parallel to the grain. *Journal of Japanese Wood Research Society*, 7, 147–150.

Schaffer, E.L. (1967) *Charring Rate of Selected Woods-Transverse to Grain*, FPL 69. Madison, WI: U.S. Department of Agriculture Forest Products Laboratory.

Schaffer, E.L. (1965) An approach to the mathematical prediction of temperature rise within a semi-infinite wood slab subjected to high temperature conditions. *Pyrodynamics*, 2, 117–132.

Schleich, J.B. (1987) Fire design of steel structures. *Steel Construction*, 3, 20–21.

Schleich, J.B. (1986) Numerical simulations: a more realistic fire safety approach in structural stability. In Grayson, S.J. and Smith, D.A., Eds., *New Technology to Reduce Fire Losses and Costs*. London: Elsevier Applied Science, pp. 203–210.

Schneider, U. (1988) Concrete at high temperatures: a general review. *Fire Safety Journal*, 13, 55–68.

Schneider, U. (1986a) *Properties of Materials at High Temperatures: Concrete*, 2nd ed. RILEM Report. Kassel: Gesamthochschule.

Schneider, U. (1986b) Modelling of concrete behaviour at high temperatures. In Anchor, R.D. et al., Eds., *Design of Structures against Fire*. London: Elsevier Applied Science, pp. 53–70.

Schneider, U. (1982) Creep effects under transient temperature conditions. In Wittmann, F.H., Ed., *Fundamental Research on Creep and Shrinkage of Concrete*. The Hague: Martinius Nijhoff, pp. 193–202.

Schneider, U. (1976) Behaviour of concrete under thermal steady state and non-steady state conditions. *Fire and Materials*, 1, 103–115.

Schneider, U. and Nägele, E., Eds. (1989) *Repairability of Fire Damaged Structures.* Publication 111. Kassel: Gesamthochschule.

Schneider, U., Schneider, M. and Franssen, J.M. (2008) Consideration of nonlinear creep strain of siliceous concrete on calculation of mechanical strain under transient temperatures as a function of load history. In Tan, K.H. et al., Eds., *Proceedings of Fifth International Conference on Structures in Fire.* Singapore: Nanyang Technical University, pp.463–476.

Schneider, U. and Schneider, M. (2009). An advanced transient concrete model for the determination of restraint in concrete structures subjected to fire. *Journal of Advanced Concrete Technology, 7,* 403–413.

SCI (2003) AD 269- *The use of intumescent coatings for the fire protection of beams with circular web openings.* New Steel Construction, 11/12, 33–34.

SCI (1993) *Building Design Using Cold-Formed Steel Sections: Fire Protection,* Publication, P129. Ascot: Steel Construction Institute.

SCI (1991) *Structural Fire Engineering: Investigation into the Broadgate Phase 8 Fire.* Ascot: Steel Construction Institute.

Selih, J. and Sousa, A.C.M. (1995) Heat transfer and thermal stress calculation in fire-exposed concrete wall. *Advanced Computing Methods in Heat Transfer IV, 12,* 265–274.

Selih, J., Sousa, A.C.M. and Bremmer, T.W. (1994) Moisture and heat flow in concrete walls exposed to fire. *Journal of Engineering Mechanics, 120,* 2028–2043.

Selvaggio, S.L. and Carlson, C.C. (1963) *Effects of Restraint on the Fire Resistance of Prestressed Concrete,* Research Bulletin 164. Skokie, IL: Portland Cement Association.

Shahbazian, A. and Wang, Y. (2012) Direct strength method for calculating distortional buckling capacity of cold-formed thin-walled steel columns with uniform and non-uniform elevated temperatures. *Thin-Walled Structures, 53,* 188–199.

Shields, T.J. (1993) *Fire and Disabled People in Buildings,* Report BR 231. BRE, Garston.

Shields, T.J., Silcock, G.W.H. and Flood, M.F. (2001a) Performance of single glazing assembly exposed to enclosure fires of increasing severity. *Fire and Materials, 25,* 123–136.

Shields, T.J., Silcock, G.W.H. and Flood, M.F. (2001b) Performance of single glazing assembly exposed to a fire in the centre of in enclosure. *Fire and Materials, 25,* 137–152.

Short, N.R. and Purkiss, J.A. (2004) Petrographic analysis of fire-damaged concrete. In *Fire Design of Concrete Structures: What Now? What Next?* Proceedings of Workshop, Politechnico di Milano, pp. 221–230.

Short, N.R., Purkiss, J.A. and Guise, S.E. (2002) Assessment of fire damaged concrete using crack density measurements. *Structural Concrete, 3,* 137–143.

Short, N.R., Purkiss, J.A. and Guise, S.E. (2001) Assessment of fire damaged concrete using colour image analysis. *Construction and Building Materials, 15,* 9–15.

Short, N.R., Purkiss, J.A. and Guise, S.E. (2000) Assessment of fire damaged concrete. In *Proceedings of 10th Concrete Communications Conference,* pp. 245–254.

Shorter, G.W. and Harmathy, T. Z. (1965) Moisture clog spalling. *Proceedings of Institution of Civil Engineers, 20,* 75–90.

Sidey, M.P. and Teague, D.P. (1988) *Elevated Temperature Data for Structural Grades of Galvanised Steel.* British Steel (Welsh Laboratories) Report.

Silva, J.C. and Landessmann, A. (2013) Performance-based analysis of steel–concrete composite floor exposed to fire. *Journal of Constructional Steel Research*, 83, 117–126.

Simms, W.I. and Newman, G.M. (2002) *Single Storey Steel Frame Buildings in Fire Boundary Conditions*, Publication P313. Ascot: Steel Construction Institute.

Smith, C.I., Kirby, B.R., Lapwood, D.G. et al. (1981) The reinstatement of fire damaged steel framed structures. *Fire Safety Journal*, 4, 21–62.

Smith, L.M. and Placido, F. (1983) Thermoluminescence: a comparison with the residual strength of various concretes. In Abrams, M.S., Ed., *Fire Safety of Concrete Structures*, Publication SP-80. Detroit: American Concrete Institute, pp. 293–304.

Somerville, G. (1996) Cement and concrete as materials: changes in properties, production and performance. *Structures and Buildings*, 116, 335–343.

Song, T.Y., Han, L.H. and Yu, H.X. (2010) Concrete-filled steel tube stub columns under combined temperature and loading. *Journal of Constructional Steel Research*, 66, 369–384.

Sterner, E. and Wickström, U. (1990) *TASEF: Temperature Analysis of Structures Exposed to Fire*, Report 1990:05. Borås: Swedish National Testing and Research Institute.

Stiller, J. (1983) Berechnungsmethode für brandbeanspruchte Holzstützen und Holzbalken aus brettschichtverleimtem Nadelholz. In *Arbeitsbericht 1981–1983*, SFB 148. Braunschweig: Technische Universität, pp. 219–276.

Stirland, C. (1981) *Sprinklers and Building Regulations: The Case for Trade-Offs*, Report T/RS/1189/22/81/C. Middlesborough: Tata Steel Teesside Laboratory.

Stirland, C. (1980) *Steel Properties at Elevated Temperatures for Use in Fire Engineering Calculations*, Report T/RS/11/80C. Middlesborough: Tata Steel Teesside Laboratory.

Sullivan, P.J.E., Terro, M.J. and Morris W.A. (1993/4) Critical review of fire-dedicated thermal and structural computer programs. *Applied Fire Science*, 3, 113–135.

Tan, K., Ting, S. and Huang, Z. (2002) Visco-elasto-plastic analysis of steel frames in fire. *Journal of Structural Engineering*, 128, 105–114.

Tata Steel Construction (2012) *Advance™ Sections*. www.tatasteelconstruction.com

Tenchev, R. and Purnell, P. (2005) An application of a damage constitutive model to concrete at high temperature and prediction of spalling. *International Journal of Solids and Structures*, 42, 6550–6565.

Tenchev, R.T., Li, L.Y. and Purkiss, J.A. (2001a) Numerical analysis of temperature and pore pressure in intensely heated concrete. In Chan, A.H.C., Ed., *Proceedings of Ninth Annual ACME Conference*, Birmingham University, pp. 5–8.

Tenchev, R.T., Li, L.Y. and Purkiss, J.A. (2001b) Finite element analysis of heat and moisture transfer in concrete subject to fire. *Numerical Heat Transfer*, 39, 685–710.

Tenchev, R.T., Li, L.Y. Purkiss, J.A. et al. (2001) Finite element analysis of coupled heat and moisture transfer in concrete when it is in fire. *Magazine of Concrete Research*, 53, 117–125.

Tenchev, R.T., Purkiss, J.A. and Li, L.Y. (2001) Numerical analysis of thermal spalling in a concrete column. In Ivanov, Y.A. et al., Eds., *Proceedings of Ninth National Congress on Theoretical and Applied Mechanics*, Vol. 1, Varna, pp. 604–609.

Tenning, K. (1969) Glued laminated timber beams: fire tests and experience in practice. In *Symposium 3 on Fire and Structural Use of Timber in Buildings*. London: Her Majesty's Stationary Office, pp. 1–6.

Terro, M.J. (1998) Numerical modeling of the behaviour of concrete structures in fire. *ACI Structural Journal*, 95, 183–193.

Thelandersson, S. (1982) On the multiaxial behaviour of concrete exposed to high temperature. *Nuclear Engineering and Design*, 75, 271–282.

Thomas, F.G. and Webster, C.T. (1953) *Fire Resistance of Reinforced Concrete Columns*, National Building Studies Research Paper 18. London: Her Majesty's Stationery Office.

Thomas, P.H., Ed. (1986) Design guide: structural fire safety. *Fire Safety Journal*, 10, 77–137.

Thomas, P.H. and Heselden, A.J.M. (1972) *Fully Developed Fires in Single Compartments*, CIB Report 20. London: Her Majesty's Stationery Office.

Tomacek, D.V. and Milke, J.A. (1993) A study on the effect of partial loss of protection on the fire resistance of steel columns. *Fire Technology*, 29, 3–21.

Touloukian, Y.S. and Ho, C.Y., Eds. (1973) *Properties of Aluminum and Aluminum Alloys*, Report 21. Thermophysical Properties Research Center, Purdue University.

Türker, P., Erdoğdu, K. and Erdoğan, B. (2001) Investigation of fire-damaged concrete with different types of aggregate, In *Proceedings of 23rd International Conference on Cement Microscopy*. International Cement Microscopy Association, pp. 193–212.

Twilt, L. and Witteveen, J. (1986) Calculation methods for fire engineering design of steel and composite structures. In Anchor, R.D. et al., Eds., *Design of Structures against Fire*. London: Elsevier, pp. 155–176.

Ulm, F.J., Coussy, O. and Bažant, Z.P. (1999) The "Chunnel" fire. II: analyses of concrete damage. *Journal of Engineering Mechanics*, 125, 283–289.

Van Acker, A. (2003) Shear resistance of pre-stressed hollow core floors exposed to fire. *Structural Concrete*, 2, 6.

Van Herberhen, P. and Van Damme, M. (1983) Fire resistance of post-tensioned continuous floor slabs with unbonded tendons. *FIP Notes*, 3–11.

Varley, N. and Both, C. (1999) Fire protection of concrete linings in tunnels. *Concrete*, 5, 27–30.

Wainman, D.E. and Kirby, B.R. (1989) *Compendium of UK Standard Fire Test Data 2: Unprotected Structural Steelwork*, Report RS/R/S1198/8/88/B. Rotherham: Tata Steel Swinden Technology Center.

Wainman, D.E. and Kirby, B.R. (1988) *Compendium of UK Standard Fire Test Data 1: Unprotected Structural Steelwork*, Report RS/RSC/S10328/1/87/B. Rotherham: Tata Steel Swinden Technology Center.

Wang, Y. (2000a) A simple method for calculating the fire resistance of concrete-filled CHS columns. *Journal of Constructional Steel Research*, 54, 365–386.

Wang, Y. (2000b) An analysis of the global structural behaviour of the Cardington steel-framed building during the two BRE fire tests. *Engineering Structures*, 22, 401–412.

Wang, Y., Burgess, I., Wald, F. et al. (2013) *Performance-Based Fire Engineering of Structures*. Boca Raton: FL: CRC Press.

Wang, Y., Davison, J.B., Burgess, I.W. et al. (2010) The safety of common steel beam/column connections in fire. *The Structural Engineer*, 88, 26–34.

Wang, Y. and Orton, A.H. (2008) Fire resistant design of concrete filled tubular steel columns. *The Structural Engineer*, 86, 40–45.

Wardle, T.M. (1966) *Fire Resistance of Heavy Timber Construction*. Wellington: New Zealand Forest Service.

Welch, S. (2000) Developing a model for thermal performance of masonry exposed to fire. In *Proceedings of First International Workshop on Structures in Fire*, pp. 117–134.

White, R.H. and Schaffer, E.L. (1978) Application of CMA program to wood charring. *Fire Technology*, 4, 279–290.

Wickström, U. (1986) A very simple method for estimating temperatures in fire-exposed structures. In Grayson, S.J., Eds., *New Technology to Reduce Fire Losses and Costs*. London: Elsevier Applied Science, pp. 186–194.

Wickström, U. (1985a) Application of the standard fire curve for expressing natural fires for design purposes. In Harmathy, T.Z., Ed., *Fire Safety: Science and Engineering*, STP 882. Philadelphia: American Society for Testing and Materials, pp. 145–159.

Wickström, U. (1985b) Temperature analysis of heavily-insulated steel structures exposed to fire. *Fire Safety Journal*, 9, 281–285.

Wickström, U. (1981/2) Temperature calculation of insulated steel columns exposed to natural fire. *Fire Safety Journal*, 4, 219–225.

Williams-Lier, G. (1973) Analytical equivalents of standard fire temperature curves. *Fire Technology*, 9, 132–136.

Witteveen, J. (1983) Trends in design methods for structural Fire Safety. In Woob B., Three Decades of Structural Fire Safety. Garston: BRE, pp. 21–30.

Witteveen, J. and Twilt, L. (1981/2) A critical view on the results of standard fire resistance tests on steel columns. *Fire Safety Journal*, 4, 259–270.

Wong, P. (2003) Performance of GRP composite structures at ambient and elevated temperatures. *The Structural Engineer*, 81, 10–12.

Wood, M. (2004) Fire-resistant glazed systems. *Fire Protection Data*, 12, 61–63.

Xu, Y., Wong, Y.L., Poon, C.S. et al. (2001) Impact of high temperatures on PFA concrete. *Cement and Concrete Research*, 31, 1065–1073.

Yin, J., Zha, X.X. and Li, L.Y. (2006) Fire resistance of axially loaded concrete filled steel tube columns. *Journal of Construction Steel Research*, 62, 723–729.

Yu, M., Zha, X.X., Ye, J.Q. et al. (2013) A unified formulation for circle and polygon concrete-filled steel tube columns under axial compression. *Engineering Structures*, 49, 1–10.

Zacharia, T. and Aidun, D.K. (1988) Elevated temperature mechanical properties of Al-Li-Cu-Mg Alloy. *Welding Journal*, 67, 281–288.

Zienkiewicz, O.C. and Taylor, R.L. (2000) *The Finite Element Method*. 5th ed. Oxford: Butterworth Heinemann.

Author Index

Subject index

Milton Keynes UK
Ingram Content Group UK Ltd.
UKHW031138141024
449569UK00024B/1229